(Eds.)
R.J. Adrian · D.F.G. Durão · F. Durst
M.V. Heitor · M. Maeda · J.H. Whitelaw

Developments in Laser Techniques
and Applications to Fluid Mechanics

Springer
Berlin
Heidelberg
New York
Barcelona
Budapest
Hong Kong
London
Milan
Paris
Santa Clara
Singapore
Tokyo

(Eds.)
R.J. Adrian · D.F.G. Durão · F. Durst
M.V. Heitor · M. Maeda · J.H. Whitelaw

Developments in Laser Techniques and Applications to Fluid Mechanics

Proceedings of the 7th International Symposium
Lisbon, Portugal, 11-14 July, 1994

With 288 Figures and 26 Tables

Springer

R. J. ADRIAN
Department of Theoretical and Applied Mechanics; University of Illinois
at Urbana - Champaign; Urbana, Illinois
USA

F. DURST
Lehrstuhl für Strömungsmechanik; University of Erlangen-Nürnberg;
Egerlandstraße 13; D-91058 Erlangen
GERMANY

M.V. HEITOR; D. F. G. DURÃO
Department of Mechanical Engineering; Instituto Superior Técnico;
Av. Rovisco Pais; 1096 Lisboa Codex
PORTUGAL

M. MAEDA
Department of Mechanical Engineering; Keio University;
1-14-1 Hiyoshi, Kohuku Yokohama 223
JAPAN

J. H. WHITELAW
Imperial College of Science, Technology and Medicine; Department of
Mechanical Engineering; Exhibition Road; London SW7 2BX
ENGLAND

ISBN 3-540-60236-4 Springer-Verlag Berlin Heidelberg New York

Cataloging-in-Publication Data applied for

Die Deutsche Bibliothek - CIP-Einheitsaufnahme
Developments in laser techniques and applications to fluid mechanics : proceedings of the 7th
international symposium, Lisbon, Portugal, 11-14 July, 1994 / R. J. Adrian ... (ed.). - Berlin ; Heidelberg ;
New York ; Barcelona ; Budapest ; Hong Kong ; London ; Milan ; Santa Clara ; Singapore ; Tokyo :
Springer, 1996
    ISBN 3-540-60236-4
NE: Adrian, Ronald J. [Hrsg.]

© Springer-Verlag Berlin Heidelberg 1996
Printed in Germany

The use of general descriptive names, registered names, trademarks, etc. in this publication does
not imply, even in the absence of a specific statement, that such names are exempt from the relevant
protective laws and regulations and therefore free for general use.

Typesetting: Camera ready copy by authors
Production: PRODUserv Springer Produktions-Gesellschaft, Berlin
SPIN: 10488771   61/3020 - 5 4 3 2 1 0 - Printed on acid-free paper

Preface

This volume comprises a selection of the best papers presented at the *Seventh International Symposium on Applications of Laser Techniques to Fluid Mechanics* held at The Calouste Gulbenkian Foundation in Lisbon, during the period of July 11 to 14, 1994. The papers describe *Applications to Fluid Mechanics, Applications to Combustion, Instrumentation for Velocity and Size Measurements* and *Instrumentation for Whole-Field Velocity* and demonstrate the continuing and healthy interest in the development of understanding of the methodology and implementation in terms of new instrumentation.

The prime objective of this Seventh Symposium was to provide a forum for the presentation of the most advanced research on laser techniques for flow measurements, and communicate significant results to fluid mechanics. The applications of laser techniques to scientific and engineering fluid flow research was emphasized, but contributions to the theory and practice of laser methods were also considered where they facilitate new improved fluid mechanic research. Attention was placed on laser-Doppler anemometry, particle sizing and other methods for the measurement of velocity and scalar, such as particle image velocimetry and laser induced fluorescence.

We would like to take this opportunity to thank those who participated. The assistance provided by the Advisory Committee, by assessing abstracts was highly appreciated. The companies who took part in the exhibition and in the manufacturers' technical presentation are also acknowledged. In addition, thanks go to the participants who contributed actively in discussions, learned from the presentations and were essential to the success of this symposium. And last, but not least, we are greatly indebted for the financial support provided by the Sponsoring Organizations that made this Symposium possible.

Sponsoring Organizations

ADIST, Banco Comercial Português, Câmara Municipal de Lisboa, Centro de Termodinâmica Aplicada e Mecânica dos Fluídos da Universidade Técnica de Lisboa, CTAMFUTL, DHL, Direcção Geral de Turismo, European Research Office: United States Army, Navy and Air Force Departments, FLAD, Fundação Calouste Gulbenkian, Instituto Superior Técnico (IST), Instituto Tecnológico Para A Europa Comunitária (ITEC), Junta Nacional de Investigação Científica e Tecnológica (JNICT), TAP - Air Portugal.

The Editors   London, March 1995

Table of Contents

Part I
Applications to Fluid Mechanics

Low Reynolds Number Effects on the Inner Region of a Turbulent Boundary Layer

C. Y. Ching, L. Djenidi and R. A. Antonia

Department of Mechanical Engineering, University of Newcastle, N.S.W, 2308, Australia

Abstract. Low-Reynolds-number effects on the inner region of a zero pressure gradient turbulent boundary layer have been investigated using a two-component laser-Doppler velocimetry (LDV) system. The momentum thickness Reynolds number R_θ is in the range 400 to 1320. The wall shear stress is determined from the mean velocity gradient close to the wall, allowing scaling on wall variables to be examined unambiguously. The results indicate that, for the present R_θ range, this scaling is not appropriate.

Keywords. Low Reynolds Number, Turbulent Boundary Layer

1 Introduction

A characteristic feature of wall-bounded turbulent flows is the existence of inner and outer flow regions, without necessarily any real demarcation between the two regions. The outer region is dominated by the Reynolds shear stress, while in the inner region, the viscous stress and the Reynolds shear stress are of the same order. The inner and outer regions are generally described by the "law of the wall" and the "defect law" respectively (e.g. Sreenivasan, 1990). However, these descriptions are strictly valid at large Reynolds numbers. At low Reynolds numbers, their justification is rather tenuous since theoretical arguments require the separation between large and small scales. Yet, low Reynolds number effects, especially in the inner region, need to be assessed since they may provide further insight into the physics of the organised motion. Also, from a turbulence modelling viewpoint, it is highly desirable that models correctly mimic the low Reynolds number effects in the inner region.

Direct numerical simulation (DNS) data for both the turbulent boundary layer (Spalart, 1988) and the channel flow (Kim et al., 1987) show that there are significant low Reynolds number effects on several turbulence quantities. In particular, the data in the inner region indicate that scaling on wall variables u_τ and ν (u_τ is the friction velocity and ν is the kinematic viscosity) is, in general, not appropriate. For the channel flow, the limiting values of the ratios u'^+/y^+, v'^+/y^{+2}, w'^+/y^+ and $-\overline{u^+v^+}/y^{+3}$ (u, v, w denote the velocity fluctuations in the streamwise x, wall-normal y and spanwise z directions respectively; the prime denotes an rms value, and the superscript $+$ denotes normalization by the wall variables) as y^+ approaches zero increase by 11, 29, 29 and 36 percent respectively when h^+ (h is the channel half width) increases from 180 to 400 (Antonia and Kim, 1994). In the boundary layer, the corresponding increases are 8, 15, 19 and 16 percent when R_θ increases from 300 to 1410. The wall-normalized Reynolds stresses exhibit different Reynolds number dependences: while u'^+ has a distinct peak at about $y^+ = 15$, the peaks in v'^+, w'^+, and $-\overline{u^+v^+}$ are much broader, and their y^+ locations increase with Reynolds number. Also, the increase in v'^+, w'^+ and $-\overline{u^+v^+}$ with Reynolds number is relatively more significant (in the inner region) than u'^+.

There is as yet no fully satisfactory explanation for these effects. Spalart (1988) attributed the effect to the inactive motion (Townsend, 1961; Bradshaw, 1967). Wei and Willmarth (1989) suggested two possible reasons: an increased vortex stretching with Reynolds number and a direct interaction between inner-region structures from opposite channel walls. Antonia et al. (1992), by introducing a small amount of heat at one of the channel walls (in experiments and simulations), dismissed any direct interaction as unlikely. They also found no significant evidence for attributing low Reynolds number effects to the inactive motion. Instead, they argued that an intensification of the active motion (which contributes practically all of $-\overline{uv}$ and v') was the most likely cause for these effects. Antonia and Kim (1994) explained this in terms of an intensification of the vortices in the wall region, although this increase in vortex strength was not accompanied by any variation in vortex size or location, when the latter are expressed in wall units.

Low Reynolds number effects in a fully developed turbulent channel flow have been confirmed by LDV measurements (Wei and Willmarth, 1987) and hot-wire measurements (Antonia et al., 1992). There have been no corresponding wall-region measurements in a turbulent boundary layer. A reason for this may be the possible ambiguity associated with the measurement of u_τ in a boundary layer (in fully developed pipe and channel flows, u_τ may be inferred directly from the pressure gradient). Most turbulent boundary layer studies have focused mainly on the log and outer regions (Erm and Joubert, 1991; Antonia et al., 1990; Murlis et al., 1982; Purtell et al., 1981). The main aim of this study is to investigate low Reynolds number effects on the inner region ($y^+ < 50$) of a turbulent boundary layer by exploiting the advantages of LDV in this region (hot wire data of v are particularly unreliable near a wall, e.g Antonia, 1993). The reliability of near-wall LDV data can be ascertained from the boundary layer measurements ($R_\theta = 2400$) of Karlsson (1993). LDV allows the wall shear stress to be estimated from the mean velocity gradient close to the wall (Karlsson and Johansson, 1988; Djenidi and Antonia, 1993; Durst et al., 1993). This is particularly important when enquiring into the relevant scaling for the turbulence quantities in the inner region.

Another motivation for the present wortk is to determine the capability of LDV for measuring terms in the transport equations that are usually modelled (e.g. the turbulent energy $\overline{q^2} \equiv \overline{u^2} + \overline{v^2} + \overline{w^2}$). The impetus has been largely provided by DNS, which have yielded estimates of terms, especially those involving spatial derivatives, which have not been generally accessible by measurement (e.g. dissipation rates). It is important that the necessary experimental techniques continue to evolve in parallel with numerical simulations.

2   Experimental Details

The experiments were performed in a closed circuit constant-head vertical water tunnel (Ching et al., 1994). The vertical 2 m high working section (250 mm square cross section) is made of 20 mm thick clear perspex. One of the walls of the working section, which is removable, was used as the smooth wall. A roughness strip, which was made up of 4.5 mm high pebbles glued onto a 30 mm wide perspex strip, was recessed into a groove about 100 mm downstream from the exit of the contraction and used to trip the boundary layer. The measurement station was located about 1 m downstream of the roughness strip. Flow visualization performed by injecting dye through a hole in the wall indicated the boundary layer to be turbulent at this station for a freestream velocity, U_1, as small as 0.08 m/s. The maximum value of U_1 is about 0.5 m/s. The freestream turbulence intensity is less than 1% for all values of U_1. The pressure gradient was checked by measuring U_1 at several axial locations and found to be negligibly small ($U_1 dU_1/dx = 5.5 \times 10^{-4}$ ms^{-2} for a freestream velocity of about 0.22 m/s). Measurements were performed at eight values of R_θ in the range 400 to 1320, corresponding to a range of 0.08 to 0.45 m/s for U_1. The stability of the tunnel during the relatively long experimental times ($\approx$ 8 hours for one profile) was confirmed by continuously monitoring the mean centerline velocity using a 15 mW He-Neon one-component TSI system.

For the boundary layer measurements, a three-component fibre optic LDV system (5W Ar-Ion) was used in forward scatter mode. Only two-component measurements (for $u - v$ or $u - w$) were performed. Since refractive index effects are wavelength dependent, the two pairs of beams with the closest wavelengths (in this case, blue and violet with wavelengths 488 and 476.5 nm respectively) were used. For the $u - v$ measurements, the measuring volumes (0.04 mm $\times$ 0.5 mm for the blue beams and 0.04 m $\times$ 0.9 mm for the violet beams) had their largest dimension oriented along the spanwise direction in order to optimize the spatial resolution (in the range 0.16 to 0.8 wall units) in the wall-normal direction. In this configuration, the beam closest to the wall (used for the wall-normal component) was centered using a pair of prisms, and the probe was slightly tilted ($\approx 2°$) with respect to the z-direction in order to obtain measurements very close to the wall. For the $u - w$ measurements, the length of the measuring volume (0.5 mm) was perpendicular to the wall, resulting in a spatial resolution (in the wall-normal direction) in the range 2 to 10 wall units. The probe was traversed using a three-axis computer-controlled traversing system, with a minimum step length of 0.025 mm in all three directions.

Enhanced Burst Spectrum Analyzers (BSA) were used for processing the photo multiplier signals. The two-component measurements were made in the coincidence mode, except very close to the wall where the data rates fall off quite steeply. In the coincidence mode, the two BSAs process the electrical signals only when the two signals are within the set coincidence time interval of about 1.5 times the transit time, allowing a more reliable measurement of the Reynolds shear stress $-\overline{u^+v^+}$. Very close to the wall, operation in the coincidence mode was not feasible because of the very low data rates. In this case, in order to improve the data rates, the BSAs were operated in the private mode, whereby the signals are processed independently. Typical data rates in the outer part of the boundary layer were about 200 Hz, falling off to about 10 Hz very close to the wall. In the outer part of the boundary layer, 20000 samples were collected at each measurement point and this was reduced to 5000 samples very close to the wall.

To obtain correct values of $-\overline{uv}$, it is important that the velocity components that are being measured are associated with the same particle. Ideally, this is achieved by ensuring that the measuring volumes from the two pairs of transmitting beams overlap completely. However, because of the different sizes of the two measuring volumes and the orientation of the two pairs of beams, this is not easy to achieve in practice. It was determined that even a slight misalignment in any of the transmitting beams resulted in a large decrease in $-\overline{uv}$ (by as much as 35%). To ensure good overlap, alignment of the two measuring volumes was performed after taking into account the refractions of the transmitting beams due to the perspex and the water. The probe was set up in the same configuration as that used in the experiments, and a water filled container fabricated from the same thickness perspex as the test section was placed in front of the probe (this was necessary because of the lack of access to the test section itself). A 30 μm pinhole was placed in the water at the focusing point of the receiving optics, which was determined by guiding part of the laser light through the receiving fibre. The four transmitting beams were then adjusted so that their beam waists intersected at this point.

3   Determination of u_τ and Mean Velocity Profiles

To examine the inner scaling laws unambiguously, an accurate estimate of the friction velocity u_τ is of paramount importance. At low R_θ, there is no rigorous basis for the log region, and the Clauser-plot technique would therefore be inaccurate for determining u_τ. The measurement of the mean velocity gradient, $d\overline{U}/dy$, at the wall should provide a more reliable estimate of u_τ. Djenidi and Antonia (1993) determined u_τ, with an estimated uncertainty of $\pm 3\%$, by fitting a straight line to their near-wall LDV data for $y^+ < 2.5$. This latter choice was based on the observation that the value of $d\overline{U}^+/dy^+$ (from the DNS data of Kim et al., 1987 and Spalart, 1988) decreases to about 0.97, i.e. 3% below the wall value, at $y^+ \approx 2.6$.

For all R_θ, it was possible to obtain sufficient data in the region $y^+ < 2.5$ to allow u_τ to be estimated using the method of Djenidi and Antonia (1993). Close to the wall (roughly $y^+ < 12$), y-steps of 0.025 mm ($\Delta y^+ = 0.5$ for $R_\theta = 1316$) were used; this ensured that there were at least four data points for the linear fit. An initial estimate of u_τ was obtained by fitting a least squares straight line to the near-wall data for $\overline{U}$. The origin for the velocity profile was estimated by extrapolation of this line to $\overline{U} = 0$; this location was always within ± 0.075 mm of the initial estimate of the origin (inferred by visual inspection of the intersection of the laser beams as they approached the perspex wall and of the analogue signals from the BSAs). From the initial estimate of u_τ, the physical values of y were converted to y^+ and a new linear fit was applied to all the data in the region $y^+ < 2.5$. The second (and final) estimate of u_τ was typically 5% larger than the first estimate; further iterations showed no change in the value of u_τ.

The error due to the velocity bias (McLaughlin and Tiederman, 1973) was corrected by weighting the individual velocity realizations with their arrival times. An additional source of error is that due to the non-uniformity of the mean velocity distribution within the measuring volume. Durst et al. (1992) showed that, in the case of $\overline{U}$, this error is proportional to the square of the diameter of the measuring volume and to the second derivative $d^2\overline{U}/dy^2$. For the present beam configuration, the correction for this effect on $\overline{U}$ is less than 0.1 percent. The corrections for the rms quantities are also negligible. Very close to the wall, truncation of the measuring volume by the wall can introduce an additional error (Kried, 1974). Djenidi and Antonia (1993) determined that this was not significant for the present set-up.

Although measurements were performed at eight different values of R_θ, results are presented only for three representative Reynolds numbers (for clarity) in the region $y^+ < 50$ to highlight the effects in this region. The normalized mean velocity profiles (Fig. 1) reflect the Reynolds number dependence displayed by the DNS data; a decrease in $\overline{U}^+$ with increasing R_θ is discernible down to about $y^+ = 10$. This is also evident in the profiles of

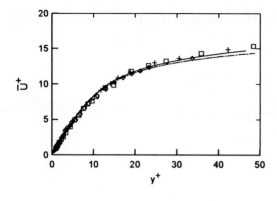

Figure 1: Mean velocity. $\diamond$, $R_\theta = 1316$; $+$, 765; $\square$, 400; — $\cdot$ —, DNS 1410; ——, 300.

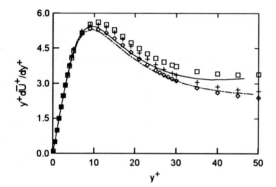

Figure 2: Distribution of $(y^+ d\overline{U}^+/dy^+)$. Symbols and lines as for Fig. 1.

$(y^+ d\overline{U}^+/dy^+)$ (Fig. 2), which accentuates the R_θ effects on $\overline{U}^+$ in the inner region. This figure suggests that the effect of R_θ on $\overline{U}^+$ extends into the sublayer. Note that the derivative $d\overline{U}^+/dy^+$ was determined after first applying a curve fit to the $\overline{U}^+$ data using the method of Bisset and Antonia (1991); the assumptions on which the calculation is based are not of particular relevance in the present context since the main interest here is to obtain a reliable fit to $\overline{U}^+$ in the inner region in order to estimate $d\overline{U}^+/dy^+$.

4 Turbulence Measurements

One measure of the uncertainty in the Reynolds stress measurements in the near-wall region was provided by the scatter inferred by repeating measurements at the same y^+ location a number of times. As the wall is approached, there is a marked decrease in the signal-to-noise ratio because of the large amount of extraneous light (reflected from the wall) collected by the receiving optics and the decrease in the velocity itself. The uncertainties (95% confidence) in u'^+, v'^+ and $-\overline{u^+v^+}$ were estimated to be $\pm4\%$, $\pm9\%$ and $\pm12\%$

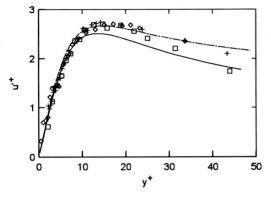

Figure 3: rms longitudinal turbulence intensity. Symbols and lines as for Fig. 1.

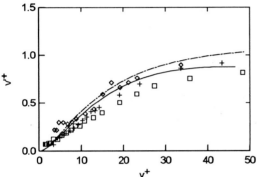

Figure 4: rms normal turbulence intensity. Symbols and lines as for Fig. 1.

respectively.

The u'^+ profiles (Fig. 3) display a discernible peak at about $y^+ = 15$. This location is consistent with that found for the DNS data of Spalart (1988), but the nearly constant peak magnitude of about 2.75 is at variance with

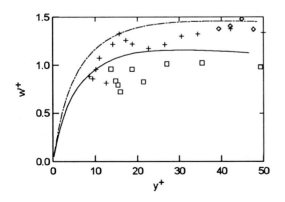

Figure 5: rms spanwise turbulence intensity. Symbols and lines as for Fig. 1.

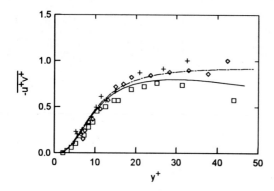

Figure 6: Reynolds shear stress. Symbols and lines as for Fig. 1.

the systematic increase observed in the DNS data. While an increase in u'^+ with R_θ is clearly discernible in the region $y^+ > 15$, it is difficult to discern a Reynolds number effect from this figure in the region $y^+ < 15$. The DNS data show only a small increase in u'^+ with R_θ in this region, and it is likely that the experimental uncertainty masks any R_θ effect for the u'^+ measurements. The present measurements for v'^+ (in the inner region) are slightly lower than the DNS data (Fig. 4). Nonetheless, like the DNS data, a systematic increase in v'^+ with R_θ is observed well into the sublayer. In contrast to u'^+, the peak in v'^+ (notwithstanding the difficulty of defining this peak) is much broader, and the magnitude and y^+ location of v'^+_{max} increase with R_θ (this is not seen in Fig. 4 since the peak locations occur in the range $60 < y^+ < 130$). The profiles of w'^+ (Fig. 5) also show a systematic increase with R_θ in the inner region. Relative to u'^+ and v'^+, the increase in w'^+ with R_θ is much larger. However, the orientation of the measuring volume (recall that for the w'^+ measurements, the length of the measuring volume was perpendicular to the wall) did not allow w'^+ measurements very close to the wall. The large scatter in w'^+ (relative to u'^+ and v'^+) can be attributed to the increased noise associated with the refraction of the laser beams through the wall (this is far more intense than the light scattered from the measuring volume). Distributions of the Reynolds shear stress $-\overline{u^+v^+}$ are in good agreement with the DNS data (Fig. 6), and the increase in $-\overline{u^+v^+}$ with R_θ extends into the sublayer.

The skewness (S) and flatness (F) factors of u (Figs. 7 and 8) are generally in agreement with the DNS data. Both S_u and F_u increase as the wall is approached, and the local DNS peak at $y^+ \approx 2$ appears to be captured by the measurements. The scatter in the data makes it difficult to comment on the behaviour of S_v and F_v (Figs. 9 and 10) in the inner region. At the two lower Reynolds numbers, F_v deviates significantly from the DNS data in the near-wall region. Instead of increasing as y^+ goes to zero, F_v reaches a maximum at about $y^+ = 10$ before decreasing towards the wall. Durst et al. (1993) observed a similar trend for their F_v data in a turbulent pipe

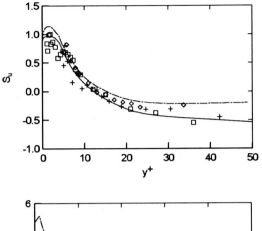

Figure 7: Skewness of the longitudinal velocity fluctuation. Symbols and lines as for Fig. 1.

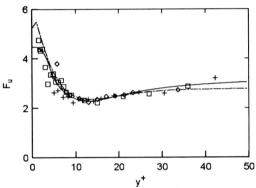

Figure 8: Flatness of the longitudinal velocity fluctuation. Symbols and lines as for Fig. 1.

flow. This seems to reflect the inability of the LDV technique to resolve the limiting behaviour of v as $y^+ \to 0$. The primary difficulty appears to be the increase in the noise due to the scattering of light from the wall. This is reflected in Fig. 11, where the limiting behaviours of u'^+/y^+, v'^+/y^{+2} and

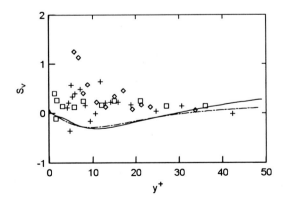

Figure 9: Skewness of the normal velocity fluctuation. Symbols and lines as for Fig. 1.

12

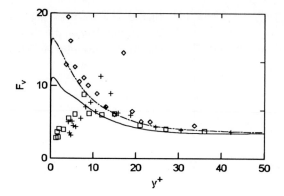

Figure 10: Flatness of the normal velocity fluctuation. Symbols and lines as for Fig. 1.

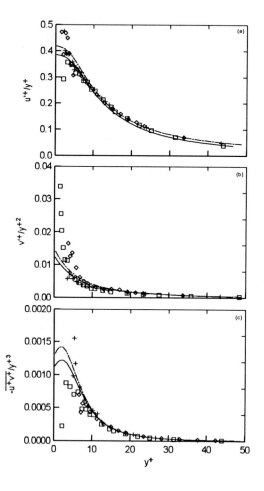

Figure 11: Limiting behaviour of turbulence intensities and Reynolds shear stress. Symbols and lines as for Fig. 1.

$-\overline{u^+v^+}/y^{+3}$ deviate significantly from the DNS data. This is not unique to the present experiments and certainly requires further study. We are currently investigating methods of enhancing the signal-to-noise ratio in the near-wall region by improving the surface finish of the wall to reduce speckle formation and using anti-reflective wall coatings. Durst et al. (1994) also suggest using a refractive index matched test section and operating with a low frequency shift.

5   Conclusions

The present LDV investigation indicates that there are significant low Reynolds number effects on the mean velocity and Reynolds stresses in the inner region of a turbulent boundary layer. The effect extends well into the sublayer, indicating that, in the range $400 < R_\theta < 1320$, scaling on wall variables is, in general, not appropriate. This agrees with Spalart's (1988) conclusions on the basis of a direct numerical simulation of this flow. It also agrees with the experimental results of Wei and Willmarth (1989) and Antonia et al. (1992) in a fully developed turbulent channel flow.

Although there is little doubt that the LDV technique represents a considerable improvement over the hot-wire technique in connection with the measurement of the Reynolds stresses (especially $\overline{v^2}$) in the near-wall region, the limiting behaviour of the stresses does not seem to be reliably captured by the LDV data. This is particularly true for $\overline{v^2}$, although the results for $\overline{u^2}$ are encouraging. It is important that the correct limiting wall behaviour is achieved as this would further increase our confidence in the use of LDV.

Acknowledgements

The support of the Australian Research Council is gratefully acknowledged.

References

Antonia, R. A., 1993 : Direct Numerical Simulations and Hot Wire Experiments: A Possible Way Ahead?. In *New Approaches and Concepts in Turbulence*, (eds. T. Dracos and A. Tsinober), Birkhäuser Verlag, Basel, Switzerland, 349-365.

Antonia, R. A., Bisset, D. K. and Browne, L. W. B., 1990 : Effect of Reynolds Number on the Topology of the Organized Motion in a Turbulent Boundary Layer, *J. Fluid Mech.*, **213**, 267-286.

Antonia, R. A. and Kim, J., 1993 : Low Reynolds Number Effects on Near-Wall Turbulence, *J. Fluid Mech*, **276**, 61-80.

Antonia , R. A., Teitel, M., Kim, J. and Browne, L. W. B., 1992 : Low-Reynolds-Number Effects in a Fully Developed Turbulent Channel Flow, *J. Fluid Mech.*, **236**, 579-605.

Bisset, D. K. and Antonia, R. A., 1991 : Mean Velocity and Reynolds Shear Stress in a Turbulent Boundary Layer at Low Reynolds Numbers, *Aeron. Quart.*, **95**, 244-247.

Bradshaw, P., 1967 : 'Inactive' Motion and Pressure Fluctuations in Turbulent Boundary Layers, *J. Fluid Mech.*, **30**, 241-258.

Ching, C. Y., Djenidi, L. and Antonia, R. A., 1994 : Low Reynolds Number Effects in a Turbulent Boundary Layer, *Proc. Seventh International Symposium on Applications of Laser Techniques to Fluid Mechanics*, Lisbon, Portugal, 4.1.1-4.1.8.

Djenidi, L. and Antonia, R. A., 1993 : LDA Measurements in a Low Reynolds Number Turbulent Boundary Layer, *Expts. in Fluids*, **14**, 280-283.

Durst, F., Jovanovic, J. and Sender, J., 1993 : Detailed Measurements of the Near Wall Region of Turbulent Pipe Flows, *Ninth Symposium on Turbulent Shear Flows*, Kyoto, Japan, 2.2.1-2.2.6.

Durst, F., Jovanovic, J. and Sender, J., 1994 : LDA Measurements in the Near-Wall Region of a Turbulent Pipe Flow (personal communication).

Durst, F., Martinuzzi, R., Sender, J. and Thevenin, D., 1992 : LDA Measurements of Mean Velocity, RMS-Values and Higher Order Moments of Turbulence Intensity Fluctuations in Flow Fields with Strong Velocity Gradients, *Sixth International Symposium on Applications of Laser Techniques to Fluid Mechanics*, Lisbon, Portugal, 5.1.1-5.1.6.

Erm, L. and Joubert, P. N., 1991 : Low Reynolds Number Turbulent Boundary Layers, *J. Fluid Mech.*, **230**, 1-44.

Karlsson, R. I., 1993 : Near-Wall Measurements of Turbulence Structure in Boundary Layers and Wall Jets. In *Near-Wall Turbulent Flows*, (eds. R. M. C. So, C. G. Speziale and B. E. Launder) Elsevier Science Publishers, 423-432.

Karlsson, R. I. and Johansson, T. G., 1988 : LDV Measurements of Higher Order Moments of Velocity Fluctuations in a Turbulent Boundary Layer. In *Laser Anemometry in Fluid Mechanics III* (eds. R. J. Adrian et al.), Ladoan - Instituto Superior Tecnico, Portugal, 273-289.

Kim, J., Moin, P. and Moser, R., 1987 : Turbulence Statistics in Fully Developed Channel Flow at Low Reynolds Number, *J. Fluid Mech.*, **177**, 133-166.

Kried, D. K., 1974 : Laser Doppler Velocimeter Measurements in Nonuniform Flow: Error Estimates, *Appl. Optics*, **13**, 1872-1881.

McLaughlin, D. K. and Tiederman, W. G., 1973 : Biasing Correction for Individual Realization of Laser Anemometer Measurements in Turbulent Flows, *Phys. Fluids*, **16**, 2082-2088.

Purtell, L. P., Klebanoff, P. S. and Buckley, F. T., 1981 : Turbulent Boundary Layers at Low Reynolds Numbers, *Phys. Fluids*, **24**, 802-811.

Spalart, P. R., 1988 : Direct Simulation of a Turbulent Boundary Layer up to $R_\theta = 1410$, *J. Fluid Mech.*, **187**, 61-98.

Sreenivasan, K. R., 1990 : The Turbulent Boundary Layer. In *Frontiers in Experimental Fluid Mechanics* (ed. M. Gad-el-Hak), Berlin, Springer-Verlag, 159-210.

Townsend, A. A., 1961 : Equilibrium Layers and Wall Turbulence, *J. Fluid Mech.*, **11**, 97-120.

Wei, T. and Willmarth, W. W., 1989 : Reynolds-Number Effects on the Structure of a Turbulent Channel Flow, *J. Fluid Mech.*, **204**, 57-95.

CIRCULAR COUETTE FLOW AND TAYLOR VORTICES IN SHEAR-THINNING LIQUIDS

M P Escudier, I W Gouldson and D M Jones

University of Liverpool, Department of Mechanical Engineering,
P O Box 147, Liverpool, L69 3BX, England.

Abstract The flow structure in a concentric annular geometry with a large aspect ratio and a radius ratio of 0.506 has been investigated for inner cylinder (centrebody) rotation for one Newtonian and two non-Newtonian liquids: glucose, Xanthan gum and a blend of Laponite and CMC. Tangential velocity distributions have been measured for circular Couette flow (i.e. sub-critical Taylor numbers) and found to be in excellent agreement with theory. Tangential and axial velocity components have been measured for high-Taylor numbers to reveal the internal structure of the Taylor vortices. The axial-velocity data have been used to construct streamline plots of the radial-axial flowfields. The shear-thinning nature of the two non-Newtonian fluids is shown to have a strong influence on both the Couette flow and also the Taylor cell structure. Axial drift of the Taylor cells for the two non-Newtonian liquids is attributed to differences in their rheological characteristics (viscoelasticity versus thixotropy).

1.   Introduction

The cellular motion that develops in the fluid contained between two concentric cylinders when the rotation speed exceeds a critical value, strongly dependent upon the radius ratio, was discovered and analysed theoretically and experimentally by Taylor in 1923. The review of Stuart (1986) lists nearly a hundred subsequent papers concerned in one way and another with Taylor-vortex flows whilst Koschmieder (1993) puts the total at about three hundred. Although some quite remarkable experiments have been reported, starting with those of Taylor himself and including the definitive work of Donnelly (1958) and Coles (1965), almost all have been limited to either torque measurements or visualisation of flow structures through the outer cylinder wall with considerable emphasis on transitions between flow states rather than the internal details of a particular flow state. Hot-film, laser Doppler anemometer and electrochemical current measurements of power spectra have also been employed to detect departures from the axisymmetric structure of the primary Taylor state to increasingly complex wavy vortices as the Taylor number is increased. In view of the interest in Taylor vortex flow which has continued for more than 70 years, it seems remarkable that only limited measurements of the internal structure of Taylor vortices have been reported hitherto.

So far as non-Newtonian fluids are concerned, the Taylor-vortex literature is very limited. Recent theoretical work includes the finite-element study of the stability of inelastic non-Newtonian fluids in Couette flow by Lockett et al (1992) based upon the far more extensive numerical work of Lockett (1992) which also includes the influence of an imposed axial flow and eccentricity. The latter topics are also being investigated experimentally in the wider research programme of which this study of Taylor vortices is a part (Escudier et al 1994a, Escudier and Gouldson 1994). Previous experimental work on non-Newtonian liquids has been limited to either detecting the critical rotation speed for the onset of Taylor motion or using the critical speed to determine rheological properties.

As was Lockett's (1992) work, the present study was motivated by the need for a more complete understanding of the flow of drilling fluids in the annulus created between drillpipe and wellbore during oil- and gas-well drilling operations. The selection of the two non-Newtonian fluids used for the work described here was strongly influenced by the viscometric characteristics of a typical drilling fluid: shear thinning, thixotropic, viscoelastic and formulated to gel below a critical shear (yield) stress. To permit the use of laser Doppler anemometry for the detailed flow measurements, it was essential that the fluids were also optically transparent. Of the two fluids selected, one, Xanthan gum, is shear thinning and elastic whilst the other, a Laponite/CMC blend, is shear thinning with a low yield stress and is also thixotropic. A second aspect of the present work which was influenced by the relevance to drilling situations is the geometry of the apparatus. The radius ratio is 0.506 and the aspect ratio (i.e. annulus length/gap width) is 233:1, which is considerably higher than any previous wide-gap Taylor-vortex apparatus. Such a high aspect ratio was essential since the main aim of the research programme was the investigation of fully developed flow through the annulus.

The present measurements reveal the influence of fluid rheology on circular Couette flow and on the internal structure of Taylor cells. Variations in the internal vortex structure with increasing Taylor number are investigated by determining the changes in the radial distributions of the maximum and minimum axial velocities within a Taylor cell.

2.   Experimental Rig and Instrumentation

The annular test section comprises six precision-bore borosilicate glass tubes (ID 100.4 ± 0.1mm) with a 50.8mm diameter thin-wall stainless steel inner tube giving a radius ratio of 0.506. There are five modules each of 1.027m length and one of 0.64m, which gives an overall length of 5.775m and a length-to-hydraulic diameter ratio of 116 (i.e. an aspect ratio of 233). The centrebody rotation speed may be infinitely varied up to a maximum of 126rpm, the speed being measured by means of a slotted disk and optical encoder arrangement giving a resolution of 0.1 rpm. Detailed measurements were made at a location 600 mm (24 gap widths) from the endhousing furthest away from the centrebody drive. A more complete description of the flow facility is given by Escudier et al (1994a).

Flow velocities were determined using a Dantec Fibreflow laser Doppler anemometer (LDA) system comprising of a 60X10 probe and 55X12 beam expander in conjunction with a Dantec BSA 57N10 Burst Spectrum Analyzer signal processor and Hewlett Packard 286/12 microcomputer. The length of the principal axis of the LDA measurement volume is estimated as 0.193 mm in water. The traversing system was controlled by a microcomputer (IBM XT PS2 model 30) and had a spatial resolution of 0.015mm. A flat-faced optical box containing castor oil (R_n = 1.478) was positioned over the pipe at the measurement location to minimise refraction of the laser beams.

Rheological measurements of the non-Newtonian liquids were made on a CarriMed controlled-stress rheometer (CSL 100) using a combination of cone-and-plate and parallel-plate geometries and employing CarriMed's flow equilibrium software for the Laponite/CMC blend. Fluid refractive indices were determined using an ABBE 60/ED high accuracy refractometer.

3.   Test Fluid Preparation and Rheology

The first (purely polymeric) non-Newtonian fluid was an aqueous solution of food-grade Xanthan gum (Keltrol TF supplied by the Kelco Division of Merck and Co Inc) and the second a blend of Laponite RD (supplied by Laporte Inorganics), a synthetic clay, and high-viscosity carboxymethylcellulose (CMC), sodium salt (supplied by BDH Laboratory Supplies), also a polymer. For comparison purposes, a series of experiments was carried out for a representative Newtonian fluid, a mixture of glucose syrup and water.

A working quantity of each fluid was prepared by filtering tap water prior to the addition of the glucose or polymer and, for the blend, the Laponite powder. To retard bacteriological degradation, 80-200 ppm of formaldehyde was added. Seeding particles (Timiron MP-1005, mean diameter ~20 μm) at a concentration of 1ppm were added to improve the LDA signal/noise ratio and the data rate.

Viscometric measurements were made on the controlled-stress rheometer at temperatures bracketing those observed in the experimental rig and the viscosity values used in data reduction arrived at by linear interpolation. The rheological data given in this section are representative: Complete details of the fluids used are given by Escudier et al (1994b).

The Newtonian fluid used was a 2:1 w/w glucose syrup (Cerestar) and water mixture with an initial viscosity μ = 0.065 Pa.s at 20°C. Due to water evaporation, the viscosity gradually increased to a value of 0.08 Pa.s with no measurable change in either density or refractive index, which were 1,265 kg/m^3 and 1.436, respectively.

The purely polymeric non-Newtonian fluid was a 0.15% aqueous solution of Xanthan gum, for which the viscosity data (**Figure 1**) are well represented by the Sisko model:

$$\mu = \mu_{ref}(\lambda_S \dot{\gamma})^{n-1} + \mu_\infty$$

with λ_S = 11.0 s, n = 0.437, μ_∞ = 0.001 Pa.s and μ_{ref} = 1 Pa.s. For shear rates between about 1 and 1000 s^{-1}, Xanthan gum (at the concentration used here) essentially behaves as a slightly elastic power-law fluid.

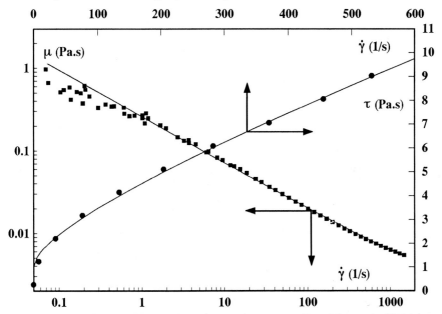

Figure 1. Viscometric characteristics of Xanthan gum and Laponite/CMC.

The second non-Newtonian liquid was a blend of aqueous solutions of 0.5% Laponite and 0.07% CMC which produces a thixotropic, shear-thinning fluid with a low yield stress (**Figure 1**). In spite of the known difficulties of characterizing thixotropic fluids (Alderman et al, 1988), the CarriMed flow-equilibrium procedure, whereby a shear stress is maintained and the shear rate monitored until steady-state conditions are achieved, led to excellent reproducibility of viscometric data for this fluid. The data are well modelled over the shear-rate range considered by the Casson equation:

$$\tau^{1/2} = \tau_y^{1/2} + \tau_{ref}^{1/2} \, (\lambda_C \dot{\gamma})^{1/2}$$

with $\tau_y = 0.99$ Pa, $\lambda_C = 7.55$ ms and $\tau_{ref} = 1$ Pa.

Curves of viscosity versus time for the Laponite/CMC blend reveal timescales for both buildup and breakdown of fluid structure of about 3,000 s, so that the time to establish an equilibrium flow for this fluid is very long.

4. Experimental Results

4.1 Glucose

For the glucose solution, monitoring the development of an axial velocity component at a fixed radial location close to the centrebody ($\xi = 0.8$) as the centrebody rotation rate was progressively increased proved to be a very sensitive and repeatable indicator for the onset of Taylor flow: the results of two separate runs are shown in **Figure 2**. For a radius ratio of 0.506, a critical Taylor number of 4,500 would be consistent with the experiments of Donnelly (1958) and with the theoretical work of several workers listed by Di Prima and Swinney (1981). The lower value of 3,700 found here is thought to be a consequence of slight departures from straightness of the centrebody geometry.

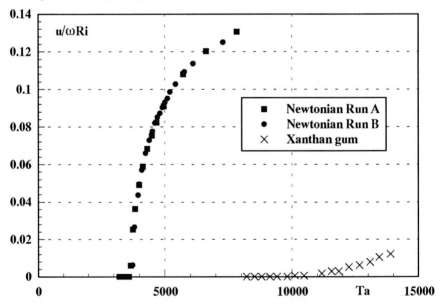

Figure 2. Axial velocity at $\xi = 0.8$ versus Taylor number.

Once the transition point had been identified, tangential velocity distributions were measured at sub-critical Taylor numbers of 430 and 2900 and, as seen from **Figure 3** found to be in excellent agreement with the theoretical profile for circular Couette flow of a Newtonian fluid

$$\text{i.e. } wr = \frac{\omega R_i^2}{1 - \kappa^2} \left[1 - \left[\frac{r}{R_o} \right]^2 \right]$$

wherein κ is the radius ratio R_i/R_o.

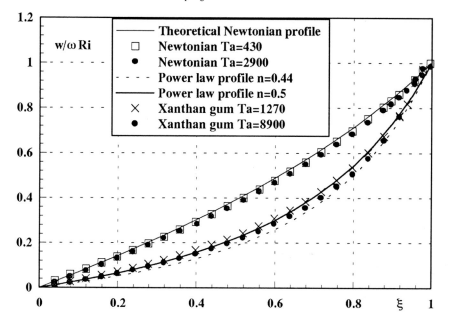

Figure 3. Tangential velocity profiles for Newtonian fluid and Xanthan gum.

A detailed mapping of the axial and tangential velocity components over an adjacent pair of counter-rotating vortices was carried out for a supercritical Taylor number of 15,400 which is well below the second critical Taylor number at which the axisymmetric Taylor vortex flow becomes unstable: for a radius ratio of 0.5, Snyder and Lambert (1966) put the second critical value at at least an order of magnitude higher than the first. The mapping was performed as a succession of traverses along the axis of the geometry at fixed radial locations. For the glucose solution, the vortices were found to be stable over the timescale of the experiment and to exhibit no drift along the axis of the geometry. The axial-velocity data are presented in **Figure 4** with a schematic diagram indicating the sense of axial-radial circulation within the vortices. The tangential velocity component was found to exhibit an axial periodicity of the same frequency as the axial velocity component but to lag the axial component by 90° (i.e. the maximum and minimum tangential velocities correspond to zero axial velocity). From the radial distributions of the tangential velocity at a series of fixed axial locations (**Figure 5**), obtained by interpolation of the measurements at fixed ξ, it is seen that the tangential velocity at any radial location is highest at the boundary between adjacent Taylor cells where there is radial outflow ($\zeta = 1$) and lowest on the inflow boundary ($\zeta = 0$). This observation is consistent with considerations of angular-momentum transport with reference to the base circular Couette flow: In contrast with a potential vortex, fluid transported radially outward carries an excess of angular momentum leading to a velocity overshoot, whilst the reverse is true for fluid transported radially inward. It may also be remarked that at any radial location the departure of the tangential velocity from the theoretical Couette-flow variation generally exceeds the axial velocity at that location: it is clearly inappropriate to regard the Taylor vortex motion as a minor perturbation of the underlying Couette flow.

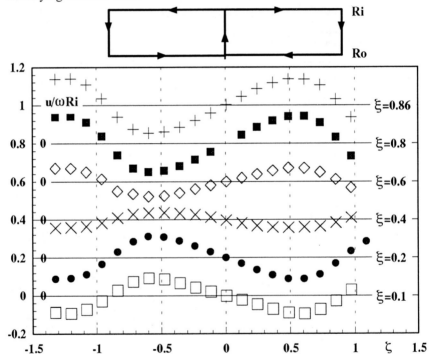

Figure 4. Axial velocity component within Taylor cells for glucose solution at a Taylor number of 15400. The schematic diagram shows the sense of the radial/axial circulation.

The overall structure of the motion within a Taylor cell is most clearly revealed by the streamlines (**Figure 6**) computed from the measurements shown in **Figure 5**. The axial asymmetry, with the eye of the vortex closest to the

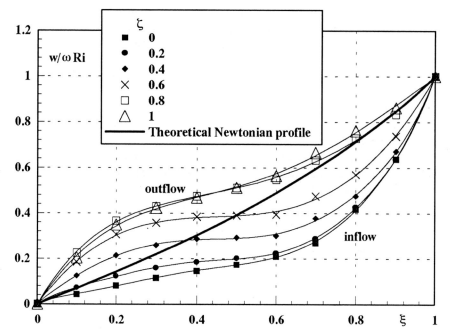

Figure 5. Tangential velocity distributions within a Taylor cell for glucose (Ta=15400).

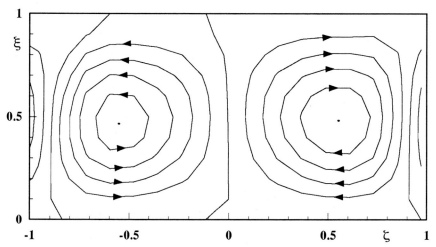

Figure 6. Streamline patterns within Taylor cells for glucose solution at a Taylor number of 15400 (the stream function Ψ varies from 0 on the boundaries to a maximum of 1 in the cell interior).

radial outflow boundary, would seem to be consistent with the asymmetry in the tangential velocities commented on above. No measurements of comparable detail have been reported to date, but there is clear evidence of asymmetry in the measurements of Pfister and Gerdts (1981) and to a lesser degree in the data reported by Kusnetsov et al (1977).

The maximum axial velocity components at Taylor numbers of 4,460, 15,400 and 25,100, presented in **Figure** 7, confirm that both velocities and velocity gradients are higher in the vicinity of the centrebody than at the outer wall of the annulus, and that with increasing Taylor number the position of zero velocity (the eye of the vortex) moves towards the outer wall.

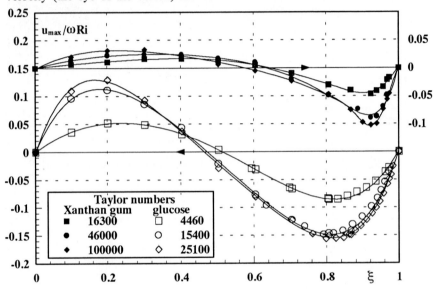

Figure 7. Radial variation of maximum axial velocity within a Taylor cell for glucose and Xanthan gum.

4.2   Xanthan Gum

Since Xanthan gum may be considered to be essentially a power-law fluid, the Taylor number was defined as suggested by Sinevic et al (1986) with the effective viscosity determined from the theoretical shear rate prevailing at the rotating surface for circular Couette flow of a power-law fluid, i.e. $\dot{\gamma}_C = (2\omega/n)$ $[1 - \kappa^{2/n}]^{-1}$ with values of n in the range 0.44 to 0.47 determined from the viscometric data.

The critical Taylor number was again determined by monitoring the development of an axial velocity component at a fixed radius with changing centrebody rotation rate. The data in **Figure 2** reveal that the onset of Taylor flow is much more gradual than was the case for glucose. For a shear-thinning fluid the apparent viscosity is evidently lowest close to the centrebody and it is entirely plausible that small disturbances initiated at the centrebody are suppressed in the higher viscosity fluid a short distance away. Nevertheless, our observations are at variance with the conclusion of Green and Jones (1982) who found that the increase in torque which accompanies the onset of Taylor flow was just as abrupt for Xanthan gum and polyacrylamide as for a Newtonian fluid. From the present work it may be concluded that the critical Taylor number for 0.15% Xanthan gum is about 10,000.

The two tangential-velocity profiles measured in circular Couette flow for Xanthan gum (**Figure 3**) correspond to sub-critical Taylor numbers of 1,270 and 8,900. Agreement with the theoretical profile appropriate for a power-law fluid of index 0.44 is good, although the data for the lower Taylor number fall slightly closer to the curve for n = 0.5. In this case, for circular Couette flow

$$w = \frac{\omega\, r}{1-\kappa^{\frac{2}{n}}} \left[\left(\frac{R_i}{r}\right)^{\frac{2}{n}} - \kappa^{\frac{2}{n}} \right]$$

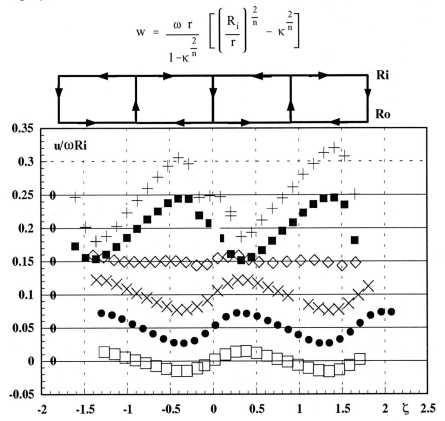

Figure 8. Axial velocity component within Taylor cells for Xanthan gum (Ta=46000). ξ values as for Figure 5.

26

Mappings of the Taylor vortex structure in the axial and tangential planes were conducted at a supercritical Taylor number of 46,000 after a stabilization time of about 240 minutes. It became apparent during the mapping that the vortices were drifting at approximately 3-4 μm/s along the axis of the geometry opposite to the direction of the centrebody rotation vector. As for the Newtonian case, the tangential velocity component was found to exhibit a periodicity of the same frequency as the axial velocity component (**Figure 8**) but to lag the axial component by 90°. The tangential velocity distributions (**Figure 9**) show qualitatively the same characteristics as for glucose but with smaller departures from the base Couette flow as represented by the theoretical power-law profile for n = 0.44. It should be noted that for both Xanthan gum and Laponite/CMC, $\zeta = 0$ corresponds to the outflow boundary whereas for glucose it corresponds to the inflow boundary. This difference is a consequence of the arbitrary location of the origin for the axial location (i.e. the vortex cells select their own axial location).

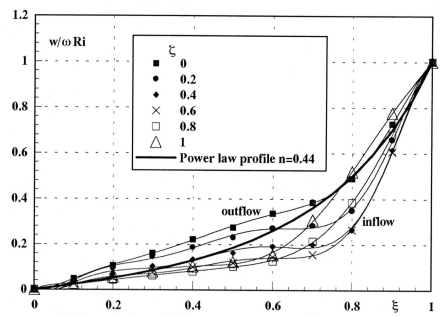

Figure 9. Tangential velocity distributions within a Taylor cell for Xanthan gum (Ta=46000).

Streamline patterns, (**Figure 10**) determined from the axial velocity measurements again reveal the overall structure of the interior motion. The essential difference compared with the observations for glucose is the radial shift of the vortex eye towards the centrebody coupled with an axial shift towards the radial outflow boundary, i.e. in the direction of higher shear stress and lower viscosity. The maximum axial velocity components, also measured at two additional Taylor numbers (**Figure** 7), confirm the increased radial asymmetry compared with the Newtonian fluid with significantly higher velocities and velocity gradients in the near centrebody region and also the lower amplitude of the motion compared with glucose.

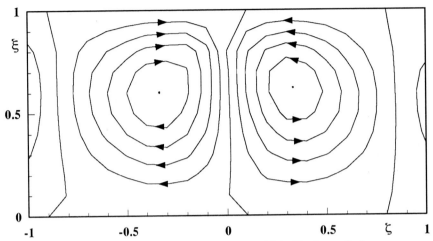

Figure 10. Streamline patterns within Taylor cells for Xanthan gum (Ta=46000).

4.3   Laponite/CMC

Given the long equilibrium times associated with the thixotropic character of the Laponite/CMC blend, it was not practical to investigate the precise onset of Taylor flow by progressively increasing the rotation speed. Three profiles were measured in circular Couette flow together with one detailed vortex mapping at the highest possible centrebody rotation rate (126 rpm). As for Xanthan gum, the Taylor number has been defined by evaluating the apparent viscosity at the centrebody surface at a shear rate $\dot{\gamma}_C$ corresponding to circular Couette flow using the Casson model for the fluid rheology.   At low rotation rates, the theoretical velocity variation is given by

$$w = \omega R_i \alpha \left[\frac{1}{2} \frac{R_i}{r} \left[\frac{1}{\kappa_y} - 3 \frac{r}{R_i} \right] \left[\frac{1}{\kappa_y} - \frac{r}{R_i} \right] - \frac{r}{R_i} \ln \left[\frac{1}{\kappa_y} \frac{r}{R_i} \right] \right]$$

where $\alpha = \tau_y/(\omega \tau_{ref} \lambda_C)$ and κ_y ($\equiv R_i/R_y$) is the location of the yield surface given by the condition $w = \omega R_i$ at $r = R_i$. The corresponding value of $\dot{\gamma}_c$ is $\alpha\omega$ (1 - $1/\kappa_y)^2$. At higher rotation rates where the fluid is ungelled ($\tau > \tau_y$) throughout the annulus, the velocity variation for Couette flow is given by

$$w = \alpha\omega r \left[\frac{\sigma^2}{2} \left[\frac{R_i^2}{r^2} - \kappa^2 \right] - 2\sigma \left[\frac{R_i}{r} - \kappa \right] - \ln \left[\kappa \frac{r}{R_i} \right] \right]$$

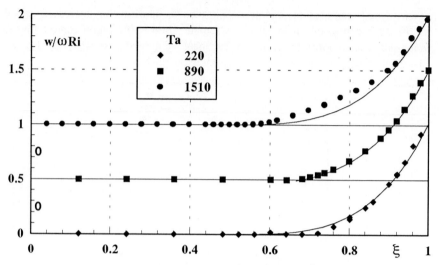

Figure 11. Tangential velocity distributions for Laponite/CMC.

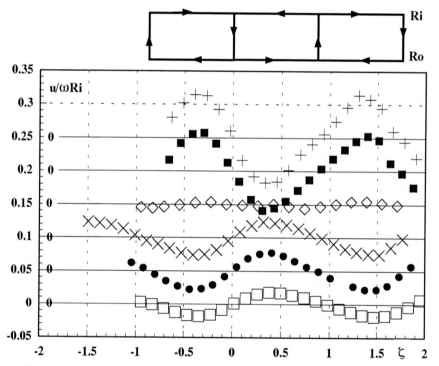

Figure 12. Axial velocity component within Taylor cells for Laponite/CMC (Ta=54000). ξ values as for Figure 5.

where the non-dimensional torque σ is given by the condition $w = \omega R_i$ at $r = R_i$ and the corresponding value of $\dot{\gamma}_C$ is now $\alpha\omega\,(\sigma - 1)^2$. The three velocity distributions in **Figure 11** for Ta = 220, 890 and 1510 are well represented by the first of the two equations given above with flow restricted to the inner region of the annulus, stationary fluid in the outer annular region and the yield surface located progressively closer to the outer wall as the centrebody rotation rate increases.

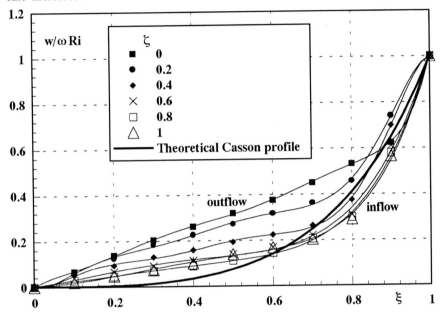

Figure 13. Tangential velocity distributions within a Taylor vortex for Laponite/CMC (Ta=54000).

The vortex mapping conducted at a Taylor number of 54,000 (**Figures 12, 13**) shows similar trends to those seen previously for the Newtonian fluid and Xanthan Gum, with no region of gelled fluid in evidence. The detailed differences in the tangential velocity distributions for the two non-Newtonian liquids are attributed to the much longer timescales required for the fluid structure to change for Laponite/CMC compared with Xanthan gum. Again the vortices were found to exhibit a slow drift (3-4 μm/s) along the axis of the geometry but now in the direction of the centrebody rotation vector. This contrast with the behaviour for Xanthan Gum suggests the drift is a consequence of the differences in fluid rheology, since all other influences were the same for the two fluids. The streamline patterns (**Figure 14**) are remarkably similar to those for Xanthan gum suggesting that the global structure of the Taylor vortices is largely determined by the shear-thinning characteristics of these two fluids and not significantly influenced by either elasticity or thixotropy.

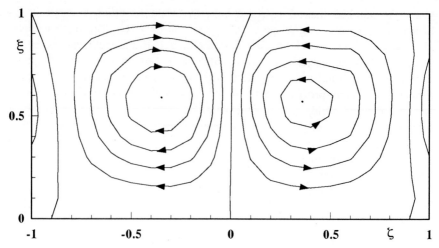

Figure 14. Streamline patterns within Taylor vortices for Laponite/CMC (Ta=54000).

5.   Conclusions

For circular Couette flow of a Newtonian fluid, the measured tangential velocity distribution is in excellent agreement with the theoretical prediction and the transition from Couette flow to a flow exhibiting Taylor vortices is clearly identified by an abrupt increase from zero in the axial velocity. Detailed velocity mapping of a vortex pair showed the tangential velocity to exhibit a periodic structure of the same frequency as the axial component but out of phase by 90°. Higher values of the maximum axial velocity and velocity gradient are observed closer to the centrebody than the outer wall with the eye of the vortex moving towards the outer wall with increasing Taylor number. Streamline patterns determined from the measured axial velocities show that there is an appreciable axial asymmetry in the vortex structure, with the eye of the vortex closest to the radial outflow boundary where the tangential velocity was found to lie above the Couette-flow profile in contrast to the inflow boundary where it was lower.

Measurements for Xanthan gum were again in close agreement with the theoretical tangential velocity profile for a power-law fluid in circular Couette flow although there was evidence of sensitivity to the power-law index, n. The transition to Taylor flow was marked by a more gradual increase from zero in the value of the axial velocity component. A slow drift of the vortices along the axis of the annulus was attributed to the elastic character of the fluid. The tangential and axial velocity components exhibited similar periodicity to that observed for the Newtonian fluid. The radial asymmetry in the vortex structure was more marked than for glucose with the eye of the vortex closer to the centrebody.

The thixotropic nature of the Laponite/CMC blend did not permit precise identification of the transition point from circular Couette flow to Taylor flow. Tangential velocity profiles measured in circular Couette flow are characteristic of a yield-stress fluid, with flow being restricted to the inner region of the annulus. No region of gelled fluid was observed for rotation rates well above the critical Taylor number and the vortex cell structure was found to be remarkably similar to that for the Xanthan Gum suggesting that the shear-thinning aspect of the fluid rheology is far more important than either thixotropy or viscoelasticity whereas the axial drift, in the opposite direction to that of Xanthan gum, is more likely to be associated with thixotropy.

Acknowledgments

The work reported here represents part of a programme of research which has received financial support from SERC (GR/F 87813), BP Exploration Company Ltd, Shell Research BV and AEA Petroleum Services. This support is gratefully acknowledged. Frequent meetings with Professor J H Whitelaw, Imperial College of Science, Technology and Medicine, Dr C F Lockyear and Dr D Ryan, BP Research, Ms B Kampman, Shell Research BV and Dr W J Worraker, AEA Technology, were of considerable benefit to the research.

Nomenclature

n	power-law exponent for fluid viscosity
r	radial distance from pipe centreline (m)
R_i	outer radius of centrebody (m)
R_n	refractive index
R_o	inner radius of outer tube (m)
R_y	radial location of yield surface (m)
Ta	Taylor number $(\rho\omega/\mu)^2\, R_i\, (R_o - R_i)^3$
Ta_c	critical Taylor number
u	axial component of velocity (m/s)
u_{max}	maximum axial velocity (m/s)
w	tangential component of velocity (m/s)
z	axial distance along annulus (m)
$\dot{\gamma}$	shear rate (s^{-1})
$\dot{\gamma}_C$	shear rate at centrebody (s^{-1})
ζ	non-dimensional axial location $z/(R_o - R_i)$
κ	radius ratio R_i/R_o
λ_C	constant in Casson model (s)
λ_S	constant in Sisko model (s)

μ	dynamic viscosity (Pa.s)
μ_{ref}	reference viscosity (1 Pa.s)
μ_∞	infinite shear-rate viscosity (Pa.s)
ξ	non-dimensional radial location $(R_o - r)/(R_o - R_i)$
τ	shear stress (Pa)
τ_{ref}	reference shear stress (1 Pa)
τ_y	fluid yield stress (Pa)
ψ	non-dimensional stream function
ω	angular velocity of centrebody (rad/s)

7.   References

Alderman N.J., Ram Babu D., Hughes T.L. and Maitland G.C., 1988, The rheological properties of oil well drilling fluids, Proc. Xth Int. Cong.Rheology; Sydney, pp 140-142.

Coles, D., 1965, Transition in circular Couette flow. J. Fluid Mech., Vol 21, pp 385-425.

Di Prima, R.C. and Swinney, H.L., 1981, Instabilities and transition in flow between concentric rotating cylinders, in Hydrodynamic Instabilities and the Transition to Turbulence, H.L. Swinney and J.P. Gollub, eds., Springer, New York, 139-180.

Donnelly R.J., 1958, Experiments on the stability of viscous flow between rotating cylinders. 1. Torque measurements. Proc. R. Soc. Lond. A, Vol 246. pp 312-325.

Escudier M. P., Gouldson I. W. and Jones D. M., 1994a, Flow of shear-thinning fluids in a concentric annulus. Experiments in Fluids (in press).

Escudier M. P., Gouldson I. W. and Jones D. M., 1994b, Taylor vortices in Newtonian and shear-thinning liquids. Proc. Roy. Soc. A (in press).

Escudier M. P. and Gouldson I. W., 1994, Concentric annular flow of shear-thinning liquids with centrebody rotation. University of Liverpool, Department of Mechanical Engineering, Report No TF/038/94.

Green, J. and Jones, W.M., 1982, Couette flow of dilute solutions of macromolecules : embryo cellos and overstability. J. Fluid Mech., Vol 119, pp 491-505.

Koschmeider, E.L., 1993, Bénard cells and Taylor vortices. Cambridge University Press.

Kusnetsov, E.A., Lvov, V.S., Nesterikhin, Y.E., Shmojlov, Y.F., Sobolev, V.S., Spector, M.D., Timokhin, S.A., Utkin, E.N. and Vasilenko, Y.G., 1977, About turbulence arising in Couette flow, Institute of Automotion and Electrometry, Siberian Branch, USSR Academy of Sciences, Preprint No. 58.

Lockett, T.J., 1992, Numerical simulation of inelastic non-Newtonian fluid flows in annuli. PhD Thesis. Imperial College of Science, Technology and Medicine.

Lockett T.J., Richardson, S.M. and Worraker W.J., 1992, The stability of inelastic non-Newtonian fluids in Couette flow between concentric cylinders: a finite-element study. J. Non-Newtonian Fluid Mech., Vol 43, pp 165-177.

Pfister G. and Gerdts U., 1981, Dynamics of Taylor wavy vortex flow. Phys. Lett., 83A, 23-25.

Sinevic V, Kuboi R and Nienow A.W., 1986, Power numbers, Taylor numbers and Taylor vortices in viscous Newtonian and non-Newtonian fluids. Chem. Eng. Sci., Vol 41. No 11, pp 2915-2923.

Snyder, H.A., 1969, Change in waveform and mean flow associated with wavelength variations in rotating Couette flow. Part 1. J Fluid Mech 35. 337-352.

Snyder H.A. and Lambert R.B., 1966, Harmonic generation in Taylor vortices between rotating cylinders. J. Fluid Mech. Vol 26, pp 545-562.

Stuart, J.T., 1986, Taylor-vortex flow: A dynamical system. SIAM Review, Vol 283, pp 315-342.

Taylor, G.I., 1923, Stability of a viscous liquid contained between two rotating cylinders. Phil. Trans. Roy. Soc., London, A., Vol 223, pp 289-343.

LDV Measurements of Confined Vortex Breakdown Generated by a Rotating Cone

D.F.G. Durão, J.C.F. Pereira and J.M.M. Sousa

Instituto Superior Técnico/Technical University of Lisbon
Mechanical Engineering Department, Av. Rovisco Pais
1096 Lisbon Codex, Portugal

Abstract. LDV measurements of the three-dimensional velocity flowfield in the confined swirling flow generated by a rotating cone are presented. The reported measurements covered two particular regimes exhibiting bubble-type of vortex breakdown. The corresponding Reynolds numbers were $Re = 2200$ and 2570, with values of the gap ratio $H/R = 2$ and 3, respectively. The former regime displayed single breakdown, while the latter regime was characterized by double breakdown. The present measurements and flow visualization have corroborated the approximate axisymmetry of the flowfield associated to bubble-type of vortex breakdown already observed in different flow geometries.

Keywords. Swirling flows, vortex breakdown, laser-Doppler velocimetry

1 Introduction

The drastic effects of imparting swirl to fluid flows, giving rise to simultaneous axial and vortex motions, have received eminent attention since many years. A comprehensive description of the mechanisms and general characteristics associated to swirling flows can be found in Gupta et al. (1984). Most of those characteristics present favorable features which made the utilization of swirl attractive for practical applications. Hence, it has been widely used for the stabilization of high-intensity combustion processes (Syred and Beér, 1974). For the case of atomizers (e.g., Carvalho and Heitor, 1993) and combustors (e.g., Bach and Gouldin, 1982; Rhode et al., 1982; Rhode et al., 1983), the consequent regions of turbulent recirculating flow are beneficial, providing suitable zones for flame stabilization with increased mixing. In contradistinction, for the case of water turbines (e.g., Gerich and Raabe, 1975), the vortex flow has been identified as responsible for the draft-tube surge, due to unsteady vortex breakdown (VB) or precession of vortex core, as observed by Cassidy and Falvey (1970). The resulting helical and mostly cavitating "dead water" core beginning at the runner cone originates undesirable low-frequency vibrations affecting the turbine.

In this context, the present work will focus on the phenomenon of VB. The words "vortex breakdown" are nowadays assentingly used to describe an abrupt change in the structure of the core of a swirling flow. When a certain critical state, determined by the Reynolds number and the geometry of the swirl generator, is reached, the flow in the axis undergoes a drastic deceleration, leading to the occurrence of a stagnation point with the concomitant enlargement of the core, often associated to regions of recirculating flow. The phenomenon may, however, present distinctive features, depending upon the case, as it has been reported to occur in a panoply of circumstances (see Leibovich, 1984; Escudier, 1987; Escudier, 1988, for excellent reviews).

The major general classification of the different forms of VB divides them into two categories: the "bubble-type" and the "spiral-type", the former usually displaying a high degree of axisymmetry and the latter being intrinsically non-axisymmetric. Nevertheless, since the first recognized observation of VB, by Peckham and Atkinson (1957), in the tip vortices of a delta-winged aircraft, the problem has received the attention of many other researchers, giving rise to the discovery of a number of sub-categories. The complexity of VB flows and its sensitivity to external disturbances advised the implementation of experimental apparatus where the phenomenon might be accurately investigated under controlled conditions. The tube flow with swirl generated by an upstream guidevane arrangement was widely used, first by Harvey (1962) and later by several others, e.g., Sarpkaya (1971), Faler (1976), Faler and Leibovich (1977), Escudier and Keller (1983), and Bornstein and Escudier (1984). Escudier and his co-workers have also used another apparatus (Escudier et al., 1980 and Escudier et al., 1982), the slit-tube arrangement, exhibiting the advantage of a resulting vortex core which was relatively thin, although the swirling flow has been found to be slightly asymmetrical. A third experimental device which has been employed in many investigations was the cylindrical container, where the fluid motion is produced by a rotating endwall (e.g., Escudier, 1984 and Spohn et al., 1993). Mainly due to the fact that for this configuration the boundary conditions are precisely specified, successful numerical simulations have been recently also performed for this geometry. These were reported by Lugt and Abboud (1987) and in a series of papers by Lopez (1990), Brown and Lopez (1990), and Lopez and Perry (1992). Despite the fact that the flow visualization studies have probably played the dominant role in this line of research, detailed and very reliable measurements of bubble-type breakdown have also been carried out (Faler and Leibovich, 1978 and Bornstein and Escudier, 1984) using a non-intrusive technique, laser-Doppler velocimetry (LDV). It is undeniable that the complete and accurate mapping of the internal structure of VB flows represents an invaluable contribution towards the understanding of the underlying mechanisms, allowing a more efficient design of flow systems incorporating swirl.

The present work describes an experimental study of the swirling flow generated by a rotating cone in a sealed rotor-stator system. The main objective of the reported investigations is to apply LDV techniques to accurately characterize the core structure of the flow, with special emphasis put on the occurrence of VB.

Herein, detailed measurements of the three-dimensional velocity field associated to bubble-type VB are complemented with flow visualization. The next section describes the experimental arrangement, presents details of the flow configuration, of the laser-Doppler velocimeter and flow visualization. In Section 3, the results of the present work are shown and discussed. The last section summarizes the main findings of this study.

2 Experimental Arrangement

2.1 Flow Configuration

The essential features of the experimental setup are schematically shown in Fig.1. The swirling flow is generated inside a DURAN glass cylinder with an inner radius $R = 50$ mm and the length of the gap H may be adjusted, up to $H/R = 4$, using the screw fixed to the top wall. The rotating cone, $(H/R)_{cone} = 1$, is driven by an electronically controlled motor which velocity is adjustable in a continuous range up to 1500 rpm ($\Omega = 157$ rad/s). The working fluid employed for the LDV measurements was "baby oil", a colorless oil mixture, aiming to avoid the loss of applicability of LDV due by changes of position of the measuring volume which occur when the light passes through solids and fluids of different refractive index. By refractive index matching (see, e.g., Pereira, 1989), this fluid allowed to eliminate the problem of distortion of the optical paths through the curved surfaces in the liquid flow, so that reliable measurements close to the walls could be

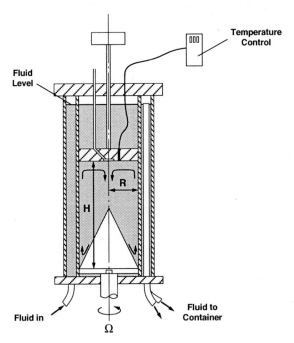

Fig. 1 Geometrical flow configuration

carried out. In order to ensure stable operating conditions (constant viscosity and refractive index) it was necessary to immerse the cylindrical container in a thermostatically controlled bath of the same working fluid, with temperature adjusted up to ± 0.1 °C, using a GRANT TD immersion thermostat and water cooling. The viscosity values have been obtained employing a digital viscometer

BROOKFIELD model DV-II. Due to the pureness of the working fluid, it was further necessary to add polystyrene spherical particles ($d \approx 1$ μm) to the oil, so that a reasonable signal-to-noise ratio could be obtained.

After flow visualization two regimes were selected to be studied in detail, covering the gap ratios $H/R = 2$ and 3. The Reynolds numbers characterizing these cases were $Re = \Omega R^2/\nu = 2200$ and 2570, respectively.

2.2 Instrumentation

The measurements were obtained using a two-component LDV system. A 3W (nominal) Ar-ion laser was employed as laser source for the velocimeter, which was operated in the dual-beam (two green and two blue) backward-scatter mode and included acousto-optic modulator to provide sensitivity to the flow direction. A shift frequency of 200 kHz was used. The principal characteristics of the velocimeter are summarized in Table 1.

Table 1 Principal characteristics of the laser-Doppler velocimeter

3W (nominal) Ar-ion laser — wavelengths:	514.5 nm (green)
	488.0 nm (blue)
Virtual beam waist diameter, @ $1/e^2$ points:	1.4 mm
Focal length of focusing lens:	350 mm

Half-angle of beam intersection:
    measured (in air):
    calculated (in working fluid):
	
	4.086°
	2.774°

Fringe separation in working fluid:	5.315 μm (green)
	5.042 μm (blue)
Number of fringes within the control volume:	31
Calculated dimensions of measuring volume, @ $1/e^2$ points, in working fluid: major axis of the ellipsoid:	 3.384 mm (green) 3.210 mm (blue)
minor axis of the ellipsoid:	0.164 mm (green) 0.155 mm (blue)
Velocimeter transfer constants:	0.188 MHzm^{-1}s (green) 0.198 MHzm^{-1}s (blue)

The backward-scatter light (green and blue) was collected through the transmitting and receiving lenses and focused into the multimode receiving fiber.

The received light from the multimode fiber was collimated with the double-input coupler. The collimated received light was directed into a color separator to separate the light of the two velocity components. Then, the photomultiplier output signals were monitored in a two-channel oscilloscope, filtered and processed using a double counter (TSI 1990C) interface with a digital computer, as shown in Fig. 2. The velocity values were evaluated by ensemble averaging, calculated from 5000 samples, using the TSI FIND Software. This software was also used to control an ISEL three-dimensional traversing system, where the fiberoptic probe containing the transmitting and receiving optics was mounted. The test section was held stationary. In order to map the three-dimensional velocity flowfield, two traversings through centerplanes of the test section were carried out for stations along the axial direction. In the first traversing, the axial and radial components (u and v) were measured, while during the second traversing, performed in a direction normal to the previous one, data to compute the swirl component (w) was acquired.

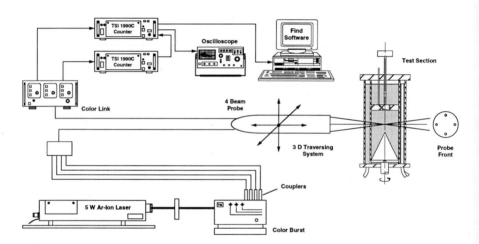

Fig. 2 Optical arrangement and signal processing

The velocity measurements may be affected by non-turbulent Doppler broadening errors, due to gradients of mean velocity across the measuring volume (see, e.g., Durst *et al.*, 1981). The maximum error in the mean velocity value was estimated to be normally at least four orders of magnitude smaller than the maximum velocity in the flow. On the other hand, random errors associated to the statistical evaluation of the mean velocity values were found to be below 3%, according to the analysis referred by Yanta and Smith (1978) for a 95% confidence level.

2.3 Flow Visualization

The fluid motion inside the cylinder was made visible by light reflecting the seeding particles in the oil. A lateral laser light illuminated the cylinder centerplane and photographs were taken at right angle to the sheet of light. Still photographs were obtained with 3200 ASA film, using exposure times of 1/15 to 1/2 seconds.

3 Results

The angular speed Ω of the cone was gradually increased, starting from rest. This originated the development of a rotating boundary layer on the cone surface which caused a centrifugal pump effect, sending the fluid radially outwards in spiral trajectories. Due to the presence of the cylinder walls, the fluid was then forced to spiral up them, giving rise to the establishment of a sidewall boundary layer. Except for the losses caused by viscous dissipation, the fluid conserved its angular momentum, until reaching the top stationary wall. Then, another boundary layer was formed, decelerating the fluid in the vicinity of the wall. As the radial pressure forces had to be compensated by the centrifugal forces, the fluid trajectory radius was reduced, directing the flow towards the axis. Then, by continuity, the spinning cone further sucked the fluid from the axis of the top wall, again in a spiral motion downwards to the tip of the cone. These features were observed during the flow visualization that was carried out prior to the detailed measurements. It guided the selection of the regimes to study and also the choice of the measurement locations.

Figures 3a and 3b show the flow patterns inside the cylinder, for $H/R = 2$, $Re = 2200$ (case 1), and for $H/R = 3$, $Re = 2570$ (case 2), respectively. The photographs reveal fully developed recirculation bubbles in the core flow, indicating single breakdown for case 1 and double breakdown for case 2. It was possible to observe that, close to $Re = 1350$, in case 1 (single breakdown), and $Re = 2240$ in case 2 (double breakdown, almost simultaneous), spiral structures were formed. With a small increase in Ω, these structures were radially stretched, giving place to the aforementioned recirculation bubbles, as a central stagnation point appeared.

Figures 4a and 4b display the velocity vectors formed by the measured axial and radial velocity components, corresponding to case 1 and case 2, respectively. In agreement with the flow visualization study, case 1 was characterized by the presence of a significantly stretched (in radial direction) single central stagnant region, while case 2 exhibited a first central stagnant region near the top wall and a second much more elongated one (in longitudinal direction) closer to the rotating cone. Throughout the recirculation zones interior it can be seen that the fluid velocities are very low, which implies that these are regions of practically uniform pressure. The vector plots further illustrate the locations where the two sets of velocity measurements have been taken, aiming to provide a reliable characterization of the vortex breakdown structures.

A better understanding of the internal flow structure can be obtained from a streamline map. Figures 5a and 5b show interpretive sketches, representing case 1 and case 2 respectively, constructed from the "streamline" contours, corresponding to the intersection of stream surfaces with a meridional plane. These contours have been computed using the axial velocity measurements but only the flow region above the cone is shown. The well-defined recirculation bubbles observed in the constructed streamline maps now reveal the details of the vortex breakdown flow structure. Both cases show approximately axisymmetric closed bubbles between two stagnation points. The minimum (outer flow) and maximum (bubble flow)

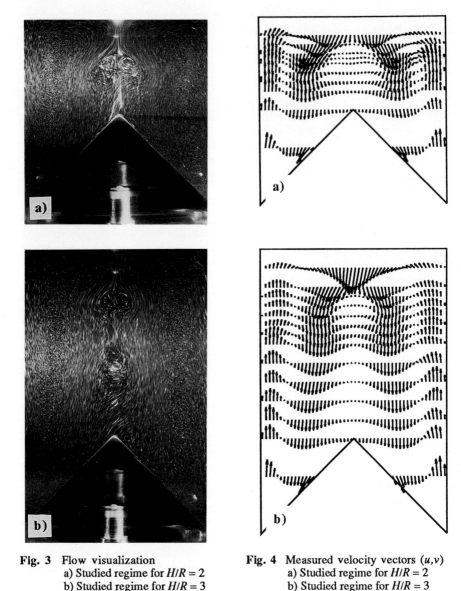

Fig. 3 Flow visualization
    a) Studied regime for $H/R = 2$
    b) Studied regime for $H/R = 3$

Fig. 4 Measured velocity vectors (u,v)
    a) Studied regime for $H/R = 2$
    b) Studied regime for $H/R = 3$

values of the non-dimensional stream function contour levels in Fig. 5a are -7.3×10^{-3} and 1.2×10^{-5}. In Fig. 5b, the value is -7.5×10^{-3} in the outer flow, while 1.0×10^{-5} and 2.5×10^{-6} are the values corresponding to main and secondary bubbles, respectively. Therefore, the computed streamlines indicate that the recirculating flow in the secondary bubble is considerably slower than in the primary bubble. Further, the above values evidence the large magnitude difference between the volume of fluid recirculated in the vortex breakdown bubbles and in the main flow.

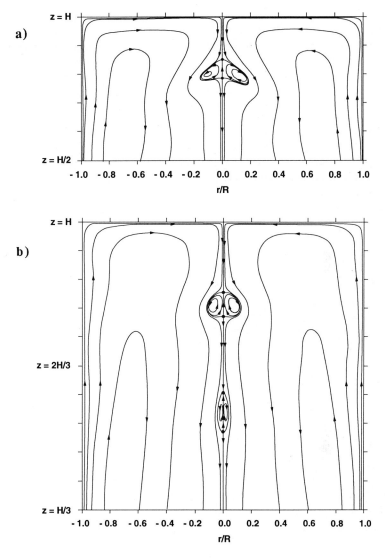

Fig. 5 Streamline maps constructed from the axial velocity measurements
a) Studied regime for $H/R = 2$
b) Studied regime for $H/R = 3$

Lopez (1990) has reported similar findings in his numerical calculations of confined swirling flow generated by a rotating flat lid. The contraction of the streamlines to the outer edge of the bubbles, while spreading out in the interior region, also referred by the aforementioned author, is clearly observed for the primary bubbles in Figs 5a and 5b as well. Additionally, one may infer from the streamline patterns in the vicinity of the centerline, the decelerations (particularly

42

dramatic before the first stagnation point) and accelerations experienced by the fluid in the central core.

The evolution of the swirl component along a centerplane is portrayed in Figs. 6a and 6b. The most remarkable features are the high gradients in the cone region and, moving upwards into vortex breakdown locations, the development of double-zoned velocity profiles. It is discernible that the presence of a recirculation bubble yields a region of low swirl velocity in solid-body rotation, while an outer region with higher and approximately constant swirl can also be identified in these profiles. The magnitude of this effect was observed to be correlated with the strength of the recirculation bubble. Still worth noting is the fact that the location of the maxima of the swirl velocity component moves continuously inwards as the outer fluid approaches the top wall.

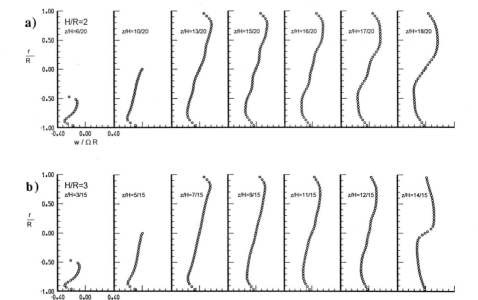

Fig. 6 Profiles of w-velocity component
a) Studied regime for $H/R = 2$
b) Studied regime for $H/R = 3$

4 Concluding Remarks

Flow visualization and detailed LDV measurements of the three velocity components in the confined swirling flow generated by a rotating cone have been presented for two distinct flow conditions. These were characterized by the presence of bubble-type vortex breakdown. For $H/R = 2$ and $Re = 2200$, single breakdown was observed, while for $H/R = 3$ and $Re = 2570$ the flow displayed double breakdown. Both the flow visualization and the streamline maps constructed from the axial velocity measurements have shown that the flowfields were very close to axisymmetric.

Large efforts have been invested to assure the reliability of the reported measurements. However, improved spatial resolution could still be obtained by reducing the dimension of the measuring control volume of the LDV-optical system. Future work should be directed towards the investigation of the unsteady behavior of vortex breakdown and precession of vortex core.

References

Bach, T. Vu and Gouldin, F.C. 1982, Flow Measurements in a Model Swirl Combustor, AIAA J., vol. 20, no. 5, pp. 642-651.

Bornstein, J. and Escudier, M.P. 1984, LDA Measurements within a Vortex Breakdown Bubble, Laser Anemometry in Fluid Mechanics, pp. 253-263, Ladoan-Instituto Superior Técnico.

Brown, G.L. and Lopez, J.M. 1990, Axisymmetric Vortex Breakdown. Part 2. Physical Mechanisms, J. Fluid Mech., vol. 221, pp. 553-576.

Carvalho, I.S. and Heitor, M.V. 1993, Vortex Stability and Breakdown in Coaxial Turbulent Swirling Jets, Ninth Symposium on Turbulent Shear Flows, Kyoto, Japan, Paper 9-3.

Cassidy, J.J. and Falvey, H.T. 1970, Observations of Unsteady Flow Arising After Vortex Breakdown, J. Fluid Mech., vol. 41, part 4, pp. 727-736.

Durst, F., Melling, A. and Whitelaw, J.H. 1981, Principles and Practice of Laser-Doppler Anemometry, 2nd ed., Academic Press, New York.

Escudier, M. 1987, Confined Vortices in Flow Machinery, Ann. Rev. Fluid Mech., vol. 19, pp. 27-52.

Escudier, M. 1988, Vortex Breakdown: Observations and Explanations, Prog. Aerospace Sci., vol. 25, pp. 189-229.

Escudier, M.P. and Keller, J.J. 1983, Vortex Breakdown: a Two-Stage Transition, AGARD CP-342, Paper 25.

Escudier, M.P. 1984, Observations of the Flow Produced in a Cylindrical Container by a Rotating Endwall, Exp. Fluids, vol. 2, no. 4, pp. 189-196.

Escudier, M.P., Bornstein, J. and Maxworthy, T. 1982, The Dynamics of Confined Vortices, Proc. R. Soc. Lond., Ser. A, no. 382, pp. 335-360.

Escudier, M.P., Bornstein, J. and Zehnder, N. 1980, Observations and LDA Measurements of Confined Turbulent Vortex Flow, J. Fluid Mech., vol. 98, no. 1, pp. 49-63.

Faler, J.H. 1976, Some Experiments in Swirling Flows: Detailed Velocity Measurements of a Vortex Breakdown using a Laser Doppler Anemometer, NASA CR-135115.

Faler, J.H. and Leibovich, S. 1977, Disrupted States of Vortex Flow and Vortex Breakdown, Phys. Fluids, vol. 20, no. 9, pp. 1385-1400.

Faler, J.H. and Leibovich, S. 1978, An Experimental Map of the Internal Structure of a Vortex Breakdown, J. Fluid Mech., vol. 86, no. 2, pp. 313-335.

Gerich, R. and Raabe, J. 1975, Measurement of the Unsteady and Cavitating Flow in a Model Francis Turbine of High Specific Speed, Trans. ASME, J. Fluids Eng., vol. 97, pp. 402-411.

Gupta, A.K., Lilley, D.G. and Syred, N. 1984, Swirl Flows, Abacus Press.

44

Harvey, J.K. 1962, Some Observations of the Vortex Breakdown Phenomenon, J. Fluid Mech., vol. 14, no. 4, pp. 585-592.

Leibovich, S. 1984, Vortex Stability and Breakdown: Survey and Extension, AIAA J., vol. 22, no. 9, pp. 1192-1206.

Lopez, J.M. and Perry, A.D. 1992, Axisymmetric Vortex Breakdown. Part 3. Onset of Periodic Flow and Chaotic Advection, J. Fluid Mech., vol. 234, pp. 449-471.

Lopez, J.M. 1990, Axisymmetric Vortex Breakdown. Part 1. Confined Swirling Flow, J. Fluid Mech., vol. 221, pp. 533-552.

Lugt, H.J. and Abboud, M. 1987, Axisymmetric Vortex Breakdown With and Without Temperature Effects in a Container With a Rotating Lid, J. Fluid Mech., vol. 179, pp. 179-200.

Peckham, D.H. and Atkinson, S.A. 1957, Preliminary Results of Low Speed Wind Tunnel Tests on a Gothic Wing of Aspect Ratio 1.0, Aero. Res. Counc. CP-508.

Pereira, J.C.F. 1989, Refractive Index Matching for LDV Measurements Near Walls and in Complex Geometries, in Instrumentation for Combustion and Flow in Engines, pp. 267-284, Kluwer Academic Publishers.

Rhode, D.L., Lilley, D.G. and McLaughlin, D.K. 1982, On the Prediction of Swirling Flowfields Found in Axisymmetric Combustor Geometries, Trans. ASME, J. Fluids Eng., vol. 104, pp. 378-384.

Rhode, D.L., Lilley, D.G. and McLaughlin, D.K. 1983, Mean Flowfields in Axisymmetric Combustor Geometries With Swirl, AIAA J., vol. 21, no. 4, pp. 593-600.

Sarpkaya, T. 1971, Vortex Breakdown in Swirling Conical Flows, AIAA J., vol. 9, no. 9, pp. 1792-1799.

Spohn, A., Mory, M. and Hopfinger, E.J. 1993, Observations of Vortex Breakdown in an Open Cylindrical Container with a Rotating Bottom, Exp. Fluids, vol. 14, no. 1/2, pp. 70-77.

Syred, N. and Beér, J.M. 1974, Combustion in Swirling Flows: A Review, Combust. Flame, vol. 23, pp. 143-201.

Yanta, Z. and Smith, R.A. 1978, Measurements of Turbulent-Transport Properties with a Laser-Doppler Velocimeter, AIAA-73-169, Eleventh Aerospace Science Meeting, Washington.

LDV–measurements on wide gap instabilities in spherical Couette flow

Christoph Egbers and Hans J. Rath

Center of Applied Space Technology and Microgravity (ZARM),
University of Bremen, 28359 Bremen, F.R.G.

Abstract. A new type of instability during the laminar-turbulent transition of a viscous incompressible fluid flow in the gap between two concentric spheres, where only the inner sphere rotates (spherical Couette flow), was detected. In case of two relatively wide gap widths (β = 0.33 and β = 0.5) it was found that the well-known Taylor-instability does not exist. At the stability threshold, where the laminar basic flow loses its stability, the first instability manifests itself as a break of the spatial symmetry and non-axisymmetric secondary waves with spiral arms appear. They spread from the pole to the equator. With increasing the Reynolds number above the critical one, the number of secondary waves with spiral arms decreases. Flow visualization studies and simultaneously laser-Doppler-velocimeter measurements show that the transition of the secondary wave flow with spiral arms is periodic and quasi-periodic before small scale turbulent structures occur.

1 Introduction

The subject of hydrodynamic instabilities and the transition to turbulence has an importance for the understanding of non-linear dynamics systems. Progress in understanding instabilities and the onset of turbulence has been made primarily by focussing attention on a small number of relatively simple hydrodynamic systems like Rayleigh-Bénard convection, the flow between two concentric rotating cylinders (Taylor-Couette flow) and parallel shear flows. Furthermore, a considerable progress in understanding the first instability in the form of Taylor vortices of a viscous incompressible fluid flow between two concentric rotating spheres for small and medium gap widths has been achieved over the last two decades. The three examples just reviewed are examples of transition to turbulence through a repeated finite number of symmetry-breaking bifurcations. Especially the study of instabilities and turbulence in spherical Couette flow is of basic importance for the understanding of global astrophysical and geophysical motions. Much of the universe is filled with fluids in turbulent motion, and instabilities are quite common in planetary atmospheres. But the study of spherical Couette flow is also important for general theory of hydrodynamic stability since this flow is a natural combination of circular Couette flow at the equator and the flow between rotating disks at the poles. Another important feature of the spherical geometry is that the basic flow involves two types of symmetry, the reflection symmetry with respect to the equator and the translational symmetry with respect to the axis of rotation. Depending on the aspect ratio, both types of symmetry-breaking bifurcations can exist in the spherical Couette flow.

In this work, we consider the flow between two concentric spheres with the inner sphere rotating and the outer one at rest as illustrated in figure 1. This flow can be characterized by the following control parameters: The aspect-ratio $\beta = (R_2-R_1)/R_1$ and the Reynolds number $Re = (R_1^2 \cdot \omega_1)/\nu$, where R_1 and R_2 are the inner and outer radii, ω_1 is the angular velocity of the inner sphere and ν is the kinematic viscosity. Another control-parameter, which comes into account, is the acceleration rate $\partial\omega_1/\partial t$, where t is time, because the arising flow pattern during the transition to turbulence are also determined by the history of the flow, i.e. it depends on whether the Reynolds number is increased or decreased, quasi-stationary or fast.

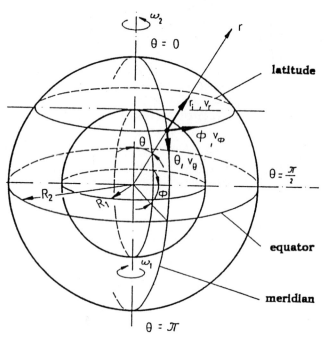

Fig. 1: Principle sketch of the spherical Couette flow

Most of previous experimental investigations on spherical Couette flow were restricted to small and medium gap widths, where Taylor vortices exist as reported by Sawatzki & Zierep (1970), Munson & Menguturk (1975), Wimmer (1976), Yavorskaya et al. (1980), Wimmer (1981), Bühler (1985) and Bühler & Zierep (1984, 1986).

Although spherical Couette flow is more relevant to astrophysical and geophysical applications, in comparison with the Taylor-Couette flow system (e.g. Fenstermacher et al. 1979, Andereck et al. 1986, Buzug et al. 1992, 1993, v. Stamm et al. 1993), the dynamic behaviour during the laminar-turbulent transition of the flow between two concentric spheres has been studied less (e.g. Belyaev et al. 1984, Nakabayashi & Tsuchida 1988).

Some new aspects of the dynamic behaviour of the spherical Couette flow during the transition to turbulence for a wide range of Reynolds numbers and for the case of wide gap widths are presented in this study. The instabilities arising are in contrast to Taylor-instabilities (Taylor 1923). They occur in the form of non-axisymmetric secondary waves with spiral arms, which break the spatial symmetry-behaviour of the basic flow. With increasing the Reynolds number, the number of secondary waves with spiral arms decreases, before the flow loses its stability to turbulence (Egbers 1994). In earlier works, this transition was thought to be a direct transition into turbulence without the existence of instabilities (e.g. Munson & Menguturk 1975).

In this work we want to give some additional quantitative estimates for the existence of a transition region during the laminar-turbulent transition and for the onset of turbulence in wide gap widths.

Experimental results connected with the problems of stability, bifurcation, non-axisymmetry, periodicity and turbulence in the spherical Couette flow will be discussed. These experiments can serve on one hand as a quantitative improvement of our previous more qualitative experiments (Egbers 1994) and on the other hand as a test for future quantitative measurements connected with the characterization of period-doubling phenomena.

The results of previous experimental, analytical and numerical work on the stability in the spherical Couette flow as a function of aspect ratio and Reynolds numbers are summarized in § 2. The experimental apparatus and experimental methods used for this investigation are described in § 3. Experimental results on simultaneous flow visualization studies and LDV-measurements are presented in § 4. In § 5 we give some conclusions.

2 Summary of previous investigations

In this chapter, we summarize previous results on the critical Reynolds numbers as a function of the aspect ratio and we compare some fits of it with our results in the form of a stability diagram as depicted in figure 2.

Frequent experimental investigations on spherical Couette flow with the inner sphere rotating and the outer sphere at rest were carried out in the region of small and medium aspect ratios, where Taylor vortices exist like in circular Couette flow (Taylor 1923). The existence of Taylor vortices in spherical gaps was first discovered experimentally by Khlebutin (1968). For aspect ratios $\beta \leq 0.19$, he calculated a good fit of the critical Reynolds number for the onset of Taylor instability to give $Re_{crit} = 49 \cdot \beta^{-3/2}$. He carried out flow visualization experiments and torque measurements in the range of $0.037 < \beta < 1.515$, but for $\beta \geq 0.44$, he did not find Taylor vortices in his experiments.

Further investigations of Taylor instability in spherical Couette flow were carried out by Sawatzki & Zierep (1970), Yakushin (1970), Munson & Men-

48

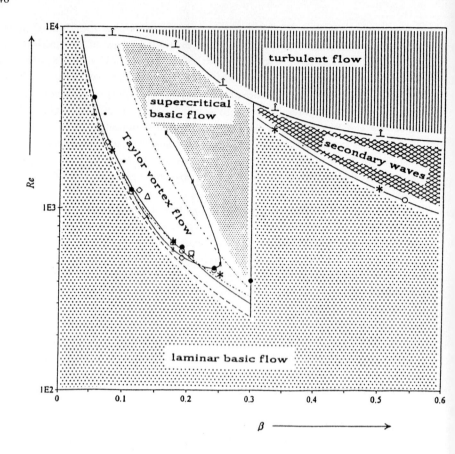

Theory

+ linear theory. Brathukin (1961)
· linear theory. Yakushin (1970)
■ energy theory. Munson & Joseph (1971)
▲ linear theory. Munson & Menguturk (1975)
● theory. Yavorskaya et. al. (1980)
── linear theory. Brathukin (1961)
····· Bartels & Krause (1979)
─╫─ Schrauf (1986)
─── cylindrical theory. Taylor (1923)

Experiment

◇ Khlebutin (1968)
+ Zierep & Sawatzki (1970)
△ Munson & Menguturk (1975)
○ Yavorskaya et. al. (1980)
× Nakabayashi (1983)
□ Bar-Yoseph et al. (1990, 1992)
···· Yavorskaya (1980)
─●─ present study (Re $_{krit}$)
⇧ present study (Re $_{turb}$)

Fig. 2: Stability diagram: Overview of previous and present investigations
    on the critical Reynolds number as a function of the aspect-ratio.

guturk (1975), Wimmer (1976), Yavorskaya (1980) and Nakabayashi (1983).
Bühler & Zierep (1984) found new secondary instabilities for higher Rey-
nolds numbers and medium sized gap widths. A survey of recent work on
Taylor vortex flow in small and medium sized gap widths is enclosed in the

work of Bühler (1985). For the case of large aspect ratios, however, only a few experimental investigations were carried out: Sorokin *et al.* (1966) experimentally tested the validity of Bratukhin's (1961) linear stability analysis for the case of very large gap (β = 1.0) and they found a continuous change with Reynolds number, but not a sudden transition. Munson & Menguturk (1975) and Waked & Munson (1978) reported that the laminar basic flow becomes unstable by transforming directly into turbulent flow and not by Taylor instability, while Yavorskaya *et al.* (1975) could detect a transition region for β = 0.54, where secondary waves exist before the flow becomes turbulent. But no more details about the behavior of these secondary waves were reported. Only a few investigations on spherical Couette flow were extended to the case that both spheres can rotate independently. Experiments on the stability of co- and counter-rotating spheres were carried out by Wimmer (1981) and Yavorskaya *et al.* (1980), but only for small aspect ratios. Furthermore, the flow between eccentric rotating spheres was investigated experimentally and numerically both for co- and counter-rotating spheres by Bar-Yoseph *et al.* (1990,1992).

Numerical investigations on the spherical Couette flow for Re $\geq$ Re$_{crit}$ in wide and very wide gap widths are not available, because of the difficulties arising to simulate non-axisymmetric flows. For small and medium sized gap widths, however, especially for β = 0.18, several axisymmetric flows with Taylor vortices were calculated by Bonnet & Alziary de Roquefort (1976), Bartels & Krause (1979), Dennis & Quartapelle (1984) and Markus & Tuckerman (1987 a,b). Schrauf (1986) investigated the influence of β on the first appearance of a pair of Taylor vortices in the spherical Couette flow and he obtained steady, but not necessarily stable solutions. He concluded that Taylor vortices exist even for $\beta \leq 0.48$, which is in contrast to our previous results in the form of visual observations in wide gap widths for β = 0.33 and β = 0.5 (Egbers 1994, Egbers & Rath 1995).

3 Experimental methods

3.1 Experimental apparatus

An experimental set-up for spherical Couette flow was constructed, consisting of an inner sphere rotating concentrically inside another rotating outer spherical shell. A principle sketch of the experimental apparatus is illustrated in figure 3:

The outer sphere (R_2 =40.00±0.02 mm) is composed of two transparent acrylic plastic hemispheres. The upper hemisphere has a spherical outer surface of about $0 \leq \Theta \leq 110$ $^\circ$ to investigate whether the occuring flow patterns are symmetric with respect to the equatorial plane or not. The inner sphere is made out of aluminium having the five various radii R_1 shown in table 1, to variate the aspect ratio β in a range between $0.08 \leq \beta \leq 0.5$.

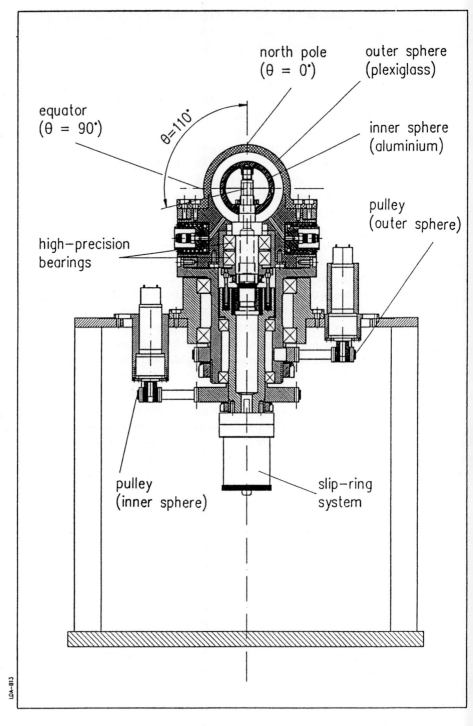

Fig. 3: Principle sketch of the experimental apparatus

No.	radius R_1 of inner spheres [mm]	aspect ratio β $\beta = (R_2 - R_1)/R_1$
1	37.00 ± 0.01	0.08
2	33.96 ± 0.02	0.18
3	32.03 ± 0.02	0.25
4	29.95 ± 0.03	0.33
5	26.70 ± 0.01	0.50

Table 1: Dimensions of inner spheres and variation of aspect ratio (radius of the outer sphere: R_2 = 40.00±0.02 mm)

Generally, both spheres can be rotated independently by means of two belt-drives. The eccentricity between outer and inner sphere could be minimized to ±0.015 mm using high precision bearings and shafts. Two different synchronous motor drives were adapted, which provide a uniform and stable rate of rotation up to n_{max}= 850 rev/min with fluctuations of less than 1.5%. They variate the Reynolds numbers of the inner and outer sphere respectively in a range from 0 to Re = 10^5. The revolutions were calibrated using an optical tracking system (optical coupling). Most of our experiments were carried out by increasing the Reynolds number quasistationary from zero. However, because the occuring flow structures depend also on initial conditions (Egbers 1994), the acceleration rate for both spheres should be variated. A schematic view of the experiment control is shown in figure 4:

Experiments with the spherical Couette flow system were carried out in a laboratory condition, where the temperature could be kept uniformly up to ± 0.3 °C. Since the viscosity of the silicone oils, which were used as working fluids, vary by approximately 2%/°C, the temperature must be precisely controlled and measured in order to have a well-defined Reynolds number. A temperature accuracy of ± 0.15 ° C was achieved for all six temperature sensors (PT 1000) used in our experiment: For measuring the fluid temperature, three temperature sensors are installed just below the outer surface of the inner sphere at latitudes Θ_{i1} = 10 ° , Θ_{i2} = 80 ° and Θ_{i3} = 160 °. These temperature data were transferred from the rotating sphere to the stationary part by a slip-ring system. For investigations with only the inner sphere rotating three other temperature sensors are installed at the inner surface of the outer sphere at latitudes Θ_{o1}=0°, Θ_{o2}=45° and Θ_{o3}=90° as illustrated in figure 4. Most of the measurements were carried out with the silicone oil M3 at a constant temperature of 25°C.

Fluid data			Baysilon-oil M3
density	ρ	[kg/m^3]	910
kin. viscosity	ν	[10^{-6}m^2/s]	3.30
refractive index	n	[-]	1.394

Table 2: List of characteristic fluid data (25°C)

52

As tracer particles small aluminium flakes were used. The concentration by weight was 0.05%. The viscosities of the working fluids with tracer particles were measured with a VOR-rheometer (BOHLIN-Reologi AB, Sweden). An effect on the viscosity was not detected. The physical properties of the working fluid M3 are listed in table 2 for the temperature of 25 °C.

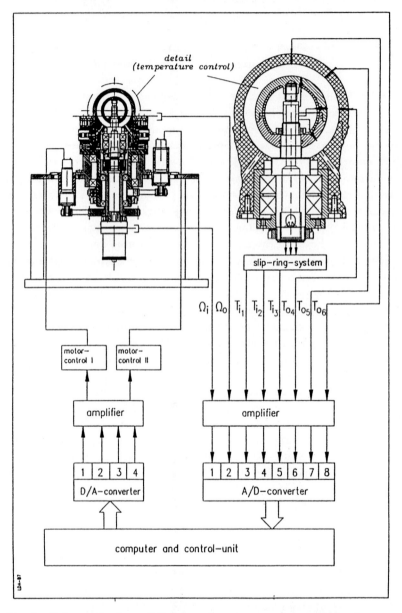

Fig. 4: Schematic view of the experiment control (principle of temperature control and rotation rate measurement)

3.2 Flow visualization technique

Because some of the observed flow structures appearing in the spherical Couette flow are non–axisymmetric and non–equatorialsymmetric, it was necessary to develop a new observation–technique, which provides a simultaneously flow visualization of both the azimuthal and the meridional flow. Therefore, a combination of the following two visualization methods was used:

To investigate the flow structures occuring in the meridional cross-section of the spherical annulus, a slit illumination technique is employed as illustrated in figure 5 (a). In addition to the slit illumination technique, a system with a fiber-optic is applied to visualize the front flow region with azimuthal waves. Furthermore, for characterizing the non–axisymmetric instabilities in wide gap widths, two cameras observe simultaneously the polar and the equatorial region as illustrated in figure 5 (b). In practice, the angle of observation of the azimuthal and meridional flows is of about 90°. In this way, the cellular structure of the occuring vortices in the meridional plane as well as the azimuthal and polar behaviour of the arising flow pattern can be obtained. Photographs or print sequences from video-records were taken.

a.) light sheet illumination   b.) frontal illumination

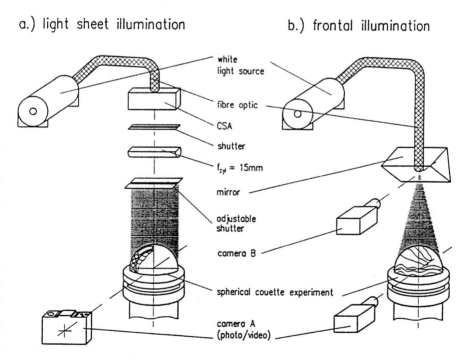

Fig. 5: Flow visualization techniques:
a.): light sheet illumination
b.): frontal illumination

3.3 Laser-Doppler-velocimeter measuring system

For the application of laser-Doppler-velocimetry (LDV) on the spherical
Couette flow experiment, a special traversing system has been constructed
to mount the optic probe on the spherical Couette flow experiment as depicted in figure 6 (a), while the laser and the transmitter are mounted
apart from the experiment on a mounting bench. The traversing system
consists of a high-precision bow with a traversing sledge and a traversing
table. The traversing sledge is capable of moving the optic probe in meridional direction ($0 \leq \Theta \leq 110^\circ$) and the traversing table is capable of moving
in radial direction over a range of 60 mm in order to determine the meridional dependence of the velocity and to obtain velocity profiles. The
laser-Doppler-velocimeter system used in our experiments, consists of a
100 mW Ar-laser, a transmitter unit and a 1-D fiber flow optic probe (DAN-
TEC-Electronics, Denmark). A frequency shift is added by the Bragg-cell to
one of the beam pair to allow for measurements of reversing flows. The
optic probe with a fibre optic cable is connected to the transmitter via mani-
pulators. The optical part of the laser-Doppler-system, which uses the dual
beam backscatter geometry, is shown schematically in figure 6(b). The back-
scattered light is focussed on a photomultiplier tube. All LDV-measurements were made with the traversing system mentioned above. Data records
of each measurement in the centre of the gap consisted of 32000 validated
Doppler bursts, while for each location along the profile, a sequence of
8000 validated Doppler bursts was accumulated in the buffer interface. A
principle sketch of the data processing technique with the laser-Doppler
signal- processor, which is based on FFT- technique, is shown in figure 6 (c).
Parameters characterizing the laser-Doppler-velocimeter and traversing
data are listed in table 3.

Ar-laser	100	[mW]
focal length	160	[mm]
wavelength	514.5	[nm]
beam spacing	38	[mm]
beam diameter	2.2	[mm]
number of fringes	22	[–]
probe volume: length	0.4	[mm]
diameter	0.05	[mm]
fringe spacing	2.18	[μm]
Bragg-cell frequency	40	[MHz]
meridional traversing angle	$0 \leq \Theta \leq 110^\circ$	
radial traversing distance	$0 \leq r \leq 60$ mm	

Table 3: Parameters characterizing the laser-Doppler-velocimeter
and traversing data

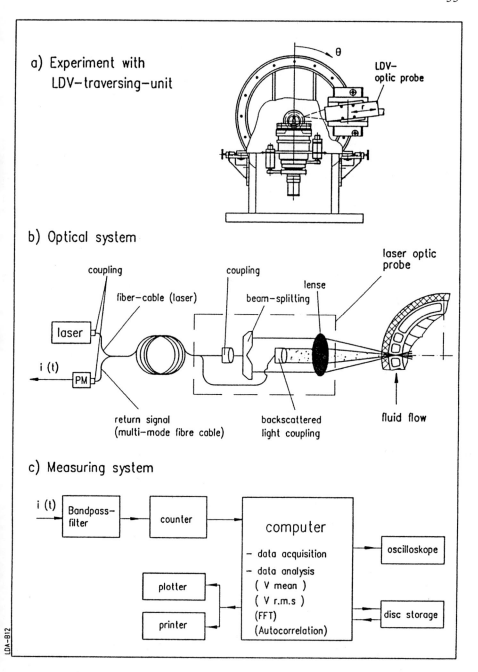

LDA-BI2

Fig. 6: Schematic diagram of the applied laser-Doppler-velocimeter (LDV):
    a.) Experiment with LDV-traversing-unit
    b.) Optical system
    c.) Measuring system

56

The application of the LDV-technique on the spherical Couette flow experiment requires an optical correction for the accurate determinations of the probe volume locations and for the interference fringe spacing due to refraction effects of the spherical outer surface. Because the probe is adjusted in radial direction, the optical axis of the front lense of the probe passes perpendicular through the spherical outer surface. Thus, the correction for the two laser beams, which are in the same plane, could be calculated for a cylindrical surface. However, the fact, that a small probing volume is needed to produce sufficient spatial resolution, which could be obtained only by a large intersection angle, the small-angle approximation cannot be used in this case. The correction method used in this work for the case of large intersection angles is illustrated in figure 7. The ray tracing begins at the front lense of the LDV optics. The equations describing the beam propagation follow immediatly from Snells law and from geometric considerations. For further details, the reader might be referred to Kekoe & Desai (1987), Gardavský & Hrbek (1989) and Meijering (1993).

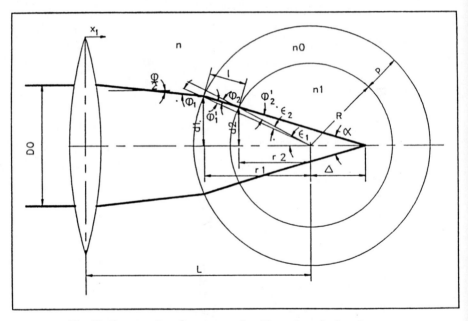

Fig. 7: Ray tracing for measuring the azimuthal and
meridional velocity component (Meijering 1993)

The LDV-technique requires the addition of particles to the working fluid. For this study of simultaneously LDV-measurements and flow visualization we were able to use the same aluminium flakes as tracer particles ranging between 10 and 50 μm by calculating the particle adaption capability, which was improved with the formulas given by Ruck (1990). For an amplitude ratio η = 99%, which defines the accuracy with which the particles follow the fluid fluctuations, the limitation frequency for the fluid/tracer combination as a function of particle diameter used in the experiments was calculated by Meijering (1993). As a result from this calculation, the frequency limit for 50 µm particles is only of about 500 Hz, which is 5 times higher as the measured frequencies of the power spectra during the transition with several instabilities. Furthermore, the position Δ of the laser beam intersection relative to the centre of the sphere in radial direction was calculated by Meijering (1993) to be

$$\Delta = d_2 / \tan(\alpha/2) - r_2.$$

The interference fringe spacing Δx at the laser beam intersection that is traversed in radial direction is

$$\Delta x = (\lambda/n_1)/ 2\sin(\alpha/2).$$

The azimuthal velocity will be

$$v = f_D \cdot \Delta x.$$

The calculated refraction corrections for the probe volume location and the fringe spacing are shown in figure 8.

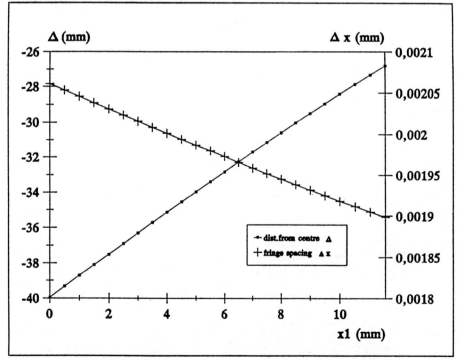

Fig. 8: Refraction correction for the optic probe volume location Δ and the interference fringe spacing Δx as a function of the radial traversing position (Meijering 1993)

58

4 Experimental results

4.1 Visual observations of wide gap instabilities

In our previous investigation on the stability of spherical Couette flow
(Egbers & Rath 1995), it was found, that the laminar basic flow between two
concentric spheres, where only the inner sphere rotates and the outer one
is at rest, loses its stability not only in the form of Taylor-instability. The
instabilities occuring depend strongly on the aspect ratio. For wide gap
widths ($\beta \geq 0.33$) considered here, the first instability manifests itself as
a break of the spatial symmetry and a new non-axisymmetric secondary
wave mode was observed, which spreads from the pole to the equator.
With increasing the Reynolds number, the number of secondary waves with
n spiral vortices decreases. For $\beta = 0.33$, secondary waves with n = six, five
and four spiral arms were found, while in the gap with an aspect ratio of
$\beta = 0.5$ waves with n = five, four or three spiral arms exist. As an example
for this type of instability, the secondary wave flow with n = five and four
spiral arms – observed from the pole – is depicted in figure 9 (a) and 9 (b).

The flow regimes, their spatial states with non-axisymmetric and non-
equatorial symmetric secondary waves were characterized by measuring
the phase velocities and the displacement between the northern and the
southern hemisphere of the arising secondary waves (Egbers 1994).

Furthermore, the spherical Couette flow in wide gap widths shows the
well-known transition phenomena of hysteresis and non-uniqueness, depending on the acceleration rate of the inner sphere (Egbers & Rath 1995).

In order to give some additional quantitative estimates on the dynamic
behaviour of these secondary waves, for a wide range of Reynolds numbers,
the azimuthal velocity components were recorded as a function of time in
the center of the gap. From these records, the mean azimuthal velocity component, velocity profiles, the turbulence intensity (r.m.s. value), the autocorrelation coefficients and the power spectra were calculated. Some of the results will be presented in the following chapters.

$\beta = 0.5$	$Re = 1357$

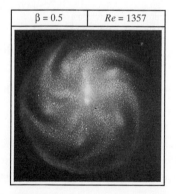

$\beta = 0.5$	$Re = 1810$

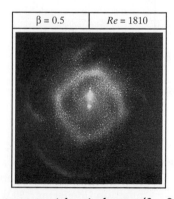

Fig. 9: Visual observations of the secondary waves with spiral arms ($\beta = 0.5$):
(a) Re = 1357, n = 5; (b) Re = 1810, n = 4

4.2 Velocity distributions

During the transition region, the azimuthal velocity profiles as a function of the dimensionless radial coordinate $d^* = (r - R_i)/(R_a - R_i)$ were measured at $\Theta = 48^0$ ($\beta = 0.33$) and $\Theta = 52^0$ ($\beta = 0.5$) with increasing the Reynolds numbers. The results are depicted in figure 10. As can be seen there for $\beta = 0.33$ and $\beta = 0.5$, the non-dimensional velocity profiles become more concave (upward) with increasing the Reynolds number in comparison with the theoretical one for the laminar basic flow calculated from Ritter (1973). Furthermore, with increasing the Reynolds number, the profiles exihibit a typical boundary layer characteristic with a minimum near the inner sphere and a maximum near the outer sphere. The velocity gradients become greater near the inner sphere in comparison with the outer sphere.

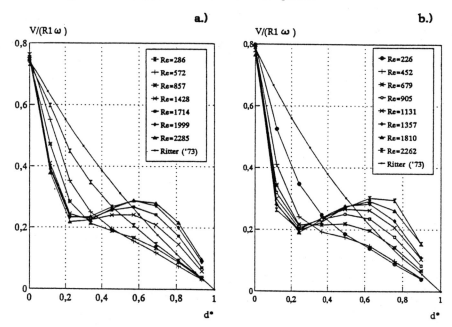

Fig. 10: The dependence of the azimuthal velocity profiles in the gap as a function of the Reynolds number during the transition region in comparison with the analytical solution (Ritter, 1973)
(a) $\beta = 0.33$, $\Theta = 48^\circ$; (b) $\beta = 0.5$, $\Theta = 52^\circ$

4.3 Velocity fluctuations

The temporal dependence of the azimuthal velocity component in the centre of the gap ($d^* = 0.5$) as a function of the Reynolds number during the laminar/turbulent transition for the flow in the gap with an aspect ratio $\beta = 0.5$ and at the latitude $\Theta = 52°$ are illustrated in figure 11:

For Re = 905 and Re = 1131 the flow is laminar stable. But with increasing the Reynolds number above the critical value ($Re_{crit} = 1244$) up to Re = 1357 (corresponding to the visual observation shown in figure 9 (a)), the velocity shows a sinusoidale fluctuation, which corresponds to the phase velocity of the secondary waves (Egbers & Rath 1995). As in this work already observed by flow visualization, the secondary waves are periodic in time (Re = 1357, Re = 1583), quasi-periodic with noise (Re = 1810, Re = 2262) and finally they are fully turbulent (Re = 4524). These results could be improved by calculating the corresponding autocorrelation coefficients as depicted in figure 12. From these records it can be seen, that the periodicity is also available for Re = 1357 and Re = 1583, but for Re $\geq$ 1810, the autocorrelation coefficients are quasi-periodic and damped, while for Re $\geq$ 2262 they are strongly damped.

4.4 Power-spectra

The calculated power-spectra of the azimuthal velocity and the corresponding simultaneously observed flow pattern are illustrated in figure 13. The Reynolds numbers are increased quasistationary from Re = 1131 just below the critical one ($Re_{krit} = 1244$) up to Re = 20.000. For Re = 1583, where the secondary wave flow with five spiral arms exist, the power-spectrum contains a single frequency peak, which corresponds to the rotation frequency of the wave mode (Egbers 1994). The spectrum for Re = 1809 contains six frequency peaks, which correspond to the quasi- periodic motion. For Re $\geq$ 2262, the frequency spectrum is filled continuously with noise and the flow becomes turbulent as it is shown in the corresponding visual observations. This noisy time-dependent behaviour is accompanied during the transition to turbulence by a spatial transition from large scale secondary waves with spiral arms into small scale structures.

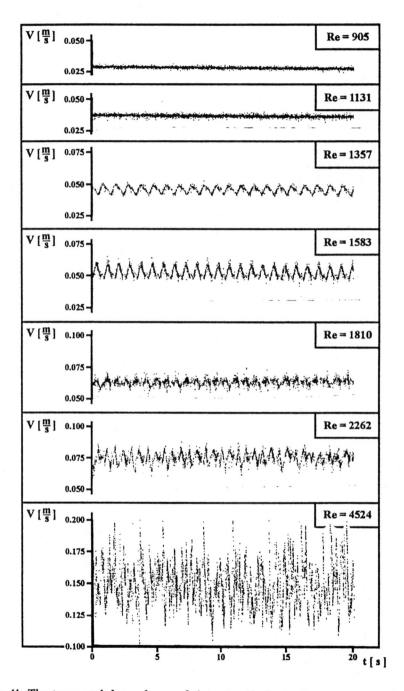

Fig. 11: The temporal dependence of the azimuthal velocity components in the centre of the gap ($d^* = 0.5$) as a function of the Reynolds number during the laminar/turbulent transition ($\beta = 0.5$, $\Theta = 52°$)

62

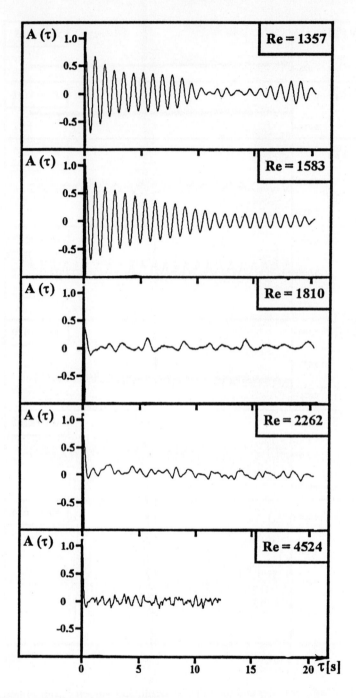

Fig.12: The autocorrelation coefficients of the azimuthal velocity components in the centre of the gap ($d^* = 0.5$) as a function of the Reynolds number during the laminar/ turbulent transition ($\beta = 0.5$, $\Theta = 52°$)

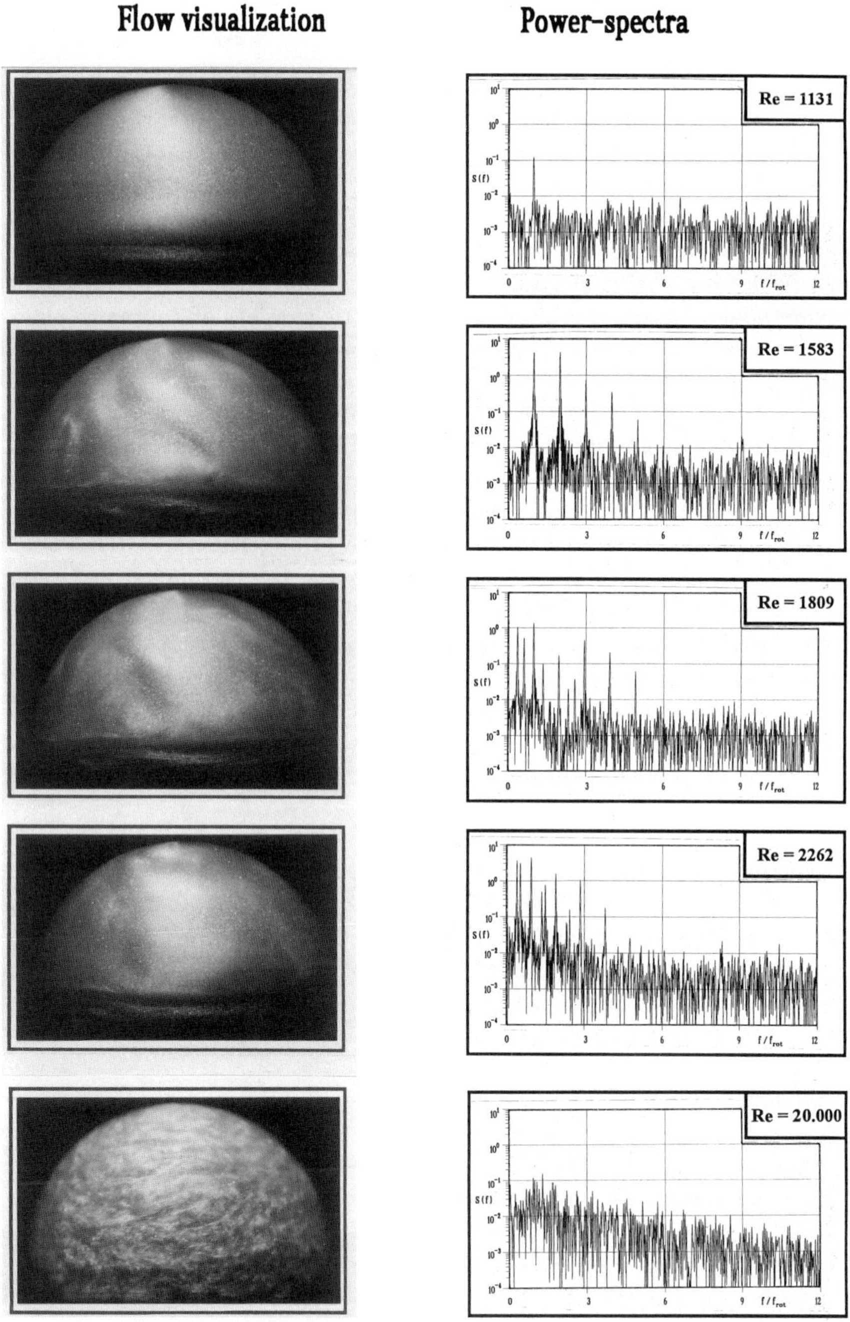

Flow visualization

Power-spectra

Fig. 13: Power-spectra of the azimuthal velocity component in the centre of the gap ($d^* = 0.5$) and corresponding observed flow regimes as a function of the Reynolds number during the laminar/turbulent transition ($\beta = 0.50$; $\Theta = 52°$)

5 Conclusions

We have considered some characteristics of the laminar/turbulent transition of the spherical Couette flow for wide gap widths. In contrast to previous investigations, we could detect a transition region, where instabilities in the form of secondary waves with spiral arms exist, which break the spatial symmetry. The LDV-technique could be applied successfully for measurements of the azimuthal velocity components. In this way, the time-dependent behaviour of the secondary waves during the transition to turbulence was obtained. The system shows the typical rich variety of routes to chaos, e.g. the periodicity, quasi-peridicity and stochastic regions. Simultaneous measurements of the azimuthal velocity components and flow visualization studies give a good impression of the spatial-temporal development of the flow during the transition to turbulence. For a complete description of these transitions, however, we need to conduct simultaneous high-resolution LDV-measurements of the fully three-dimensional velocity vector. Therefore, further investigations must be carried out to give more informations about the development of turbulence as a function of the meridional and radial coordinate. Furthermore, we need to calculate the fractal dimensions to analyze the properties of this dissipative non-linear dynamical system. As a next step we plan measurements on bifurcations with a high resolution and noise reduced LDV-measuring technique by plotting consecutive extrema of the time series, ramping the Reynolds number continuously from $Re \approx Re_{crit}$ up to Re_{turb}.

Acknowledgements. This work was supported by the Deutsche Agentur für Raumfahrtangelegenheiten (DARA) under grant number 50 QV 8898. The authors thank Prof. Zierep, Prof. Bühler and Dr. Belyaev for several discussions and Dipl.-Ing. Wolfgang Beyer and Dipl.-Ing. Alexander Meijering for their assistence during this work.

References

Andereck, C.D., Liu, S.S. and Swinney, H.L. (1986): Flow regimes in a circular Couette system with independently rotating cylinders. J. Fluid Mech., vol. 164, 155–183

Bar-Yoseph, P., Solan, A., Hillen, R. and Roesner, K.G. (1990): Taylor vortex flow between eccentric coaxialrotating spheres. Phys. Fluids A 2 (9), pp. 1564–1573

Bar-Yoseph, P., Roesner, K.G. and Solan, A. (1992): Vortex breakdown in the polar region between rotating spheres. Phys. Fluids A 4 (8), pp. 1677–1686

Bartels, F. and Krause, E. (1979): Taylor-Vortices in Spherical Gaps. in: Laminar-Turbulent Transition IUTAM-Symp. Stuttgart, Springer

Belyaev, Yu.N., Monakhov, A.A., Scherbakov, S.A. and Yavorskaya, I.M. (1984): Some routes to turbulence in spherical Couette Flow. in: Kozlov, V.V. (ed.): Laminar-TurbulentTransition. IUTAM-Symp. Novosibirsk/USSR, Springer

Bonnet, J.P. & Alziary de Roquefort, T. (1976): Ecoulement entre deux spheres concentriques en rotation. J. Mec., vol. 13, pp. 373

Bratukhin, Yu.K. (1961): On the evaluation of the critical Reynolds number for the flow of fluid between two rotating spherical surfaces. PMM, vol.25, No.5, pp. 858-866

Bühler, K. (1985): Strömungsmechanische Instabilitäten zäher Medien im Kugelspalt. VDI-Berichte, Reihe7: Strömungstechnik Nr.96

Bühler, K. and Zierep, J. (1984): New secondary instabilities for high Re-number flow between two rotating spheres. in: Kozlov,V.V. (ed.): Laminar- TurbulentTransition. IUTAM-Symp. Novosibirsk/USSR, Springer, Berlin, Heidelberg, NewYork, Tokyo

Bühler, K. and Zierep, J. (1986): Dynamical instabilities and transition to turbulence in spherical gap flows. in: Comte-Bellot, G. and Mathieu, J.: Advances in turbulence. Proc. 1st Europ. Turb. Conf., Lyon, France, 1-4 July, Springer, Berlin, Heidelberg, NewYork, London, Paris, Tokyo

Buzug, Th., v. Stamm, J. and Pfister, G. (1992): Fractal dimensions of strange attractors obtained from the Taylor-Couette experiment. Phys. A, 191, 559

Buzug, Th., v. Stamm, J. and Pfister, G. (1993): Characterization of period-doubling scenarios in Taylor- Couette flow. Physical Review E, 47, no. 2, 1054-1065

Dennis, S.C.R. and Quartapelle, L. (1984): Finite difference solution to the flow between two rotating spheres. Comp. Fluids, vol. 12, pp. 77

Egbers, C. (1994): Zur Stabilität der Strömung im konzentrischen Kugelspalt. Dissertation, Universität Bremen

Egbers, C. and Rath, H.J. (1995): The existence of Taylor vortices and wide-gap instabilities in spherical Couette flow. Acta Mech. 111, 3-4, 125-140

Fenstermacher, P.R., Swinney, H.L. and Gollub, J.P. (1979): Dynamic instabilities and the transition to chaotic Taylor vortex flow. J. Fluid Mech., 94, part1, 103-128

Gardavský, J. and Hrbek, J. (1989): Refraction Correction for LDV-Measurements in Circular Tubes within Rectangular Optical Boxes. DANTEC 8

Kekoe, A.B. and Desai, P.V. (1987): Compensation for refractive-index variations in laser-Doppler-anemometry. Applied Optics 26, 2582-2591

Khlebutin, G.N. (1968): Stability of fluid motion between a rotating and a stationary concentric sphere. Fluid Dynamics Vol.3, No.6, p.31-32

Markus, P.S. and Tuckerman, L.S. (1987a): Simulation of flow between concentric rotating spheres. Part 1. Steady states. J. Fluid Mech., vol185, 1-30

Markus, P.S. and Tuckerman, L.S. (1987b): Simulation of flow between concentric rotating spheres. Part 2. Transitions. J. Fluid Mech., vol185, 31-65

Meijering, A. (1993): Anwendung der Laser-Doppler-Anemometrie für Strömungsuntersuchungen im Spalt zwischen zwei konzentrisch rotierenden Kugeln. Diplomarbeit, Universität Bremen, FB 4, ZARM

Munson, B.R. and Joseph, D.D. (1971a): Viscous incompressible flow between concentric rotating spheres. Part1: Basic flow. J. Fluid Mech. vol. 49, part2, pp. 289–303

Munson, B.R. and Joseph, D.D. (1971b): Viscous incompressible flow between concentric rotating spheres. Part2: Hydrodynamic stability. J. Fluid Mech. vol. 49, part2, pp. 305–318

Munson, B.R. and Menguturk, M. (1975): Viscous incompressible flow between concentric rotating spheres. Part 3: Linear stability and experiments. J. Fluid Mech., vol. 69, 705–719

Nakabayashi, K. (1983): Transition of Taylor-Görtler vortex flow in spherical Couette flow. J. Fluid Mech., vol. 132, pp. 209–230

Nakabayashi, K. and Tsuchida, Y. (1988): Spectral study of the laminar-turbulent transition in spherical Couette flow. J. Fluid Mech., vol. 194, 101–132

Ritter, C.F. (1973): Berechnung der zähen, inkompressiblen Strömung im Spalt zwischen zwei konzentrischen, rotierenden Kugelflächen. Dissertation. Universität Karlsruhe

Ruck, B. (1990) Lasermethoden in der Strömungsmeßtechnik. AT-Fachverlag, Stuttgart

Sawatzki, O. und Zierep, J. (1970): Das Stromfeld im Spalt zwischen zwei konzentrischen Kugelflächen, von denen die innere rotiert. Acta Mech. 9, 13–35

Schrauf, G. (1986): The first instability in spherical Taylor-Couette flow. J. Fluid Mech., vol. 166, pp. 287–303

Sorokin, M.P., Khlebutin, G.N. and Shaidurov, G.F. (1966): Study of the motion of a liquid between two rotating spherical surfaces. J. Appl. Mech. and Tech. Phys., vol. 6, pp. 73–74

v. Stamm, J., Buzug, Th. and Pfister, G. (1993): Frequency locking in axisymmetric Taylor-Couette flow. submitted to Physical Review E.

Taylor, G.J. (1923): Stability of a viscous liquid contained between two rotating cylinders. Phil. Trans. A223, 289–293

Waked, A.M. and Munson, B.R. (1978): Laminar-turbulent flow in spherical annulus. Trans. ASME I, J. Fluids Engng., vol. 100, pp. 281

Wimmer, M. (1976): Experiments on a viscous fluid flow between concentric rotating spheres. J. Fluid Mech., vol. 78, part2, 317–335

Wimmer, M. (1981): Experiments on the stability of viscous flow between two concentric rotating spheres. J. Fluid Mech., vol. 103, 117–131

Yakushin, V.I. (1970): Instability of the motion of a liquid between two rotating spherical surfaces. Fluid Dynamics, vol. 5, no. 2, p. 660–661

Yavorskaya, I.M., Belyaev, Yu.N. and Monakhov, A.A. (1975): Experimental study of a spherical Couette flow. Sov. Phys. Dokl., vol. 20, 4, 256–258

Yavorskaya, I.M., Belyaev, Yu.N., Monakhov, A.A., Astaf'eva, N.M, Scherbakov, S.A. and Vvedenskaya, N.D. (1980): Stability, non-uniqueness and transition to turbulence in the flow between two rotating spheres. IUTAM-Symposium, Toronto

AXIAL AND TANGENTIAL VELOCITY COMPONENTS AT THE EXHAUST END OF A HIGHLY COMPLEX CYCLONIC FLOW

Mr. P. A. Yazdabadi, Dr. A. J. Griffiths, Prof. N. Syred
Division of Mechanical Engineering and Energy studies,
University of Wales, Cardiff,
PO Box 925,
Cardiff, CF2 1YF,
UK

Keywords. Precessing vortex core, swirling flow, instability

1 Introduction

1.1 Swirling Flows

Swirling flow, in which there is large scale rotation of the mean flow, is one of the well recognised configurations of flow. The use of vortex type flows are extremely widespread in a range of different devices and applications (1). The list of industrial operations in which they are used includes such diverse applications as the separation of gases of different molecular weights, Ranque-Hilsch tubes, cyclone dust separators, spray dryers, gas scrubbers, flash dryers, cyclone evaporators, combustion devices, plasma flame stabilisation, in agitators and in the piping associated with fluid turbo-equipment (2)(3)(4)(5)(6)(7)(8)(9)(10). Also fluidic vortex valves, and gas-core nuclear rockets are among the numerous practical devices, the performance of which is dominated by a confined turbulent vortex created by tangential injection of all or part of the through-flow (11)

These flows exhibit a rich variety of flow features including flow recirculation (11) and non axisymmetric velocity profiles (12). In general, experimental and theoretical results of most workers agree in demonstrating the following significant points about confined vortex flows:

(1) The confined vortex flow field is three dimensional in nature (13).
(2) The tangential component of the velocity is predominant in the entire vortex flow, with the exception of the turbulent core present in the centre (13).
(3) In a swirl flow, the radial component of velocity is very small (14)
(2) In a confined vortex flow, the predominant axial velocity is concentrated in an annular region adjacent to the wall, while in the interior of the flow the axial velocities are small (14). Actually , the axial velocity falls off sharply towards the centre of the vortex, and may exhibit a reverse flow near the centre (13).
(5) Spiral flow patterns are quite insensitive to changes in the inlet velocities (14).
(6) High turbulence is inherent in any device employing a rotating flow within a stationary container. The boundary layer on the outer wall of the container is

unstable to radially inward displacement and the separation of this boundary layer in certain locations of the apparatus is the source of high turbulence (15).

It has been reported that the shape of the tangential velocity profile of a swirling flow is not very sensitive to swirl number alterations (16). In contrast the axial velocity profile can change dramatically. Flow reversals are known to appear around the flow axis at swirl numbers greater than 0.6 (17), whereby the decay of swirl velocity profiles cause adverse pressure gradient to be set-up. As this mechanism depends on changes in swirl velocity profiles caused by sudden expansions, friction etc. the flow then becomes sensitive to changes in the downstream flow conditions, like bends, obstacles and changes in pipe diameter outside the device.

1.2   The Precessing Vortex Core

The formation of the recirculation zones in a swirling flow device depends upon the swirl number and its geometry. It is now well documented that the assumptions of axisymmetry in such flows are only true for a very low Swirl and Reynolds numbers. At a critical Swirl number of about 0.6 as the Reynolds number is increased an instability develops called the vortex breakdown phenomenon. As the Reynolds number is further increased, a large three dimensional time dependent instability, called the precessing vortex core (P.V.C.) develops. In this the central forced vortex region of flow is displaced from the axis and starts to precess about the axis of symmetry. This phenomenon is very regular in nature and can cause other instabilities to arise (18). The resultant effect of this type of flow regime is akin to a large out of balance mass of fluid/particles rotating at a high swirl velocity.

The P.V.C. appears to be a mechanism for the rapid transport of fluid from the wall opposite the exit of the swirling device to the downstream end of the reverse flow zone. For it to be a stable oscillation, there must be feedback, and it has been suggested (19) that this is provided by the reverse flow zone. It has also been suggested that the reverse flow zone displaces the central vortex core giving rise to the P.V.C (20).

It is known that the P.V.C. has a regular frequency and amplitude, simply dependent upon the system configuration and flowrate, and that it can cause many other problems by locking onto other system instabilities and resonating (or driving) with them. This can be a severe problem in large installations where low frequency high amplitude oscillations can easily arise. Beside being undesirable from a 'noise' aspect, this may easily cause sufficiently large pressure fluctuations to overstress the outer shell of both the cyclone and the downstream interconnecting systems (10).

This work, therefore, investigates the naturally occurring three dimensional time dependent instability associated with strongly swirling flows, using a laboratory scale cyclone dust separator and Laser Doppler Anemometry (L.D.A) techniques. The P.V.C is studied for a range of different parameters, and downstream pipework configurations. Clarification is given to the 'cause and effect' of the P.V.C., especially in relation to the reversed flow zone.

2 Experimental Analysis

Experiments were carried out using a laboratory scale cyclone dust separator, the set-up for the experimentation being shown in figure 1. Particular attention was paid to the exit of the cyclone and thus the effects of downstream pipe configurations on the velocity components of the swirling flow were examined. Table 1 highlights the equivalent lengths of those pipework configurations attached to the cyclone. The exit diameter was also varied by the application of inserts allowing the cyclone to have two distinct Geometric Swirl numbers, 1.790 and 3.043 respectively. The different experimental conditions are listed in Table 2.

A laser probe was carefully aligned such that the control volume was located at the central axis of the exhaust/pipe bend configurations under consideration. A hotwire anemometer probe, utilising linearising and filtering equipment, was positioned close to the outer edge of the exhaust nozzle. A clear sinusoidal P.V.C. signal was obtained, this was input into a triggering unit, which output a

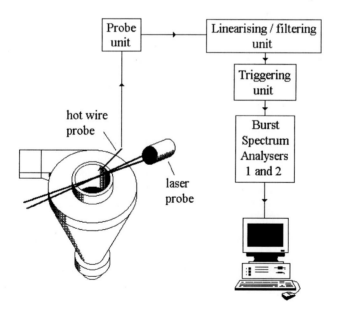

Figure 1. Schematic diagram of the experimental set-up

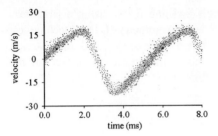

Figure 2a. Typical velocity/time scatter diagram obtained via the LDA system

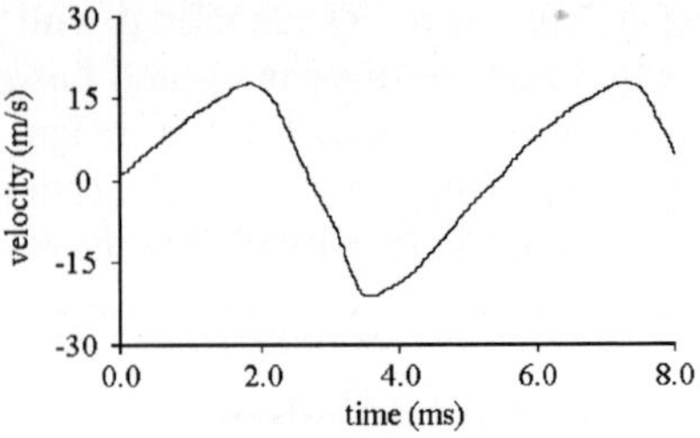

Figure 2b. Digitised mean velocity/time curve for one and a quarter P.V.C cycles

Figure 2. Velocity/time curves for one and a quarter P.V.C cycles

synchronised square wave, used to reset the L.D.A time base after each cycle of the P.V.C.

This meant that for a specific radial position the instantaneous velocity variations with time for several cycles of the P.V.C. could be superimposed over one another. Thus representative mean and r.m.s instantaneous rotating velocity values for an averaged single P.V.C. cycle could be produced. This is commonly termed phase averaging and is a technique derived from L.D.A work in reciprocating engines. This novel approach to collecting and manipulating data is used extensively in engine analysis, but it is the first time the technique has been used in swirling flows.

The laser control volume was traversed from the centre of the flow radially outwards to beyond the extremities of the cyclone exhaust/pipe bend configuration under consideration, taking axial and tangential velocity measurements every 1mm.

At each 1mm step axial and tangential velocity components of the flow were recorded, and the data was presented as a mean instantaneous velocity time curve for a complete cycle of the P.V.C, an example of which is shown by figure 2a. From this data, derivations of rotating mean axial and tangential velocities were made by digitising each velocity/time curve for a complete P.V.C. cycle, as shown in figure 2b. Nine identical radial paths were chosen and velocity data tabulated as a function of radius.

Finally, the processed data was plotted as spatial distributions of rotating mean axial and tangential velocities in a plane. Also derived were the associated r.m.s profiles, these have not been examined in detail in this paper as they are known to be prone to P.V.C. ' jitter ' causing slight changes in the trigger point from cycle to cycle. Again, this jitter is typical in engine analysis and is caused by cycle to cycle variations. Obviously this also affected the mean rotating axial and tangential profiles as did the size of the control volume (effectively 2mm long). By using methods described by Kline and McClintok (21) the accuracy of these rotating mean phased averaged velocity measurements is estimated to be ±5 % for velocities >4 m/s, ±10 % for velocities <4 m/s.

Table 1. Dimensions and equivalent lengths of downstream pipework configurations

downstream pipework configuration	Equivalent length /cyclone exit diameters
Cyclone exhaust tube	1.3
Straight pipe	6.5
60° angled bend	2.7
90° angled bend	3.2

Table 2. Test apparatus set-up and flow conditions

Sg	Re	experimental set-up
3.043	28500	cyclone only
3.043	14250	cyclone only
1.790	25600	cyclone only
3.043	42750	cyclone only
3.043	14250	cyclone + 6.5 exit diameter
3.043	28500	cyclone + 3.2 exit diameter
3.043	28500	cyclone + 2.7 exit diameter

3. Preliminary Analysis of Results

As discussed in the last section the flows at the exhaust nozzle of the cyclone dust separator freely exhausting to atmosphere have been characterised for the mean rotating axial and tangential velocity for two Swirl levels, a range of Reynolds numbers, and three different exhaust configurations as illustrated by the spatial distributions in figures 4 to 10.

In order to interpret the data it is useful to be able to postulate a model which can explain the very unusual results obtained for the mean rotating tangential velocities, figures 4a to 10a, namely the area of negative tangential velocity extending from the geometrical centre of the device, to typically 0.25R, where R is the radius of the cyclone exhaust. Careful examination of data from work reported, herein, detailed measurements at various distances downstream of the exhaust and indeed from swirl burners (where similar phenomenon occur) indicates that the problem is caused by an inappropriate choice of co-ordinate system.

Figure 3 illustrates a model for the motion of the vortex which explains the measurement of the negative tangential velocity close to the normal axis, this being due to the fact that the geometrical centre of the unit and the centre of the vortex do not coincide. It appears that the centre of the vortex is displaced to one side of the geometric centre of the system and precesses about this axis. This is

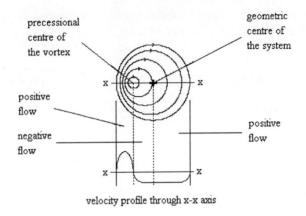

velocity profile through x-x axis

Figure 3. Schematic diagram of the off set vortex

well illustrated by figure 3, which shows mean rotating tangential velocities through the Precessional Centre Of the Vortex (P.C.O.V) and the geometric centre of the device when allowance is made for a change in the sign of velocity as one passes through the true vortex centre.

Calculation of the angular momentum flux balance along this line clearly showed that there was a balance on both sides of the precessional centre of the vortex to within ±0.2%. As discussed later this gives a clear explanation of the form of the mean rotating tangential velocity fields measured. For convenience the mean rotating tangential velocity has been collected, processed and presented using the geometric centre of the device as the centre of the co-ordinate system.

4 Discussion of Results

Using the model of figure 3 detailed interpretations of the various results obtained can be undertaken. Figures 4a and 4b illustrate results from the normal cyclone configuration, operating at Sg=3.043 and Re=28500, with a precessional frequency of 150 Hz. Figure 4a shows the mean rotating tangential velocity field. The approximate centre of the vortex has been marked and it can be seen that this displacement from the geometric centre is sufficient to accelerate the tangential velocity to a maximum of 25 m/s from a typical mean level of 10 to 11 m/s. In effect the swirling flow is accelerated as angular momentum flux is conserved by being squeezed through a narrow region between the precessional centre of the vortex and the wall. The formation of reverse flow zones in swirling flow is well known to be caused by,

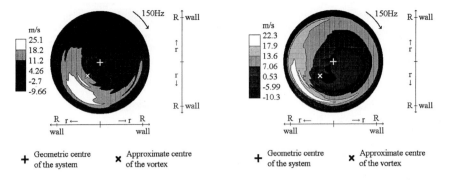

Figure 4a. Tangential velocities   Figure 4b. Axial velocities
Figure 4. Spatial distribution of rotating mean velocities for the cyclone with no
attachments: Re=28500, Sg=3.043

(a) The formation of strong radial pressure gradients due to the term
$$\frac{dp}{dr} = -\rho\frac{w^2}{r}$$ which give for cyclonic units operating at near to atmospheric
pressure sub atmospheric pressure levels on the axis
(b) Axial decay of tangential velocity level (as occur at sudden enlargements etc.
as in this work) which translates via the radial pressure gradient term in (a)
above into negative axial pressure gradients which induce reverse axial flow.

Since the P.C.O.V. is clearly shown in figure 3 for tangential velocities, it must
be logical to assume that the axial velocity profiles will have a similar centre of
flow, i.e. displaced from the geometric centre. Indeed this does infact occur, as
shown by figure 4b. The central reverse flow zone encompasses the P.C.O.V., as
to be expected, and also extends to a non dimensional radius of up to 0.25 to $0.3R_X$
in front of the P.C.O.V. Several other very important flow features are evident in
figures 4a and 4b viz.
(i) Not only are high mean tangential velocities generated by the P.C.O.V. effect
over a sector extending some 20° ahead and 60° behind the P.C.O.V., but also
are high mean axial velocities over a similar region. Clearly it is in this region
that the largest mass fluxes of flow are occurring.
(ii) Just beyond the exhaust lip of the cyclone at $R_x \geq 1.05$ a sector of reverse axial
flow extending over 340° on the outer periphery of the flow has developed as
shown in figure 4b. Flow visualisation studies have revealed that this region is
associated with the formation of engulfment vortices and high initial
entrainment rates (a well known phenomenon with swirling flows).
Indications are that these engulfment vortices are spiral or helical in nature and
break-up after about 0.25 to 0.5 exit diameters.

Figure 5 shows the effect of halving the Reynolds number on the flow. The

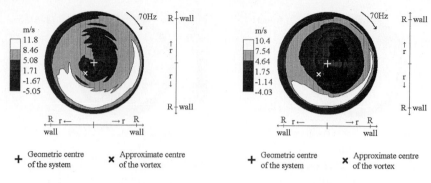

Figure 5a. Tangential velocities   Figure 5b. Axial velocities
Figure 5. Spatial distribution of rotating mean velocities for the cyclone with no
attachments: Re=14250, Sg=3.043

broad features of the flow are similar to figure 4, although some important
differences emerge:

(a) The peak tangential velocity when compared to the average over the section is
somewhat reduced, comparing figures 5a and 4a, whilst the region of negative
tangential velocity is of similar area, but of a slightly different shape.

(b) The lower Reynolds number flow produces a slightly larger area of reversed
axial flow, comparing figures 5b and 4b. For both regions it is noticeable how
the largest area of reversed axial flow occurs behind the P.C.O.V.

(c) The reverse flow zone on the outer periphery of the flow has virtually
disappeared at the lower Reynolds number, comparing figures 5b and 4b, this
is because of the reduced velocity levels.

The effect of Swirl number on the mean rotating flow fields are shown by

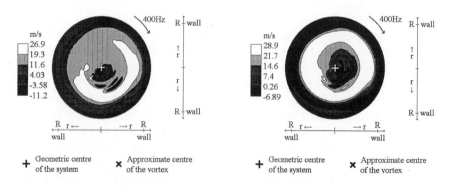

Figure 6a. Tangential velocities   Figure 6b. Axial velocities
Figure 6. Spatial distribution of rotating mean velocities for the cyclone with no
attachments: Re=25600, Sg=1.790

comparing figures 6a and 6b (Sg=1.790) with figures 4a and 4b (Sg=3.043) for similar Reynolds numbers. The region of high tangential velocity (up to 26.9 m/s) for Sg=1.790, figure 6a, can be seen to extend over more than 180° as compared to 60° for the high Swirl number result, as shown in figure 4a for Sg=3.043. The region of negative tangential velocity is reduced somewhat, but there are still clear indications of the presence of the P.C.O.V. effect. Again there is evidence of an uneven region of reverse flow in the centre of the flow, but again off axis, identified in figure 6b. The outer section of the flow between the wall and the P.C.O.V. is more uniform in the θ direction, figure 6b ($1 \leq R_x \leq 0.6$). There are again evidence of large scale engulfment vortices on the outside of the flow ($R_x \geq 1$), as shown by the intermittent (in the θ direction) reversed axial flow in this region, which are of far greater extent than with the higher Swirl number; compare figures 6b and 4b.

The effect of increased Reynolds number from 28,500 to 42,750 for Sg=3.043 is shown by comparison of figures 4a and 4b, with 7a and 7b. Figure 7a shows that a substantial enlargement of the area of negative tangential velocity has occurred at the higher Reynolds number, compared to that shown in figure 4a. The highest tangential velocity region, figure 7a, is clearly located close to the outer wall in a thin narrow sector. Evidence of a second P.C.O.V. forming can be clearly seen in a region of flow 120° in front of the main P.C.O.V., figure 7a. Spectral analysis of the P.C.O.V. signal shows this to be a higher frequency or harmonic of the main P.C.O.V. This is more representative of the true flow due to better triggering by the now sharper P.C.O.V. signal (at higher velocity levels). The extent of the influence of the P.C.O.V. in the θ direction is more limited, but its relative magnitude is much greater. The pattern of axial velocities has also altered considerably, when comparing figures 7b and 4b, with an even larger area of off

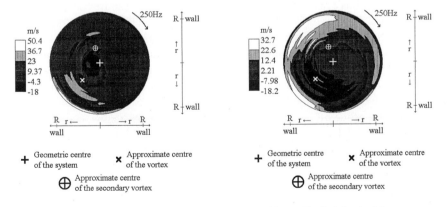

Figure 7a. Tangential velocities   Figure 7b. Axial velocities

Figure 7. Spatial distribution of rotating mean velocities for the cyclone with no attachments: Re=42750, Sg=3.043

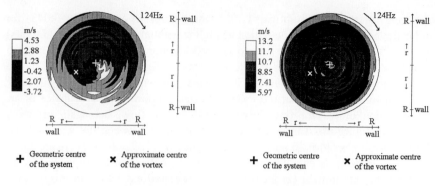

Figure 8a. Tangential velocities   Figure 8b. Axial velocities

Figure 8. Spatial distribution of rotating mean velocities for the cyclone with the 90° angled bend attached: Re=28500, Sg=3.043

centred central reverse flow. Again the squeezing effect of this reverse flow on the forward axial flow is evident, figure 7b. However the flow is complicated by a further region of high axial velocity forward flow about 70° to 90° in front of the centre of the central reverse flow zone, again indicative of the formation of a second P.C.O.V.

Figure 8a illustrates the instantaneous mean tangential rotating velocity profile taken at the exhaust of the cyclone with the 90° bend attached, operating at a Reynolds number of 28700 and a Geometric Swirl number of 3.043. For these conditions the P.V.C was found to be rotating about the exhaust at a frequency of 124 Hz. To examine the effect the 90° bend upon the velocity profile, this figure is best compared to figure 4a, as both have identical Reynolds and Geometric Swirl numbers.

Figure 8a exhibits a very rough approximation as to what can be considered the classical P.V.C flow features. There are still present the eclipse shaped region of high tangential velocity, and an off set region of negative tangential flow, both characteristics however are greatly disorganised, with the negative tangential velocity region being separated at times by positive tangential velocities. For the tangential velocities there are still, close to the outer wall, a variation of more than a factor of 2 in the θ direction over 360 degrees.

The presence of the off centred negative tangential velocity region, coupled with the presence of a detectable frequency throughout the experiment, indicate that a P.V.C is still present, however it has been greatly disrupted and decayed. This is seen as the maximum tangential velocity has been reduced from 25.1 m/s to 4.53 m/s, a drop of 82%.

Figure 8b illustrate the mean instantaneous rotating axial velocities for the same experimental set up as figure 8a, thus comparisons can be made between this figure and figure 4b to examine the effect of the 90° bend upon the axial velocity profile. The initial point that must be noted is that the 90° bend has destroyed any reversed axial flow, as the minimum axial velocity detected was 5.97 m/s. The profile is still not symmetrical about the centre of the system however, indicating that a P.V.C is still present. It could be suggested that the non-axisymmetric nature of the flow could be caused souly by the turbulence and vortices set up by the flow rounding the 90° bend, however this does not explain the presence of the already stated frequency of 124 Hz that was detected.

It can be seen that the maximum positive axial velocity has dropped from 22.3 m/s to 13.2 m/s, a reduction of 41%. This indicating again that the P.V.C has been greatly decayed by the 90° hard bend.

Figure 9a illustrates the instantaneous mean tangential rotating velocity profile taken at the exhaust of the cyclone with the 6.5 exit diameter straight pipe attached, operating at a Reynolds number of 14350 and a Geometric Swirl number of 3.043. For these conditions the P.V.C was found to be rotating about the exhaust at a frequency of 43 Hz. To examine the effect the straight pipe upon the velocity profile, this figure is best compared to figure 5a, as both have identical Reynolds and Geometric Swirl numbers. Although the maximum tangential velocity has fallen from 11.8 m/s to 5.83 m/s, a reduction of 51%, and the maximum negative tangential velocity has fallen from -5.05 m/s to -2.5 m/s, a drop of 50%, the main flow features have scarcely altered. Figure 9a clearly illustrates the classical region of negative tangential velocity emanating from the geometric centre of the system, and the outer high velocity area which encompasses the full 360 degrees of the profile in the θ direction.

These findings are a direct result of the decay that the P.V.C has been subjected to by passing through the straight pipe. The maximum and minimum tangential velocities have dropped, and the centre of the vortex has moved closer to the geometric centre of the system, this implies that with a long enough length of pipework attached to the cyclone exhaust the P.V.C effect can be completely dissipated by the decay of the swirling flow.

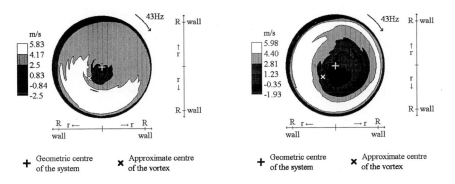

Figure 9a. Tangential velocities   Figure 9b. Axial velocities
Figure 9. Spatial distribution of rotating mean velocities for the cyclone with the straight pipe attached: Re=14250, Sg=3.043

Figure 9b is the axial velocity profile for the cyclone with the straight pipe attached, thus this figure can be compared to figure 5b. Once again the maximum and minimum velocity levels have been reduced when compared to the reference cyclone case. The maximum axial velocity has been reduced from 10.4 m/s to 5.98 m/s, a drop of 43%, and the maximum negative axial velocity has been reduced from -4.03 m/s to -1.93 m/s, a reduction of 52%.

The region of high forward axial flow has become more uniform in the θ direction, as it varies from 6 m/s down to 4 m/s, this suggesting a redistribution of the momentum flux. The area of the reversed axial flow region has been reduced in size, yet is still off centred and contains the precessing centre of the vortex Again these observations imply that the swirling flow, and thus the P.V.C, have been decayed by the presence of the straight pipe.

Figure 10a illustrates the instantaneous mean tangential rotating velocity profile taken at the exhaust of the cyclone with the 60° angled bend attached, operating at a Reynolds number of 28700 and a Geometric Swirl number of 3.043. For these conditions the P.V.C was found to be rotating about the exhaust at a frequency of 136 Hz. To examine the effect the 60° angled bend upon the velocity profile, this figure is best compared to figure 4a, as both have identical Reynolds and Geometric Swirl numbers. It has been previously shown that the introduction of the length of straight downstream pipework does not effect the shape and existence of the PVC. The introduction of the 60° angled bend however; has had a considerable effect upon the tangential velocity flow field. Both classical tangential velocity profile flow features, namely the region of high tangential velocity, and the off centred negative tangential velocity region, can not be found in this profile.

The outer edge of the profile is encompassed by a band of high velocity that is symmetrical in shape and velocity levels. Also, the region of negative tangential velocity has been centralised, and is also symmetrical about the geometric centre

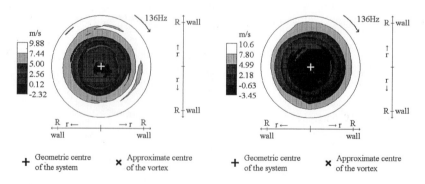

Figure 10a. Tangential velocities   Figure 10b. Axial velocities
Figure 10. Spatial distribution of rotating mean velocities for the cyclone with the 60°
angled bend attached: Re=28500, Sg=3.043

of the pipework exhaust in terms of shape and velocity levels.

The maximum and minimum tangential velocity levels have also undergone major alterations, the maximum velocity dropping from 25.1 m/s to 9.88 m/s, a reduction of 61%, and the maximum negative tangential velocity has dropped from -9.66 m/s to -2.32 m/s, a reduction of 76%.

Thus, it appears that the vortex has been centralised to the geometric centre of the pipework, and the precessing phenomenon has been destroyed. This is not in fact the case, as during the measurements a constant frequency of 136 Hz was detected at all times. Also, the presence of the negative tangential velocity region is indicative of an off centred vortex, however, the fact that this region is symmetrical about the centre of the system suggests that maybe this region is partially due to measurement error, where the control volume of the L.D.A system began its traverse beyond the centre of the system. This still does not account for the detected frequency though, and thus it must be assumed that a P.V.C is still present, although greatly decayed and centralised.

Figure 10b illustrate the mean instantaneous rotating axial velocities for the same experimental set up as figure 10a, thus comparisons can be made between this figure and figure 4b to examine the effect of the 60° angled bend upon the axial velocity profile. As with figure 10a, the profile can be seen to be symmetrical about the centre of the system. The reversed axial flow zone is now centralised in the profile, and the high forward axial velocity completely encompasses the periphery of the profile.

The levels of the maximum and minimum axial velocities have also dropped considerably when compared to figure 4b. The maximum positive axial velocity has dropped from 22.3 m/s to 10.6 m/s, a drop of 52%, and the maximum reversed axial velocity has fallen from -10.3 m/s to -3.45 m/s, a reduction of 67%.

Again this profile suggests that the application of the 60° angled bend has centralised the precessing vortex, and greatly damped the fluctuating velocities caused by the P.V.C. The velocity profiles are symmetric in both the tangential and axial planes, and the peak velocities in both cases cover a greater percentage of the flow field, again indicating a redistribution of the momentum flux. Hence a large flow reversal zone exists in the central core.

5 Conclusions

This work has provided information via phase averaging techniques of the three dimensional time dependent flows formed in the exhaust region of a swirling flow device. The work has produced a new model to describe vortex core precession, clearly showing that the phenomena is caused by the actual displacement of the centre of the vortex from and precession around the geometrical centre of the system. For tangential velocities a new rotating co-ordinate centre is needed, based on the precessional centre of the vortex. Angular momentum flux considerations based on this new co-ordinate centre clearly show why the tangential flow is accelerated by the P.V.C. phenomena, thus creating the 'classic' P.V.C. signal. The central reverse flow zone is always associated with the centre of the vortex in swirling flow and this work has been clearly shown to be correlated with the rotating precessional centre of the vortex.

The addition of external pipework has been shown to affect the P.V.C. and unique new information on the effect of a 90° hard bend produced. Axial velocity contours were shown to be much smoother in the θ direction; whilst reverse flow is destroyed. The P.V.C. phenomena propagated around the bend, albeit with much lower velocity levels.

The size and radial position of the P.V.C. is also effected by a change in Reynolds number. An increase in Reynolds number shrinking the P.V.C. size but increasing its amplitude and moving it further from the central axis of the exhaust.

Finally reverse flow on the outside of the flows points to the existence of large scale engulfment vortices on the outside of the flow, the level is dependant upon the geometrical Swirl number of the system. These can be concluded to cause vibration in the system which could increase the amplitude of the already present vibrations.

6 Nomenclature

Re - Reynolds number, defined as $\{4\,Q\,/\,(\,\pi\,D\,\nu\,)\,\}$
Sg - Geometric Swirl number, defined as $\{\,\pi\,R\,R_o\,/\,A_t\,\}$
A_t - tangential inlet area $(\,m^2\,)$
D - cyclone exhaust diameter $(\,m\,)$
Q - flowrate $(\,m^3\,/\,s\,)$
R - radius of exhaust $(\,m\,)$
R_x - non dimensional radius, defined as $\{\,r\,/\,R\,\}$
R_o - radius of cyclone main body $(\,m\,)$
r - radial position $(\,m\,)$
ν - kinematic viscosity $(\,m^2\,/\,s\,)$
P.C.O.V. - Precessional Centre Of the Vortex
P.V.C. - Precessing Vortex Core

7 References

1. O'Doherty, T., Biffin, M., Syred, N., The use of tangential offtakes for energy savings in process industries, Proc. Instn. Mech. Engineers Part E : Journal of Process Mechanical Engineering, vol. 206, 1992, pp 99-109
2. Gupta, A. K., Taylor, D. S., Beer, J. M., Investigation of combustion instabilities in swirling flows using real time LDV, Proc. Symposium on turbulent shear flows, Pennsylvania state university, April 1977
3. Beer, J. M., Chigier, N. A., Combustion aerodynamics, Applied Science

Publishers, London, 1972

4. Syred, N., Hanby, V. I., Gupta, A. K., Resonant instabilities generated by swirl burners, Journal of the institute of fuel, vol 46, no. 387, p402-407, December 1973

5. Gupta, A. K., Combustion instabilities in swirling flames, Gas Warme International, vol 28, 1979, part 1, p55-66

6. Mabrouk M Algamil, Studies of coherent structures in vortex and swirling flows, UWCC, Msc thesis, 1991

7. Ito, S., Ogawa, K., Kuroda, C., Decay process of swirling flow in a circular pipe, International Chemical Engineering, Vol. 19, no. 4, Oct 1979, p 600-605

8. Bank, N., Gauvin, W. H., Measurements of flow characteristics in a confined vortex flow, The Canadian journal of Chemical Engineering, Vol. 55, August 1977, p 397-402

9. Syred, N., Sidnu, B. S., Styles, A. C., Characteristics of swirling flow exhausing from nozzles with curved walls, UCC, dept mech eng, Wales, 1984

10. Gupta, A. K., Lilley, D. G., Syred, N., Swirl Flows, Abacus Press, Tunbridge Wells, Kent, 1984

11. Gouldin, F. C., Halthore, R. N., Vu, B. T., Periodic oscillations observed in swirling flow with and without combustion, 20th international symposium on combustion, the combustion institute, 1984, pp269-276

12. Reydon, R. F., Gauvin, W. H., Theoretical and experimental studies of confined vortex flow, Can. J. Chem. Eng., 59, 1981, pp 14-23

13. Baluev, E. D., Troyankin, Yu. V., Study of the aerodynamic structure of gas flow in a cyclone chamber, Teploenergetika, Vol. 14, No. 1, 1967, pp63

14. Schowalter, W. R., Johnstone, H. F., Characteristics of the mean flow patterns and structure of turbulence in spiral gas streams, AIChE J. 6, 1960, pp649

15. Smith, J. L., Journal of Basic Eng., Trans. A.S.M.E., Series E, 84, 602, 1962

16. Liem, T. L., van der Akker, H. E. A., LDV measurements of the turbulent flow in gas cyclones, Kramers Laboratorium voor Fysische Technologie, Delft University of Technologie, Prins Bernhardlaan 6, 2628 BW Delf, Neatherlands, 1993

17. Schetz, J. A., Injection and mixing in turbulent flows., Martin Summerfield, 1980

18. Yazdabadi, P. A., Griffiths, A. J., Syred, N., Investigation into the precessing vortex core phenomenon in cyclone dust separators, Proc. Instn. Mech. Engrs., Vol 208, 1994, pp147-154

19. Chanaud, R. C., Observations of oscillatory motion in certain swirling flows, J. Fluid Mechanics, vol 21 part 1, pp. 111-127, 1965

20. Yazdabadi, P. A., Griffiths, A. J., Syred, N., Characterisation of the PVC phenomena in the exhaust of a cyclone dust separator, Experiments in

Fluids, 17, 1994, pp84-94

21.   Kline, S. J., McClintok, I. A., Describing Uncertanties in Single-Sample Experiments, Mechanical Engineering, January, 1953, p 3-8

Mean Flow and Turbulence Measurements in a Rotating Passage Using a Stationary Fibre Optic LDA Probe

by

S C Cheah, H Iacovides, D C Jackson, H Ji and B E Launder

Mechanical Engineering Department, UMIST, Manchester, UK

ABSTRACT

This contribution describes the development of a data collection system in which a stationary fibre optic probe is used to provide LDA measurements for flows through rotating passages. The application of the technique is shown in the investigation of turbulent flow through a U-bend that rotates in orthogonal mode. A stationary 2-channel fibre optic probe is employed to collect instantaneous flow measurements from the rotating test section. The flow Reynolds number is 100,000, and the U-bend curvature ratio is 0.65. Three cases have been examined: flow through a stationary U-bend, flow through a U-bend rotating positively (the trailing side coinciding with the bend outer side) at a rotation number (Ro $\equiv \Omega D/U_m$) of +0.2, and rotation at a rotation number of -0.2. The resulting measurements produce a detailed mapping of the mean and fluctuating flow fields, which reveal the effects of strong curvature and rotation on the development on turbulent flow.

Keywords: rotating flow, curved passages, ensemble averaging

1   INTRODUCTION

Flow development within rotating, tightly curved passages, is of great interest to gas-turbine designers because of its relevance to internal blade cooling. In modern gas-turbines, serpentine passages traverse the inside of the blade and relatively cool air, extracted from the compressor stage, is circulated through them. The thermal and flow behaviour is determined by the presence of rotation as well as the passage geometry, with the former directly influencing both the mean and the fluctuating motion through the Coriolis force. In an attempt to predict the heat transfer within such passages numerical procedures have been developed (Bo et al 1991, Bo et al 1994); but it is essential to ensure that these solvers are able to produce reliable results so as to avoid potentially costly mistakes at the design stage. Numerical modellers therefore need to know how collectively, rotation, strong curvature and secondary flow influence the development of the mean and the turbulent fluid motion in a three dimensional flow.

Most experimental studies of rotating U-bend flows have so far been solely concerned with heat-transfer measurements (Guidez 1988, Wagner et al 1989). Moreover, the use of air as working fluid and the need to limit the mechanical stresses created by high rotation rates, has meant that the experimental models have had to be small in size. Consequently at best only side-averaged values have been measured. Local flow field measurements have, in fact, been obtained for flow through rotating channels of mild curvature by Kikuyama et al (1987, 1991), using hot-wire anemometry. These important studies revealed the combined effects of rotation and curvature on turbulence, but were concerned with channels of high aspect ratio in which secondary flow effects were negligible. In any case, the intrusive measuring technique employed, is not suitable for the investigation of flow through U-bends of strong curvature. In the case of turbulent flow through stationary curved ducts, detailed LDA measurements such as those of Azzola et al (1986) and Taylor et al (1982), have considerably improved our understanding of the flow behaviour and enabled turbulence modelling practices to be selected that can produce quantitatively accurate predictions. Similarly detailed measurements for flows through rotating U-bends in particular and rotating passages in general, are urgently needed if similar progress is to be made in advancing understanding and reliable simulation of flows through rotating systems. Elfert (1993) has recently applied Laser-2-Focus velocimetry (L2F) to the study of flow through a long rotating pipe. This was achieved by a system of rotating mirrors through which a stationary set of beams, aligned with the axis of rotation, was directed into the rotating test section. An important set of measurements was obtained which, though confined to one axial location, demonstrated the strong effects of the Coriolis force and of rotational buoyancy on the axial velocity and turbulence intensity.

The present work produces detailed flow-field data for rotating cases, by developing an experimental facility in which a stationary fibre optic probe is used to collect LDA measurements from a rotating test model. Our initial investigations, reported in this paper, concern flow in a U-bend of square cross-section that rotates orthogonally around an axis parallel to the axis of curvature. This test case was chosen because, despite its geometric simplicity, the resulting flow is highly complex and is also of direct relevance to blade-cooling applications. The main objective is to develop a facility that can produce flow measurements suitable for CFD code validation, over a range of Reynolds and Rotation numbers that are typical of engine operating conditions. The paper describes the experimental apparatus, the instrumentation and the data processing system that permit the collection of rotating flow data from a stationary probe and presents a representative selection of the resulting measurements. A fuller presentation and discussion of the experimental data is reported in Cheah et al (1994).

2   EXPERIMENTAL APPARATUS

In order to be able to resolve the flow field in sufficient detail, a reasonably large scale rig was necessary. As a result, rotational speeds needed to be kept at modest levels in order to avoid problems due to mechanical stresses. To reproduce engine conditions, which involve high Reynolds (Re $\equiv U_m D/v$) and rotation (Ro $\equiv \Omega D/U_m$) numbers, at modest rotational speeds, water was chosen as the working fluid because of its low kinematic viscosity. The test rig consists of a motor-driven turntable mounted in a 1.22 m diameter water tank. A U-bend of 50mm square cross-section and of curvature ratio Rc/D = 0.65, shown in Figures 1(a) and 1(b), is mounted on the turntable with the curvature axis of the duct parallel to the axis of rotation. Flow is fed to the duct through a passage that is built into the rotating turntable, as shown in Figure 1(b); the outflow is exhausted into the open water tank. The test section is made of 10mm thick perspex for optical access. The rotor turntable can be driven at any speed up to 250 rpm in either direction. The water volumetric flow rate is monitored with an orifice-plate arrangement. For all the tests, the fluid temperature was maintained at a steady level within $\pm$ 0.2 deg C, by cold water injection. At the entrance to the test section, a combination of fine wire meshes and a honeycomb section ensure that at a station five diameters upstream of the bend the flow is uniform and symmetric.

2.1   DATA ACQUISITION

The LDA system employed was a TSI two-channel, four-beam, fibre-optic probe system with frequency shifting on both the blue and green channels. A 4-watt Argon-ion laser was used to power the system and two counter processors(TSI 1980B) were used for signal validation. Subsequent data processing was done through a Zech data acquisition card on an Opus PC.

Flow measurements within the rotating U bend were made with a two channel 10 m long fibre-optic cable and probe. Because of the size and the complexity of the apparatus, a fibre optical system provided the best means of directing the laser beams into the rotating test model. The probe was clamped into a stationary X-Y-Z traversing unit mounted directly above the

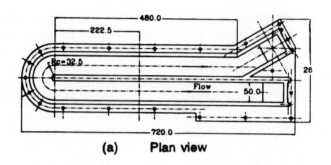

(a)   Plan view

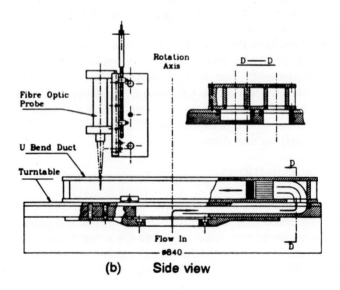

(b)   Side view

Figure 1 Experimental Apparatus

U-bend. With this arrangement, measurements could be made with the probe set at a fixed radius (with the bend rotating), or the probe could be traversed to collect data for flow in a stationary bend. In the case of the rotating U-bend, the data were collected as the duct swept past the stationary laser beams, as shown diagrammatically in Figure 2, with the laser beams entering the working section of the U-bend through the top wall. This measuring arrangement meant that the angular location at which each measurement was obtained had to be identified and recorded. The height of the measuring volume above the bottom wall was determined by establishing the intersection of the laser beams on the bottom wall as the reference point and then monitoring the height by which the probe was raised. As the probe is raised by 1 mm in air the measuring volume is raised by 1.33 mm. in water. The radial position of the probe measuring volume was determined from the radial location of the traversing unit. The laser measuring volume within the U- bend passage had a diameter of approximately 91 μm and a length of 1.8 mm, these being average dimensions for the two colours. As already mentioned, the angular position of the rotating bend was recorded with each measurement.

For the stationary case, the traversing unit was aligned with a radial line drawn from the axis of rotation to the bend axis. Traverses were made across the straight passages with the probe measuring the streamwise and transverse velocities within the passages. When radial traverses were made within the bend, the orientation of the probe relative to the traversing unit remained unchanged. Use of this method ensured that the laser probe remained undisturbed. Mean and rms quantities were determined from at least 2000 instantaneous velocity measurements at each measuring point along the traverse line.

The two orthogonal velocity measurements were obtained with the blue (488 nm) and green (514.5 nm) channels with a coincidence time window set at 20 μs for signal validation. The laser signals were filtered between high- and low-pass filters. Comparison between the periodic time averaged over 5 cycles to that averaged over 8 cycles was set so that only signals having periodic times differing by less than 1% were validated. Each validated measurement, according to its circumferential location, was stored in a corresponding angular segment bin in the form of a summed value. Each angular segment bin contained the accumulated sums of the measured instantaneous values of velocity and all velocity moments up to third order i.e. $\tilde{U}_1$, $\tilde{U}_2$, $\tilde{U_1^2}$, $\tilde{U_2^2}$, $\tilde{U_1 U_2}$, $\tilde{U_1^3}$, $\tilde{U_2^3}$, $\tilde{U_1 U_2^2}$ and $\tilde{U_1^2 U_2}$ together with the number of data points accumulated in that particular bin. For every 1000 data records collected, the acquisition program performed a statistical calculation based on all the collected data and calculated the mean velocity components, Reynolds stresses and triple correlations. This method

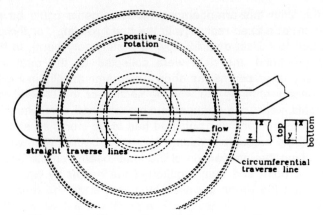

Figure 2　Circumferential traverse lines

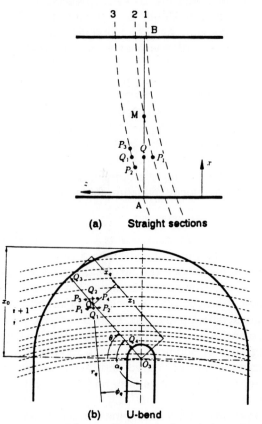

(a)　Straight sections

(b)　U-bend

Figure 3　Interpolated data lines

ensured that storage requirements were held within acceptable bounds. Each time a statistical calculation was performed by the acquisition program, the on-line PC displayed the latest velocity profiles in order to ensure that the number of measurements recorded was sufficiently high to produce a statistically independent ensemble averaged profiles. The number of measurements recorded along each traversing circle during the rotating flow traverses was between 60000 and 100000.

Seeding requirements had been determined during an earlier study of flow in a rotor-stator cavity, Cheah et al (1992), in which the same basic apparatus was used. It was found that the natural seeding material within the closed-loop water circuit produced a too low data acquisition rate. Normal hard particle seeding in the circuit was considered to be inappropriate as this could have caused damage to the large diameter seals that had been used in the construction of the turntable. Instead cornflour was selected; its introduction caused a dramatic improvement in the signal-to-noise ratio. The specific gravity of cornflour is about 1.3 and the measured diameters of the particles were in the range of 15-25 μm. Only small amounts were added so as to ensure as far as possible that there was only one particle in the measuring volume at any time. A more detailed account of the comparisons between the seeded and unseeded measurements can be found in Cheah et al (1992) and Ji (1994). Even with this seeding arrangement, however, several hours were needed to collect the quantity of data points required to generate mean values and turbulent statistics along each traverse line. Lower power levels caused by the use of optical fibres and also the use of backscatter, resulted in data collection rates, which while acceptable, were still slow. Moreover, because of the elongated control volume, at distances less than about 6mm (D/8) from the top and bottom walls, the data collection rate dropped to levels that made it impractical to carry out measurements within these regions. The inner and outer walls on the other hand, could be approached to within 1mm depending on the position of the plane of measurement within the duct.

The angular location of the rotating U-bend was obtained from a half-degree incremental encoder attached to the motor drive shaft. The motor drive shaft was connected to the turntable drive shaft by a toothed belt drive with a speed reduction of 4 to 1. With this arrangement and without any backlash in the drive, it was possible to measure the position of the U-bend to one eighth of a degree. With a pulse doubler in the LDA data acquisition interface, it was thus possible to achieve an angular resolution of 1/16 of a degree. The 5760 incremental pulses per revolution of the U bend are referenced to a reset pulse that comes from an optical switch attached to the turntable. The angular position at any time is thus determined by the pulse counter reading that is fed into the 24-bit shaft-angle input on the data acquisition system.

The rotational speed of the turntable was monitored using a counter fed with 360 pulses per motor shaft revolution from the incremental encoder. As this shaft rotated at four times the speed of the U-bend turntable this represented 1440 pulses per turntable revolution. As most of the current tests were run at 1.27 revolutions per second, this translated to a turntable speed of 1824 counts per second.

As the measurements along each circumferential line were collected over a period of several hours, it was essential that the rotational speed of the rig was maintained constant. The Ward Leonard drive system employed is an open loop system but, with the aid of an electronic speed control system that utilises the pulses from the incremental encoder, it was possible to maintain a constant rotational speed for the duration of the tests.

2.2   DATA PROCESSING

The velocities measured as the U-bend sweeps past the measuring volume are the absolute velocities for a stationary frame of reference, collected along circular arcs. It is therefore necessary to transform the measurements into a form that produces a clearer picture of the flow within the rotating passage. The resulting transformations produced a set of velocities for a rotating frame of reference, along straight traverse lines across the duct, with the two velocity components resolved along the axial and cross-duct directions.

In order to apply these transformations, two distinct regions of the rotating duct needed to be considered: the two straight sections and the bend itself. The probe collects measurements for two velocity components along the radial and circumferential directions relative to the axis of rotation. In order to transform these measurements to a rotating frame of reference, the local turntable velocity is subtracted from the measured fluid circumferential velocity. Then, at each angular location, the flow quantities that are initially measured along the circumferential and radial locations, are resolved along the axial and cross-duct directions.

Of course, it is convenient to record the data over planes normal to the duct axis. In order to achieve this for each selected plane, velocity data are successively recorded along three closely spaced circular trajectories that span the plane in question (AB in Figure 3a). The locations of these data records are denoted as points P_n in Figure 3a. Bi-linear interpolation, either between arcs 2 and 3 to the line segment AM, or between arcs 1 and 2 for the segment MB, establish the relevant data values at points such as Q along the straight line AB.

In the U-bend region data interpolation was made onto straight radial lines passing through the centre of the bend curvature in the same way that traverses were made for the stationary duct. The interpolation method employed is similar to that adopted in the straight sections. In this region the four adjacent data containing bins on the two nearest arcs were used for three linear interpolations as shown in Figure 3b. Information from bins at locations P1 and P2 and locations P3 and P4, is first interpolated to locations Q1 and Q2 respectively, both of which lie on a straight line that passes through the centre of rotation and which are on either side of the desired traverse line Q3Q4. Information from locations Q1 and Q2, is then interpolated to point Q on the selected radial straight line. Interpolations were made at 15-degree intervals around the bend. Each straight line had 49 interpolated points spaced at 1 mm intervals along the line. The 14 arcs on which data had been taken in the bend, provided more than 2500 bins of data points in the bend region from which the interpolated data were derived.

With data being stored at 1/8 degree intervals for the rotating cases, the length of the storage bins varied as the probe was set at different radii. At the smallest radius the width of the bin was 0.157 mm and at the largest 0.6 mm. Due to the difference in storage bin size, along traverse arcs of smaller radii there were more bins containing flow data, but with fewer data points stored in each bin. Within the straight sections, data from adjacent bins were therefore combined to obtain the ensemble average. The total circumferential length of the combined data bins did not however exceed 0.8 mm.

The bulk velocity, U_m, was measured through the orifice arrangement to an uncertainty less than $\pm 0.25\%$. The overall estimated uncertainties for the non-dimensionalised mean velocities in the streamwise and cross-duct directions U_z and U_x respectively, and of the corresponding turbulence quantities are:

$$U_z \le 0.02 U_m \quad ; \quad \sqrt{\overline{u_z^2}}, \sqrt{\overline{u_x^2}} \le 0.02\ U_m$$

$$U_x \le 0.03 U_m \quad ; \quad \overline{u_z^3}, \overline{u_x^3} \le 0.09 U_m^3$$

In arriving at the above figures, the following effects were considered: the uncertainty of the rotor speed, phase uncertainty of the doppler signal, uncertainty of wavelength and beam angle, velocity-gradient broadening, velocity bias and flow rate uncertainty associated with the orifice plate.

3   RESULTS

Measurements have been obtained along the symmetry plane (y/D = 0) and also along a plane near the top wall ($2y/D$ = 0.75), where y is the vertical distance from the symmetry plane.   A number of cross duct traverses have been produced covering a region from 3 hydraulic diameters (D) upstream of the bend entry to $8D$ downstream. The three cases investigated, Ro = -0.2, 0.0, 0.2, were all at a flow Reynolds number (Re = $U_m D/v$) of 100,000. A positive rotation direction, as shown in Figure 1, is defined as one in which the duct trailing (pressure) side coincides with the outer side of the U-bend. Due to space limitations only symmetry plane measurements are presented here. A more detailed account is provided in Cheah et al (1994).

The mean velocity measurements along the symmetry plane are presented in the vector plots of Figure 4. The stationary U-bend results indicate that at the bend entry there is a strong flow acceleration along the inner wall, while along the outer wall, the flow is strongly decelerated. Within the bend, a separation bubble is formed along the inner wall near the 90° location. Reattachment occurs at 1.7D downstream of the bend exit (for the stationary case). Positive rotation (Ro = 0.2) mainly affects the downstream flow. The Coriolis induced secondary motion opposes the return of the high momentum fluid to the inner side of the duct, almost doubling the length of the separation bubble.   Further downstream, the flow approaches fully developed flow conditions for a rotating straight duct. The high momentum fluid accumulates along the trailing side of the (rotating) duct, which for positive rotation, coincides with the U-bend outer wall.   Under negative rotation, the Coriolis-induced secondary motion transports the low momentum fluid to the outer side of the duct. As a result, a small region of flow reversal develops along the outer wall at the bend entry.   In the downstream region, the Coriolis-induced secondary motion begins to move the fluid across the duct towards the inner side causing reattachment to occur at around 2D downstream of the exit. After reattachment, the cross-duct movement transports the fast moving fluid to the duct inner side.

Profiles of the normal turbulent stresses along the streamwise and cross-duct directions are shown in Figures 5(a) and 5(b) respectively. In general, turbulence levels are found to be low in regions where the flow accelerates and high in regions of flow deceleration and separation. In the presence of negative rotation, turbulence levels are consequently higher than those for flow through a stationary U-bend whereas positive rotation is found to have the opposite effect.

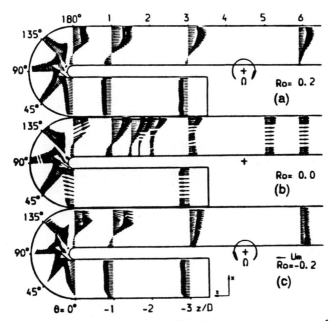

Figure 4 Measured velocity field along symmetry plane

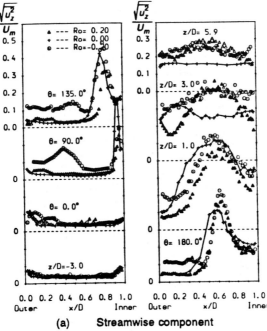

(a) Streamwise component

Figure 5 For legend see next page

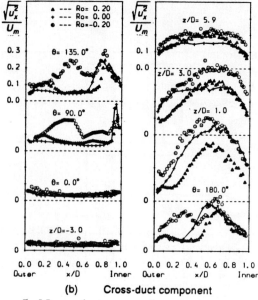

(b)   Cross-duct component

Figure 5   Normal stresses along symmetry plane

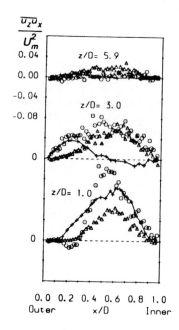

Figure 6   Profiles of shear stress $\overline{u_z u_x}$ along symmetry plane

The distribution of the turbulent shear stress component $\overline{u_z u_x}$ within the cross-duct plane is shown through the profiles of Figure 6. Since most of the significant shear stress variation occurs downstream of the bend, only the downstream profiles are shown. Within the first three downstream diameters, turbulence shear stress levels rise significantly. Evidently, the shear stress distribution is affected by rotation, negative rotation increasing stress levels and positive rotation having the opposite effect.

Finally, Figure 7 presents the distribution of triple moment components in the downstream region; again in this region the triple-moment correlations have substantially higher levels than upstream. These four triple moments reach their highest levels at 1D downstream of the U-bend. The triple moments are also affected by the sense of rotation. Negative rotation leads to higher values than positive rotation with the stationary results lying in between. Negative gradients of the triple moments imply a local diffusive gain of fluctuating energy. In the downstream region, the regions close to the inner and outer walls are consequently gaining energy at the expense of the highly turbulent fluid in the core.

4   CONCLUDING REMARKS

The work presented has shown that, with a suitable data monitoring and processing system, it is possible to collect accurate LDA measurements for internal flows through rotating ducts using a stationary, fibre-optic probe. This method provides the opportunity to generate much needed detailed information for three-dimensional flows through rotating passages and hence, to understand how the Coriolis force affects three-dimensional turbulent flows. The slow data collection rates associated with this technique do, however, limit the rate at which progress can be achieved, and prevents the resolution of the boundary layer regions along the walls normal to the laser beams. Higher data collection rates would further improve our understanding of rotating passage flows by allowing a more comprehensive examination of each case. It would also permit data to be gathered in the immediate near-wall regions which, in rotating flows, are especially influential.

Nevertheless, this method has generated the first detailed mappings of the mean and fluctuating flow fields in rotating U-bends of strong curvature. The results reveal that, within the U-bend the flow is dominated by the strong streamwise pressure gradients. At the U-bend exit these gradients cause flow separation along the duct inner wall, which starts at about the 90° location, and strong flow acceleration along the outer side. Regions of flow acceleration are also found to be regions of relatively low turbulence while, in regions of flow separation, turbulence levels are considerably higher. The triple-moment measurements indicate that in the exit region there is significant transport of turbulence from the core region to the near-wall regions through turbulence mixing. Orthogonal rotation at a rotational number of 0.2 is found to have significant effects on the flow development. Positive rotation increases the length of the separation bubble downstream of the bend and generally suppresses turbulence within the bend and in the downstream region. Negative rotation causes flow separation along the outer wall at the U-bend entry and also increases the overall turbulence levels.

ACKNOWLEDGEMENTS

The research has been supported by the SERC through Grant GR/F89039. Outstanding technical assistance has been provided by Mr J Hosker and Mr D Cooper. The camera-ready copy has been prepared by Mr Michael Newman with appreciated care. Authors' names are listed alphabetically.

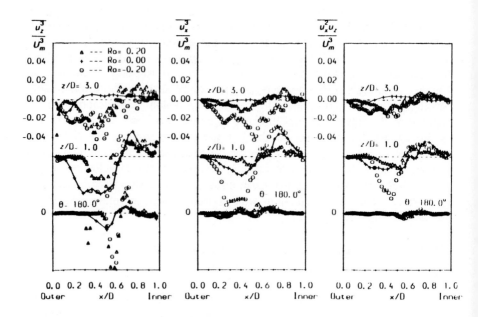

Figure 7   Profiles of triple moment components along symmetry plane

REFERENCES

Azzola J, Humphrey J A C, Iacovides H and Launder B E, "Developing turbulent flow in a U-bend of circular cross-section, measurement and computation". ASME J Fl Engrg, V108, pp 214-221, 1986

Bo T, Iacovides H and Launder B E, "Convective discretization schemes for the turbulence transport equations in flow predictions through sharp U-bends", To appear Int J Num Meth in Heat & Fluid Flow, 1994

Bo T, Iacovides H and Launder B E, "The prediction of convective heat transfer in rotating square ducts", 8th Symp on Turbulent Shear Flows, Munich, Germany, 1991.

Cheah S C, Iacovides H, Jackson D C, Ji H and Launder B E, "LDA measurements of an enclosed rotor-stator disc flow", Symp on LDA in Reciprocating, Reacting or Rotating Flows, Swansea, 1992.

Cheah S C, Iacovides H, Jackson D C, Ji H and Launder B E, "LDA Investigation of the flow development through rotating U-ducts", ASME Paper 94-57-226 Int Gas-Turbine Congress, The Hague, 1994 (To appear in ASME J Turbomachinery).

Elfert M, "The effect of rotation and buoyancy on flow development in a rotating circular channel", Engineering Turbulence Modelling and Experiments 2, Ed W Rodi and F Martelli, Elsevier Science Publishers BV, pp 815-824, 1993.

Guidez J, "Study of convective heat transfer in rotating coolant channel", ASME Paper 88-GT-33, Int Gas-Turbine Congress, Amsterdam, 1988.

Ji H, "An Experimental Investigation of two rotating turbulent shear flows with gas-turbine applications", PhD Thesis, Faculty of Technology, University of Manchester, 1994.

Kikuyama K, Nishibori K, Murakami M and Hara S, "Effects of system rotation upon turbulent boundary layer on a concave surface", 6th Symp on Turbulent Shear Flows, Toulouse, France, Sept 1987.

Kikuyama K, Nishibori K and Maeda T, "Turbulence in rotating curved channels", 8th Symp on Turbulent Shear Flows, Munich, Germany, Sept 1991.

Taylor A K M P, Whitelaw J H and Yianneskis M J, "Curved ducts with secondary motion: velocity measurements of developing laminar and turbulent flow" ASME J Fl Engrg pp 350-359, 1982.

Wagner J H, Johnson B V and Hajek T J, "Heat Transfer in rotating passages with smooth walls and radial outward flow", ASME paper 89-GT-272, Int Gas Turbine Congress, Toronto, 1989.

3-Dimensional Laser-Doppler Measurements in a Curved Flume

Robert Booij and Jan Tukker

Faculty of Civil Engineering, Delft University of Technology,
P.O. Box 5048, 2600 GA Delft, The Netherlands

Abstract. Measurements of the three Reynolds shear stresses in a fully developed flow in a mildly curved flume were executed using a 3-dimensional laser-Doppler velocimeter. An optical arrangement for turbulence measurements in shallow water flow in a flume was developed. For such flows equal eddy viscosity coefficients for the different Reynolds shear stresses are often assumed, which depend only on the main flow. However, this assumption appears not to be justified. It leads to an underestimation of the secondary flow intensity in a mildly curved flow.

Keywords. 3-dim. laser-Doppler velocimetry, shallow water flow, curved flow, turbulence, isotropic eddy viscosity

1. Introduction

The curvature of the flow in river bends gives rise to flow components perpendicular to the main flow direction, the so-called secondary flow (Bendegom, 1947; Rozovskii, 1961). This secondary flow has important consequences for the morphology of alluvial beds in river bends. Near the bottom the flow, and therewith the sediment transport, is directed towards the inside bank of the river bend. As a result the outer side of the river bend is eroded and undermining of the outside bank can occur (Odgaard, 1981).

A thorough knowledge of the secondary flow is essential for predictions about the morphology of alluvial river bends. Computed intensities and vertical distributions of the secondary flow depend strongly on the modelling of the Reynolds shear stresses, in particular the shear stress that accounts for the vertical exchange of secondary flow momentum. Generally, in turbulence models used in hydraulics gradient type momentum transport and isotropic eddy viscosity are assumed. Leschiner and Rodi (1979), for example, use a k-ϵ-model to compute the flow in a strongly curved flume. The isotropic eddy viscosity used in most estimates of the intensity of secondary flows in curved open channels is the eddy viscosity in a straight wide open channel.

To examine the correctness of the assumptions and, if not, to improve the turbulence modelling measurements of mean velocities and turbulence quantities were executed in a mildly curved flume in the Laboratory for Fluid Mechanics of the Delft University of Technology, the LFM flume, see fig. 1. (The measures of the flume are: radius of curvature, $R = 4.10$ m; width, $W = .50$ m and the water depth, $h \approx .05$ m). The cross-section of the flume is rectangular and bottom and sidewalls are transparent to allow velocity measurements by means of a laser-Doppler velocimeter (LDV).

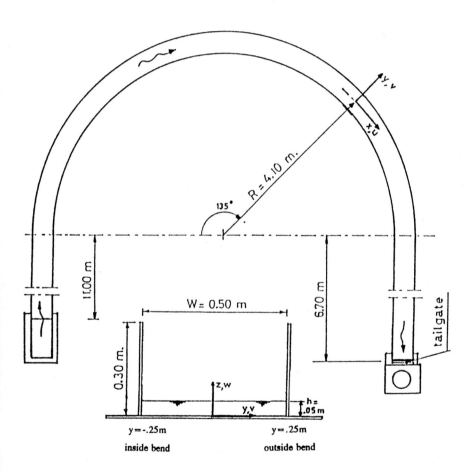

Fig. 1. The L.F.M. flume: layout and cross-section.

Previous measurements (Booij, 1985) showed that in the second part ($>90°$) of the bend the flow varied only slightly along the flume. Consequently the flow in the cross-section at 135°, where the measurements were executed, can be considered fully adapted to the flume curvature. In this part of the flume the main flow can be defined as the flow component in the direction of the flume axis and the secondary flow as the flow components perpendicular to the flume axis. Of the secondary flow components the horizontal one is the most important.

2. The experimental procedure

In the cross-section at 135° the instantaneous (local) velocity components u, v and w in longitudinal direction, x, in transverse direction, y, and in vertical direction, z, respectively, were measured (see fig. 1). The positive vertical co-ordinate is taken upward with $z = 0$ at the bottom of the flume and the positive transverse co-ordinate is taken outward with $y = 0$ at the flume axis. The instantaneous velocities (u, v, w) were decomposed into the time-averaged velocities ($\bar{u}$, $\bar{v}$, $\bar{w}$) and the turbulent velocity components (u', v', w'), e.g.

$$u = \bar{u} + u' \tag{1}$$

Cross-correlations of the turbulent velocity components yield the three Reynolds shear stresses

$$\tau_{xz} = \rho \overline{u'w'} \, , \quad \tau_{xy} = \rho \overline{u'v'} \text{ and } \quad \tau_{yz} = \rho \overline{v'w'} \tag{2}$$

and auto-correlations the three turbulence energy components $\overline{(u')^2}$, $\overline{(v')^2}$, $\overline{(w')^2}$.

Several years ago measurements were executed with a 2-dimensional reference beam mode LDV with a 10 mW He-Ne laser (Booij, 1985). A 2-dimensional LDV allows the simultaneous measurement of only 2 velocity components, from which only one shear stress can be obtained. To measure the three shear stresses several beam configurations, all with a vertical optical axis, were used. The conventional spatial beam arrangement of fig. 2a yielded simultaneous measurements of the velocity components u and v and therewith the shear stress τ_{xy}.

The LDV arrangement of fig. 2b with the beams in the cross-section plane allowed the simultaneous measurement of v and w and consequently the shear stress τ_{yz}. (The same information with a spatial beam configuration would have required the almost impossible setup with the optical axis in the flume direction.) The same arrangement rotated over 90° about the vertical yielded u, w and τ_{xz}. A drawback of this plane configuration is the small angle between the two measured velocity components. As a consequence the errors in the measured values for w and for the shear stresses τ_{yz} and τ_{xz} are relatively large. Moreover it is impossible to obtain higher order correlations of the three velocity components. The results of the 2-dimensional measurements were reported in Booij (1985).

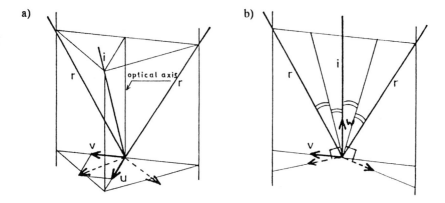

Fig. 2.   2-Dimensional LDV; laser beam configurations. (i and r are the illuminating and reference beams respectively, ---> are the measured velocity components)
a) spatial configuration, b) plane configuration

This paper describes the first results obtained by means of a 3-dimensional LDV. The measurements were executed with a DANTEC 3-colour fibre optics system in back-scatter mode with a 4 Watt Coherent Ar-ion laser and DANTEC BSA processors. The optical system used was a combination of a 2-dimensional probe and a 1-dimensional probe, see fig. 3. The velocity components measured with the 2-D probe (directions 2 and 1 in fig. 3) were the velocity component in transverse direction v and a combination of velocity components u and w. Combination of the last one with the velocity component (direction 3 in fig. 3) measured with the 1-D probe yielded u and w. The relatively large angle (60°) between direction 1 and 3 allowed a precise determination of w with this beam configuration. Identity of the measured particles was assured by requiring arrival time coincidence.

To simplify aligning of the many laser beams a water-filled prism, with sides perpendicular to the optical axes of the two probes, was attached under the measuring cross-section, see fig. 3. The refraction by the glass-plates forming the bottom of the flume and the top of the prism, which were at an angle to the optical axis, complicated the creation of overlapping measuring volumes for the different beam pairs. This refraction effect of the glass plates could be effectively compensated for by a small air-filled slit between them. The measuring system could be traversed in y- and z-direction and the two probes independently in the x-direction. This combination of 4 movements was required to be able to execute measurements over the cross-section without having to align the beams between measurements. A small measuring volume of the order of .1 mm in all three directions was realized.

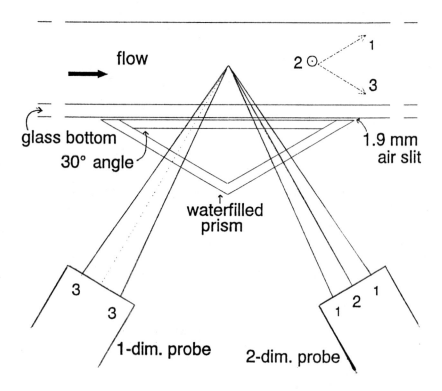

Fig. 3. The 3-dimensional LDV beam configuration.

The use of semi-automatic measuring procedures allowed measurements of 6 minutes each over a fine grid (up to 37 measuring points at each of 21 depths) to be executed within two months. The plots in this report are composed from the smoothed results of these measurements. The individual measurements are not reproduced here for clarity of the plots.

The relative shallowness of the flows in rivers, estuaries, seas, etc. indicates the importance of flow investigations in shallow flow models in laboratory flumes. The execution of 3-dimensional LDV turbulence measurements in shallow free surface flow in flumes is complicated. The obvious choice of measurement of the vertical velocity component through a (transparent) sidewall is hampered by the shallowness of the flow. The small angle between the laser beams results in a large measuring volume and in a low precision of the vertical turbulent velocity component in the centre part of the flume. Measuring in vertical direction through the free water surface in the forward or backward mode is complicated by the variable refraction of the beams at the unsteady free surface. The remaining possibility, measuring through the (transparent) bottom of the flume in backward mode allows measurements in a small measuring volume. However, the refraction index of the water requires complete re-alignment of the beams for every depth. With the system described above, including the water-filled prism and air slit, the alignment remains correct over the depth, making the execution of 3-dimensional LDV measurements in shallow water flow in flumes easier to accomplish.

3. Time-averaged velocities

Vertical profiles of the measured time-averaged velocities at different places in the cross-section are plotted in fig. 4. Fig. 4a shows an increase of the depth averaged main flow velocity $\bar{u}$ towards the outside bend. This is caused by convection of the main flow by the secondary flow.

The transverse velocity $\bar{v}$ in fig. 4b and the vertical velocity $\bar{w}$ in fig. 4c show the usual secondary flow cell, rotating clockwise. Near the water surface the flow is directed towards the outside flume wall and near the bottom towards the inside wall. The secondary cell is closed by a downward flow along the outer wall and an upward flow along the inner wall. The vertical velocity in the flume centre is almost zero. An additional small counter-rotating second cell near the upper outside flume wall can be distinguished (Vriend (1981a), see the sketch in fig. 5). This second cell stems from an instability of the flow along a concave wall (Vriend, 1981b).

The presence of a two cell system is more obvious in fig. 6, which shows a few horizontal velocity profiles of the measured secondary flow components at different depths. The distribution of the secondary flow velocity components in the lower outside bend corner appears to indicate the existence of another, very small, secondary cell in this region. This kind of secondary flow cell, which is also found in straight flumes is caused by an anisotropy of the turbulence.

104

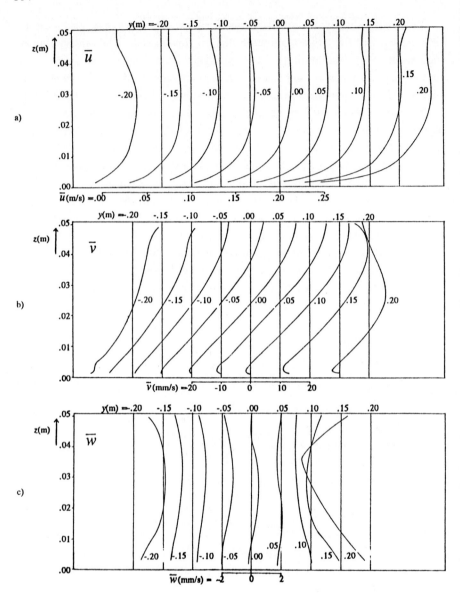

Fig. 4. Vertical profiles of the time-averaged velocity components at different places in the cross-section: a) $\overline{u}$, b) $\overline{v}$, and c) $\overline{w}$. (The velocity scale for $y = 0$ m is given at the bottom. The velocity scales for the other verticals are displaced, see top of each plot. The vertical lines indicate the characteristic velocity, e.g. $u = .2$ m/s.).

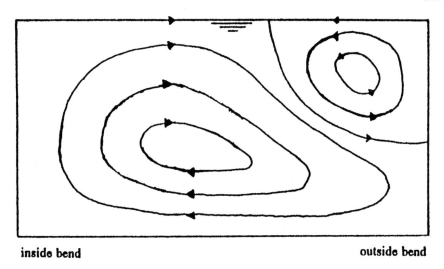

inside bend   **outside bend**

Fig. 5. Sketch of a two cell system of the secondary flow.

The main secondary flow cell has two marked effects on the distribution of the main flow. The convection of the main flow by the secondary flow distorts the nearly logarithmic velocity profiles of the main flow in free surface flow, as slower moving water from the inside bend is convected towards the outside bend in the upper part of the flow. As a result the maximum value of the main flow velocity $\bar{u}$ in the centre of the flume is reached well below the water surface. Moreover the local depth-averaged main flow velocity increases towards the outside wall.

Attention should be paid to the difference of the scales for the different velocity components in fig. 4. The scale of $\bar{v}$ is about 10% of the scale of $\bar{u}$ and the scale of $\bar{w}$ is of the order of 1% of that of $\bar{u}$, which is of importance for the relative errors in the secondary flow velocity components and in particular for the accuracy in the determination of the vertical velocity.

4. Shear stresses and eddy viscosities

The vertical profiles of the different Reynolds shear stresses obtained from the cross-correlations of the measured turbulent velocity components at different places in the cross-section are plotted in fig. 7. These shear stresses provide for the transport of main flow momentum and secondary flow momentum in

106

combination with the convective transport. Assuming gradient-type momentum transports in the various directions, eddy viscosity coefficients ν_{xz}, etc, for those transports can be derived, using

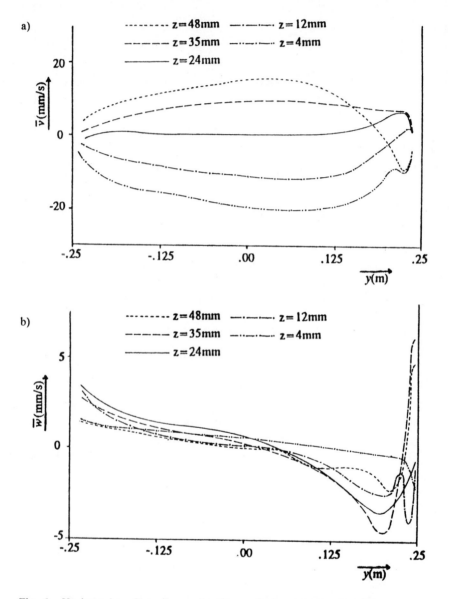

Fig. 6. Horizontal profiles of secondary flow velocity components at different depths. a) transverse velocity component, b) vertical velocity component

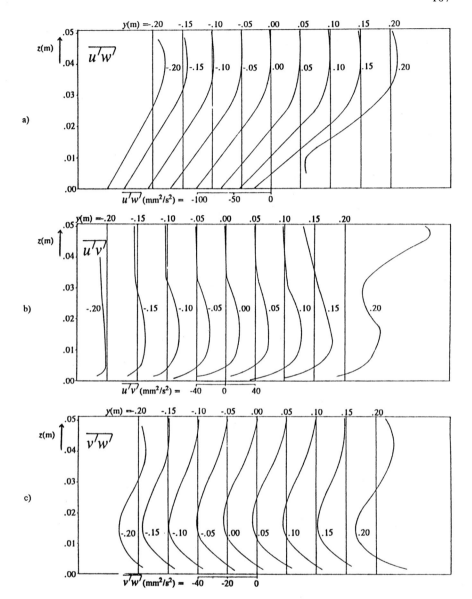

Fig. 7. Vertical profiles of the shear stresses at different places in the cross-section: a) $\overline{u'w'}$, b) $\overline{u'v'}$ and c) $\overline{v'w'}$. (The scale for $y = 0$ m is given at the bottom. The scales for the other verticals are displaced, see the top of each plot. The vertical lines indicate the characteristic shear stress value.)

$$\frac{\tau_{xz}}{\rho} = -\overline{u'w'} = -\nu_{xz}\left(\frac{\partial\overline{u}}{\partial z} + \frac{\partial\overline{w}}{\partial x}\right) \simeq -\nu_{xz}\frac{\partial\overline{u}}{\partial z} \tag{3}$$

$$\frac{\tau_{xy}}{\rho} = -\overline{u'v'} = -\nu_{xy}\left(\frac{\partial\overline{u}}{\partial y} + \frac{\partial\overline{v}}{\partial x}\right) \simeq -\nu_{xy}\left(\frac{\partial\overline{u}}{\partial y} - \frac{\overline{u}}{R}\right) \tag{4}$$

$$\frac{\tau_{yz}}{\rho} = -\overline{v'w'} = -\nu_{yz}\left(\frac{\partial\overline{v}}{\partial z} + \frac{\partial\overline{w}}{\partial y}\right) \simeq -\nu_{yz}\frac{\partial\overline{v}}{\partial z} \tag{5}$$

Here the approximations

$$\frac{\partial\overline{w}}{\partial x} \simeq 0 \; ; \quad \frac{\partial\overline{v}}{\partial x} \simeq -\frac{\overline{u}}{R} \quad \text{and} \quad \frac{\partial\overline{w}}{\partial y} \simeq 0 \tag{6}$$

are used. The assumptions for the velocity-derivatives $\partial\overline{w}/\partial x$ and $\partial\overline{v}/\partial x$ can be made as the flow varies very slowly in the longitudinal direction in this part of the flume. $\partial\overline{w}/\partial y$ is negligible except near the sidewalls. Vertical profiles of the three introduced eddy viscosities in the center of the flume are plotted in fig. 8. Apparently the three eddy viscosities are far from equal and the assumption of an isotropic eddy viscosity in analytical or numerical flow models is not justified for this kind of flows. The singular behaviour of the eddy viscosities at points with a zero velocity gradient is of minor importance as the shear stresses around those points are small too, generally.

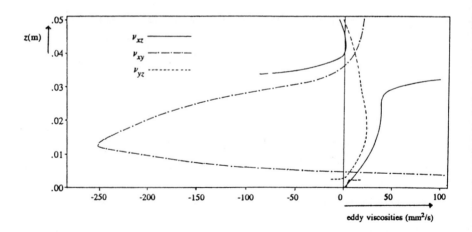

Fig. 8.   Vertical distribution of the three eddy viscosities in the centre of the flume.

The shear stress τ_{xz} transports main flow momentum in vertical direction. The measured τ_{xz} profile deviates significantly from the linear profile found in a wide straight open channel. In the central part of the flume the shear stress almost vanishes in the upper part of the flow. The corresponding eddy viscosity ν_{xz}, obtained from the measured shear stress and the vertical gradient of the main flow using equation (3) is also very small in this region (appreciably smaller than the theoretical value around half depth). To understand this behaviour of τ_{xz} and ν_{xz} it is useful to consider the distribution of the shear stress τ_{xy}.

The shear stress τ_{xy} transports main flow momentum across the flume. In a large central part of the flow the transport appears to be in the wrong direction, corresponding to a negative eddy viscosity (see fig. 8). An explanation can be found taking a closer look at the turbulence structure.

In boundary layer flow and in open channel flow an important aspect of turbulence is the movement of coherent parcels of water from the bottom in upward direction. These coherent parcels and to a lesser extent the compensating fluid moving in downward direction provide the largest contribution to the stress τ_{xz} (Brodkey, 1974). The upwards moving parcels arriving at a higher level have a positive value of w, and a relatively low value of u and v, implying a positive w' and negative u' and v'. Downward moving parcels have analogously negative w' and positive u' and v'. Both kind of parcels give a negative contribution to $\overline{u'w'}$ and a positive contribution to $\overline{u'v'}$. The positive value of $\overline{u'v'}$ measured has evidently nothing to do with a gradient of $\bar{u}$ across the flume, but stems from the mechanism described above.

The low values of τ_{xz} and τ_{xy} near the surface indicate that the upward movement of coherent parcels which is important over the whole depth in a straight flume is hindered in curved flow, especially in the upper part. This is confirmed by the distribution of the turbulence energy per unit of mass, k, see fig. 9, where

$$k = \frac{1}{2}\left(\overline{(u')^2} + \overline{(v')^2} + \overline{(w')^2}\right) \tag{7}$$

In fig. 9 the measured turbulence energy in the centre of the cross-section is compared to the corresponding turbulence energy in a straight flume. The damping of the turbulence in the central part of the curved flow increases with the distance to the bottom.

The hindrance of the vertical movement of the fluid parcels in the upper part of the curved flow is presumably caused by the typical main flow profile induced by the convection of the main flow by the secondary flow. In wide channels the convection by secondary flow is less important compared to the influence of the bed shear stress. Consequently the rate of damping may depend on the aspect ratio h/B of the channel or flume. This damping of the vertical movement of the parcels coming from the bottom region reduces the shear stresses and eddy viscosities measured in the upper part of the flow compared to a straight flume.

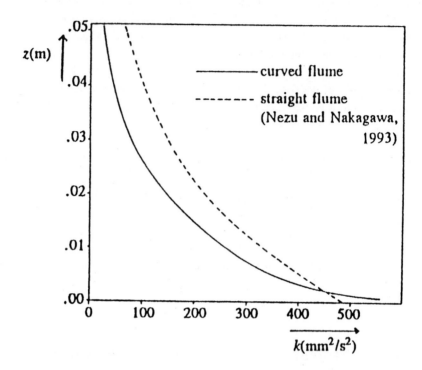

Fig. 9. Comparison between the turbulence energy in curved flow and in straight flow.

The shear stress τ_{yz} transports secondary flow momentum in vertical direction. The measured negative values of $\overline{v'w'}$ agree with the positive w' and negative v' for upward moving and the negative w' and positive v' for downward moving parcels. The upward moving parcels are created around $y^+ - yu_*/v - 30$ (Brodkey et all., 1974) or at about 1 mm from the bottom. This is near the depth where the negative secondary velocity is largest. A positive shear stress τ_{yz} that is expected below this depth is found in these measurements (see fig. 7), where in the previous 2-dimensional measurements near the bottom the noise prevented this.

The eddy viscosity v_{yz} obtained from the measurements is smaller than the measured eddy viscosity of the main flow v_{xz}. This can be understood, when it is considered that the shear stresses do not only depend on local conditions, but also on the properties transported by the parcels created in the bottom regions. This smaller eddy viscosity leads to higher secondary velocities than predicted with estimates based on the assumption of equal eddy viscosities and explains the deviations often found between measurements and theories in curved flow. In general, use of an isotropic eddy viscosity should be applied with care.

5. Estimate of the intensity of the secondary flow

Estimates of the intensity of secondary flows in curved open channels are based on the eddy viscosity in a straight wide open channel, generally. In a straight wide open channel the shear stress distribution is linear

$$\tau_{xz} = \rho \overline{u'w'} = -\rho u_*^2 (1 - \frac{z}{h})$$

(8)

with u_* the shear stress velocity. Assumption of a logarithmic velocity distribution

$$\overline{u} = \frac{u_*}{\kappa} \ln \frac{z}{z_0}$$

(9)

with κ the Von Karman constant and z_0 a measure for the roughness of the bottom, leads to a parabolic eddy viscosity distribution

$$\nu_{xz} = \kappa u_* h \frac{z}{h}(1 - \frac{z}{h})$$

(10)

To estimate the secondary flow the same main flow distribution (equations 8 and 9) in mildly curved flow is assumed generally. Assumption of an isotropic eddy viscosity, $\nu_{yz} = \nu_{xz}$, equal to the eddy viscosity in the straight flow (eq. 10), then yields the expression for the fully developed secondary flow velocity profile (Kalkwijk and Booij, 1986).

$$\overline{v} = -\frac{u_{av}}{\kappa^2} \frac{h}{R} [F_1(\frac{z}{h}) + F_2(\frac{z}{h}) - 2(1 - \frac{f}{\kappa})(1 + \frac{f}{\kappa}(1 + \ln \frac{z}{h}))]$$

(11)

where u_{av} is the local depth-averaged longitudinal velocity, f is a friction coefficient, $f = u_*/u_{av}$, and F_1 and F_2 are form functions defined by

$$F_1(\frac{z}{h}) = 2 \int_{z_0}^{z} \frac{\ln(z/h)}{z-h} dz \quad \text{and} \quad F_2(\frac{z}{h}) = \frac{f}{\kappa} \int_{z_0}^{z} \frac{\ln^2(z/h)}{z-h} dz$$

(12)

The sign in equation (11) depends on the sense of the curvature.

112

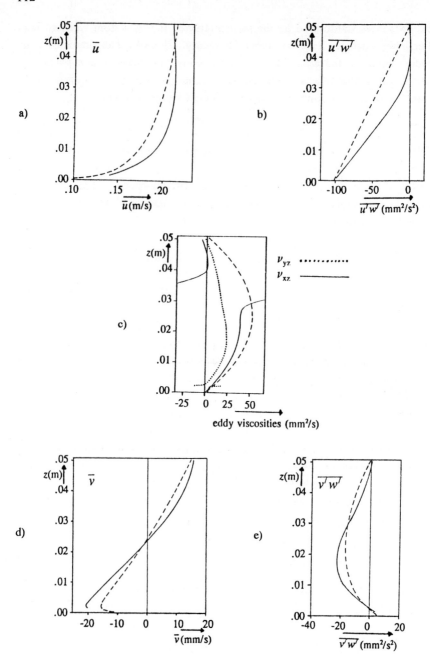

Fig. 10. Comparison between estimated and measured profiles:
a) $\overline{u}$, b) $\overline{u'w'}$, c) eddy viscosities, d) $\overline{v}$ and e) $\overline{v'w'}$.

In fig. 10 the various profiles in this estimation procedure are compared with the measured profiles in the centre of the flume. In figures 10a and 10b the profiles for the main flow velocity $\bar{u}$ and for the corresponding shear stress τ_{xz} are compared. In fig. 10c the parabolic eddy viscosity profile is compared to the two eddy viscosities ν_{xz} and ν_{yz} obtained from the measurements. In figures 10d and 10e the secondary flow velocity profiles $\bar{v}$ and the corresponding shear stress profiles τ_{yz} are compared. The relatively large velocity differences over depth in the assumed logarithmic main flow velocity profile would lead to an overestimated secondary flow. However, the parabolic eddy viscosity, which is strongly overestimated, in particular compared to the measured eddy viscosity ν_{yz}, causes a strong exchange of secondary flow momentum over the vertical and hence suppresses the estimated secondary flow intensity. This effect more than compensates the overestimation due to the assumed logarithmic main velocity profile and leads to underestimated secondary flow intensities, see fig. 10d.

6. Conclusions

- By carefully designing the optical arrangement a 3-D LDV configuration can be used for semi-automatic simultaneous measurements of various turbulence moments e.g. the three Reynolds shear stresses in a flume, even in case of shallow water flow.
- Descriptions of the flow in curved channels using gradient-type momentum transport concepts are of limited use.
- Use of equal eddy viscosities for the various momentum transports in curved channel flow is not justified, and should be applied with care in other flow conditions. The assumption leads to an underestimation of the secondary flow intensity in mildly curved flow.
- Longer measuring times may yield higher turbulence moments with sufficient accuracy to validate advanced turbulence models, for example Reynolds stress models.

References

Bendegom, L. Van, 1947, Some considerations on river morphology and river improvement, De Ingenieur, vol.59-4, pp. B1-11, in Dutch. English translation: Nat. Res. Council of Canada, Tech. Transl., 1054, 1963.

Booij, R., 1985, Eddy viscosity in channel bends, Proc. int. symp. on refined flow modelling and turbulence measurements, Iowa, E21.

Brodkey, R.S., et al., 1974, Some Properties of Truncated Turbulence Signals ri Bounded Shear Flows, J. Fluid Mech., 63, 209.

Kalkwijk, J.P.Th., and R. Booij, 1986, Adaptation of secondary flow in nearly-horizontal flow, J. Hydr. Res., 24(1), pp. 19-37.

Nezu, I., and H. Nakagawa, 1993, Turbulence in open-channel flows, IAHR Monograph, Balkema, Rotterdam.

Odgaard, A.J., 1981, Transverse bed slope in alluvial channel bends, J.Hydr.Div., ASCE, vol.107 (HY12), pp. 1677-1694.

Leschiner, M.A., and W. Rodi, 1979, Calculation of strongly curved open channel flow, J.Hydr.Div., ASCE, vol.105 (HY10), pp. 1297-1314.

Rozovskii, I.L., 1961, Flow of water in bends of open channels, Israel Progr. for Scientific Transl., Jerusalem (orig. publ. 1957).

Vriend, H.J. De, 1981a, Velocity redistribution in curved rectangular channels, J.Fluid Mech., vol.107, pp. 423-439.

Vriend, H.J. De, 1981b, Steady flow in shallow channel bends, Communications on Hydraulics, 81-3, Dept. of Civil Engrg., Delft Univ. of Technology.

APPLICATION OF LASER INDUCED FLUORESCENCE FOR MEASURING THE THICKNESS OF EVAPORATING GASOLINE LIQUID FILMS

Jean-François LE COZ and Thierry BARITAUD

Institut Français du Pétrole, BP 311, 92506 Rueil-Malmaison, FRANCE

Abstract. Laser-induced fluorescence is used to quantify the thickness of liquid films. An optimisation of the fluorescent tracer to make it evaporate at the same ratios as gasoline is performed since the final application is the study of evaporating films, as those found in intake pipes of port-injected spark-ignited engines. The liquid studied is iso-octane and the tracer is a ketone. Because of distillation, the tracer concentration in the liquid film may not remain constant. The most suitable tracers are C_6 ketones.

The optical set-up uses a single optical fibre for both laser excitation and fluorescence collection. It is designed for local measurements through transparent walls. A precise liquid wedge is used for calibration. An application is shown on a transparent pipe, with stationary air flow and pulsed low-pressure injection. The technique has proven to be able to show the unstationary phenomena. The impact of the spray, the deposition of a large quantity of fuel and the displacement of liquid waves are quantified and followed with time.

Keywords. Liquid films, evaporation, laser-induced fluorescence, gasoline engines

1. INTRODUCTION

This work is justified by the need of a characterisation of the injection of gasoline in the intake pipe of a port-injected spark-ignited engine. In these engines, injectors introduce large droplets of gasoline in the intake ports. Evaporation and convection by the air flow do not prevent a substantial part of liquid fuel from impacting the wall. Large droplets have particularly a small drag coefficient compared with their inertia and go directly to the walls. A continuous liquid film is formed. The primary atomisation (from the injector) can be practically negated by the film formation which acts as a buffer between the injector and the valve seat. This film flows towards the valve seat, where atomisation occurs during intake valve opening [1]. This secondary atomisation can be poor, so that rather large liquid drops enter the cylinder and can even reach the walls of the combustion chamber [2]. Thus, the fuel/air mixture is affected by film formation in the intake pipe [3]. This in turn determines the combustion process and the final unburned hydrocarbon emissions, which are especially important under cold conditions. There is therefore a need for a better understanding of the two-phase flow inside the intake manifold of port-injected engines.

We will first examine the different methods used for measuring the thickness of liquid films. Then the technique based on laser-induced fluorescence will be described. The choice of the fluorescent tracer will be explained in detail. The calibration procedure on a precise liquid wedge, and an application on a transparent tube with pulsed injection will end this paper.

2. LIQUID FILM DIAGNOSTIC DEVELOPMENT

2.1 Choice of measuring technique

A rather common method employed for two-phase flows uses the difference in conductivity between the liquid and the gaseous phase. Conductivity probes mounted in a wall sense the resistance of the fluid and, hence, the thickness. This method is not suited for liquids with weak conductivity, as hydrocarbons [4]. In the case of pure water, which has also a weak conductivity, a capacitance method is applicable [5]. Unfortunately, the dielectric constant of a hydrocarbon is much lower than that of water : 1.94 for iso-octane, 78.54 for water. Therefore, the capacitance method would be more difficult to apply to gasoline films.

Optical methods seem to be better adapted to the measurement of liquid films. Absorption through the film have already been used to determine the thickness of methanol films [6]. This solution requires optical access on both sides of the film. Fluorescence has been used essentially for measuring the thickness of oil films [7, 8, 9, 10, 11, 12] or the concentration of chemical solutions (e.g. [13]). With the use of laser sources and optical fibres mounted flush in opaque walls, the probe is practically non-intrusive. For transparent walls, the optical system can be very simple and easily movable for scanning large zones. This is the reason why we decided to build such a system.

2.2 Theory of fluorescence

A volume of matter illuminated can absorb a part of the incident light. The molecules from this matter are brought to an excited level. When they return to their original energy level, they can emit light with the same wavelength, that is elastic diffusion, or with a longer wavelength, that is inelastic diffusion. Among possible types of inelastic diffusion, fluorescence is characterised by an inter-mediate energy level and short lifetime, of the order of 10^{-8} seconds. Fluorescent light is omnidirectional and incoherent. The intensity of fluorescence is described by the equation : $F = \Phi I_0 (1 - e^{-\varepsilon bc})$ where Φ is the quantum efficiency, I_0 the intensity of the incident radiation, ε the molar absorptivity, b the path length in the liquid film and c the molar concentration. For small thicknesses, absorption is not very intense and the equation can be fitted by a linear function : $F = \Phi I_0 \varepsilon bc$. For these conditions, the fluorescent intensity is simply proportional to the thickness of the measured volume, to the concentration of the fluorescent compound, and to the excitation intensity.

2.3 Optical configuration

A fibre optic laser induced fluorescence technique is developed, similar to a previous one used to determine the thickness of oil films [14]. The fluid can be fluorescent itself, or a fluorescent tracer can be added to provide a controlled fluorescence. The optical arrangement is suited for transparent walls (Figure 1).

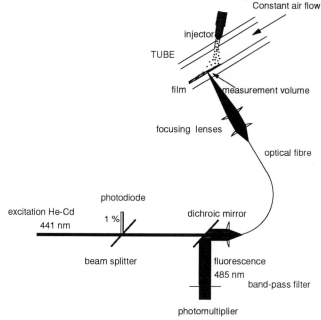

Figure 1 : Optical set-up for laser-induced fluorescence using optical fibres

Excitation is done by a continuous laser He-Cd whose wavelength (441.6 nm) is suited for the fluorescent tracer that will be selected. The laser light is carried by an optical fibre and focused on the inner surface of a transparent wall. The collection of the fluorescence light -located around 485 nm- is performed in the reverse optical path. A dichroic mirror separates the fluorescence from the excitation and directs it towards selective optical filters and a photomultiplier. The excitation intensity is monitored by a photodiode that receives about 1% of the incident light reflected by a beamsplitter.

Several filters are used to isolate the tracer fluorescence : a band-pass filter (λ=485 nm, $\Delta\lambda$=20 nm) centred about the maximum in the fluorescence spectrum of the selected tracer; a long-pass filter (λ>455 nm) for attenuating the light at the excitation wavelength, coming from reflections in the interfaces : beam splitter, transparent wall; a short-pass filter (λ<500 nm) for excluding phosphorescence from the tracer and possible fluorescence from optical components and the material of the wall.

2.4 Signal Processing

The film thickness can be calculated from the fluorescence and the excitation intensity. The following equation is used if absorption is not strong :

$$b = \frac{F}{I_0 (\varepsilon\Phi) c}$$

The requirements are the following. The background light which is different from controlled fluorescence by the tracer needs to be subtracted precisely. A precise calibration is needed, in order to determine the correspondence between the thickness and the fluorescence efficiency for a given tracer concentration : $(\varepsilon\Phi)c$. As will be shown below, a small non-linear behaviour is observed so that we will use a second order calibration function. This allows to increase the maximum measurable film thickness. The last requirement is that the tracer concentration must be constant in the liquid film.

3. CHOICE OF A FLUORESCENT TRACER

3.1 Requirements

The evaporation process gives a very restrictive condition for our application : the compounds responsible for fluorescence need to have a constant concentration in the liquid film during evaporation. This is equivalent to say that their concentrations are the same in the liquid phase and in the vapour phase with respect to the total concentration of fuel; in other words, that they follow correctly the evaporation of fuel.

Normal gasoline is not recommended since it is made of hundreds of components with very different vapour pressures and fluorescence efficiencies. A chemical analysis of gasoline has shown a molar concentration of 0.3% of naphthalene [15]. Fluorescence is generated mostly by this molecule. It cannot be taken as representative of gasoline, because its boiling point is located at 217 °C. This temperature is much higher than the 50% evaporated point of ordinary 95 octane number gasoline, which is close to 109°C.

For these reasons, we preferred to use a pure substance that is not fluorescent at all, and to add a fluorescent tracer. This way we can at least quantify our accuracy : the equilibrium between the film and the vapour phase could theoretically be determined exactly. We chose iso-octane as fuel, because it is a common reference fuel with high purity and is easy to obtain. This fuel is fluorescence free.

Ketones are widely used in studies observing mixing processes in gaseous turbulent jets [16] or fuel distribution inside the cylinder of spark-ignition engines [17, 18, 19]. Their use in engine studies is usually justified by a boiling point which is relatively close to that of iso-octane [20], or close to the 50% point of the distillation curve of normal gasoline. In all these applications they are used in the vapour phase. The use of aromatics compounds, as laser dyes, used in oil films or in diesel sprays is excluded because of their very high boiling points.

The fluorescence from ketones comes from their C=O carbonyl bond [21]. Mono-ketones have one isolated carbonyl bond. Their absorption is maximum near 270 nm, and changes slightly with the molecule. The fluorescence spectrum ranges from 330 nm to 630 nm with a maximum near 430 nm. Tests done in our laboratory showed a maximum fluorescence of 3-pentanone with an excitation around 300 nm. Fluorescence is divided by two with excitation at 275 or 330 nm, and divided by 25 with an excitation at 350 nm.

α-Diketones, as biacetyl, are molecules with two neighbouring carbonyl bonds. The conjugaison of the two bonds shifts markedly the absorption band to 440 nm, in the visible range. This allows the use of numerous materials for the walls. For our concern, plexiglas is commonly used in transparent models of pipes or cylinder heads. It does not allow transmission of UV light. This is why α-diketones are interesting. Moreover, continuous lasers near 440 nm are much more affordable than those near 300 nm.

One could classify the ketones by their boiling point and choose that with the boiling point closest to iso-octane. This would simply insure that iso-octane and the ketone have similar pressure vapours. When mixed however, the two compounds do not behave as if they were separated. Repulsive molecular interactions disturb their respective equilibrium with the vapour phase. The most visible effect is the formation of a lower azeotrope. This phenomenon can be seen as an expulsion of the minority compound from the liquid to the vapour phase by the majority compound. In principle, the tracer concentrations used are low. So, the tracer vapour pressure above a tracer/iso-octane liquid mixture is higher than with the pure liquid tracer. The practical consequence is a higher evaporation rate of the tracer than expected according to its pressure curve. Practically, the tracer whose evaporation follows iso-octane needs to have a higher boiling point than iso-octane. This results in a lower vapour pressure and tends to compensate for the repulsive forces exerted by iso-octane. A list of possible tracers is given in Table 1 with their boiling points.

Table 1 : List of possible tracers compared with iso-octane

molecule	boiling point (°C)	type
iso-octane	99	alcane
3-pentanone	102	mono-ketone
4-methyl 2-pentanone	117	"
2-hexanone	127	"
2,3 butanedione	88	α-diketone
2,3 pentanedione	108	"
2,3 hexanedione	130	"

3.2 Liquid vapour equilibriums

In the following discussion, we will use the term "discrimination coefficient" to indicate the ratio between the concentration of ketone in the vapour phase and that in the liquid phase, relatively to the concentrations of iso-octane in each phase. Figure 2 illustrates this definition :

$$\frac{dy}{dx} = \alpha \frac{y}{x},$$

where x stands for iso-octane and y for the tracer. α is the discrimination coefficient between the two phases. If α is larger than 1, the tracer concentration decreases in the liquid phase during evaporation. If α is lower than 1, the tracer concentration increases in the liquid phase.

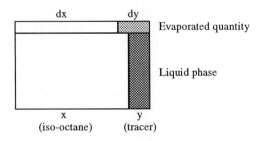

Figure 2 : Discriminated evaporation

Assuming that this coefficient is constant, as practically observed with 3-pentanone, one can integrate the equation versus x, the iso-octane quantity, and obtain the evolution of the tracer concentration versus the remaining liquid fraction :

$$\frac{y}{x} = \left[\frac{x}{x_0}\right]^{(\alpha-1)}\left[\frac{y_0}{x_0}\right].$$

We will determine the values of the discrimination coefficient for the most suitable tracers whose coefficients are probably close to unity.

Calculations of the liquid/vapour equilibrium are possible only for mono-ketones. Interaction between the carbonyl bonds in α-diketones cannot be included in the chemical models at this time. The liquid/vapour phase equilibrium was completely calculated for 3-pentanone/iso-octane mixtures [22]. The following result is obtained : at low concentrations the molar fraction of 3-pentanone in the vapour phase is *twice* that in the liquid phase (Figure 3). This is observed for concentrations of 3-pentanone lower than 5% and for temperatures between 298K and 375K. This shows the usefulness of selecting carefully another tracer.

Additional calculations were made with two hexanones : 2-hexanone and 4-methyl 2-pentanone. Figure 4 shows the equilibriums for iso-octane + 5% ketone mixtures. It is observed that the ideal behaviour is approached by 2-hexanone. One can notice that the concentration in the vapour phase changes rapidly with the boiling point.

Experiments have been conducted to validate the calculations with mono-ketones and to provide data with one α-diketone, 2,3 hexanedione, which was guessed to be a satisfactory tracer. A continuous air flow is introduced in a bubbler

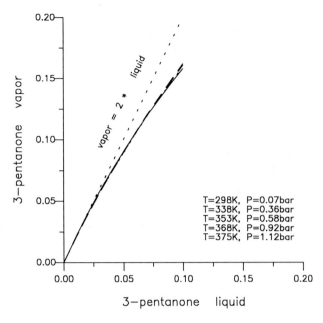

Figure 3 : Liquid/vapour equilibrium, mixture iso-octane + 3-pentanone

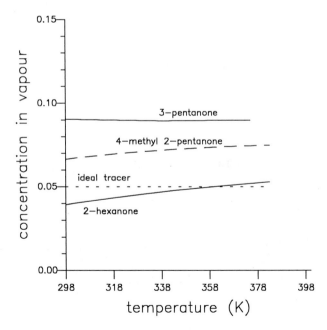

Figure 4 : Liquid/vapour equilibriums calculated for
three mixtures iso-octane + 5% ketone

containing a mixture of iso-octane with 5% ketone. Some samples are taken for analysis as evaporation goes along. Then, from the initial concentration, the remaining quantity and the new concentration, the discrimination coefficient is determined. The final results at 20°C are presented in Table 2.

Table 2 : Discrimination coefficients, 20°C, mixtures iso-octane + 5% ketone

tracer	calculated α	measured α
3-pentanone	1.8	2.19
2-hexanone	0.8	0.8
2,3 hexanedione	not available	0.75

The accuracy is about 0.05 for both calculated and experimental values. The agreement between calculations and experiments is satisfactory. The difference between hexanone and hexanedione is not important. The best tracer at this moment is 2,3 hexanedione for an application in the visible range with iso-octane. No perfect matching of a tracer can be achieved.

3.3 Error Analysis
An error analysis is now possible, based on the discrimination coefficient. If the tracer concentration decreases in the evaporating liquid film, the situation worsens as the liquid quantity decreases. Figure 5 shows that the tracer concentration diverges progressively from the original one as evaporation progresses.

If it is assumed that the concentration is constant, an error is introduced in the determination of the liquid amount, or the equivalent film thickness. The absolute error on the liquid amount is :

$$\left\{ 1 - \left[\frac{x}{x_0} \right]^{(\alpha - 1)} \right\} \cdot x$$

This error is proportional to the amount of liquid x. If we divide this error by the initial liquid amount, we obtain the following relative error :

$$\left\{ 1 - \left[\frac{x}{x_0} \right]^{(\alpha - 1)} \right\} \cdot \frac{x}{x_0}.$$

Figure 6 shows that the relative normalised error passes through a maximum during evaporation. Discrimination coefficients of 0.8 and 1.2 lead to estimated maximum errors about 7 % of the initial liquid amount. A discrimination coefficient of 2. gives an error of 25% of the initial liquid amount. From this error analysis we can conclude that 2-hexanone and 2,3 hexanedione follow correctly the evaporation of iso-octane.

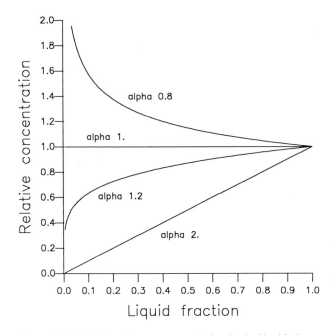

Figure 5 : Evolution of tracer concentration in the liquid phase
versus discrimination coefficient

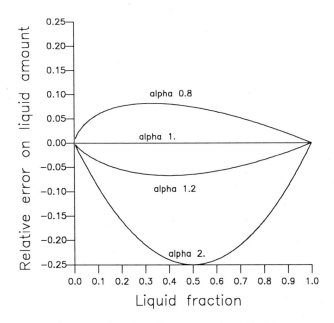

Figure 6 : Evolution of relative error made on liquid amount
versus discrimination coefficient, when assuming constant tracer concentration

4. CALIBRATION

4.1 Calibration set-up

To calibrate the technique, it is important to create very precise thicknesses of the order of those found in common situations. We use a wedge illustrated in Figure 7 that is formed by two superposed flat quartz plates separated by a known gap on one side. The flatness and the roughness of the surfaces are thus controlled. The thickness varies linearly from 0 to 500 microns from one side to the other. The displacement of the measurement volume along the horizontal plane is done by a motion controller unit whose accuracy is about 1 µm.

The influence of temperature on fluorescence was not investigated in this work. It could be introduced as a factor changing the gain of the calibration curve.

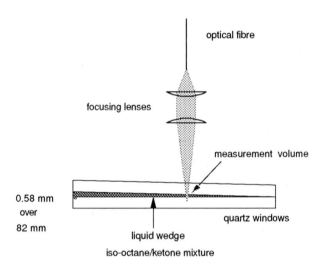

Figure 7 : Calibration wedge

4.2 Calibration procedure

Measurements are done at locations corresponding to increasing thicknesses with a step of 20 microns, from 20 to 540 microns. The acquisition lasts for 1 second without excitation and 1 second with excitation. The mean level before excitation is an offset reference for the two signals. Then, the maximum in the photomultiplier signal is only kept, and divided by the excitation intensity. The following expression summarises the data processing :

$$s_{signal} = \frac{\max(signal_2) - \overline{signal_1}}{laser_2 - laser_1}$$

'signal' represents the P.M. signal, 'laser' the laser intensity signal. Index 1 indicates the first part of the record, without excitation, that is the offset. Index 2 indicates the second part with excitation. Despite the filters, background light is always present, probably additional fluorescence from optical components made of glass, and excitation light reflected on the optical interfaces. A second test is done with pure iso-octane, which allows quantification of the background solely.

$$S_{background} = \frac{\overline{background_2} - \overline{background_1}}{\overline{laser_2} - \overline{laser_1}}$$

The final normalised useful fluorescence is the normalised fluorescence signal corrected by the normalised background :

$$S_{fluorescence} = S_{signal} - S_{background}$$

After a complete path on the quartz window and after processing, the data are plotted in a graph versus film thickness.

4.3 Calibration results

One tracer was tested : 2,3 hexanedione. Calibrations with 2% and 4% mass concentrations are shown in Figure 8. A very linear response is shown below 200 microns. A concentration of 2% gives a rather low fluorescence intensity so that a concentration of 4% was selected. We recall that a second order function is fitted to the calibration curve.

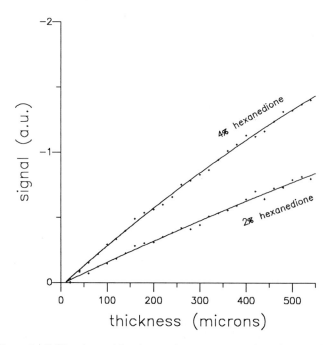

Figure 8 : Calibrations with mixtures isooctane + 2 and 4% hexanedione

We verified that either quartz or plexiglas can be used in a model of intake pipe or cylinder head. Using an upper window made of plexiglas instead of quartz in the calibration wedge does not modify the fluorescence signal significantly. Figure 9 shows calibrations made with the same mixture iso-octane + 4%

126

hexanedione with a quartz and a plexiglas window. No additional background is added by using plexiglas. The results are encouraging and show the possibility of using plexiglas instead of quartz.

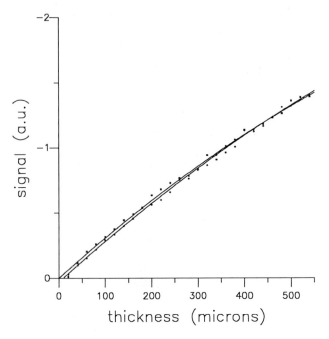

Figure 9 : Calibrations with a mixture isooctane + 4% hexanedione
with a plexiglas wall and a quartz wall

5. APPLICATION TO A MODEL INTAKE MANIFOLD

5.1 Description
The goal is to test the measurement technique in a situation similar to the intake pipe of an engine under cold conditions. The test pipe is cylindrical and made of quartz. Its inner diameter is 36 mm. It is inclined 30 degrees with respect to the horizontal. A constant air flow is introduced in the pipe at atmospheric pressure. A standard injector BOSCH EV 1.3 A is fixed to it, directing the spray with an angle of 45 degrees with respect to the tube axis. Injection is pulsed as in a port-injected engine. The fuel is supplied at 3 bars. The Sauter mean droplet diameter in the spray is estimated around 200 microns. The monitoring system controls the injection duration and frequency. A PC synchronisation board generates coding signals and injection control pulses, which makes possible the simulation of engine conditions.

5.2 Operating condition

One condition is reported to assess the performance of the measuring technique. The air flow rate, the injection frequency and the injected quantity are representative of the simulated engine condition reported in Table 3.

Table 3 : Simulated engine condition

displacement volume	441 cm^3
engine speed	1200 rpm = 10 Hz
volumetric efficiency	0.8
air flow rate (0°C, 10^5 Pa)	0.211 m^3/min
fuel/air equivalent ratio	1
injection duration	94 CAD = 13 ms
injected quantity per shot	43 mm^3

5.3 Experimental procedure

The measuring probe attached to the optical fibre is simply moved from the calibration wedge to the pipe. We observed that background light changes with time for these unstationary measurements. Probably feedback occurs between reflections and liquid film because we have an open film. Thus, a correction is necessary for each measurement location versus time. For this purpose, each test is repeated with pure iso-octane, which allows quantification of the background solely. The acquisition and processing procedures are the same as those used for the calibration except that the signal and the background are functions of time. The fluorescence signal is obtained by normalising by the laser power and by subtracting the normalised background. This signal is transformed to a thickness signal by inverting the equation determined by the calibration.

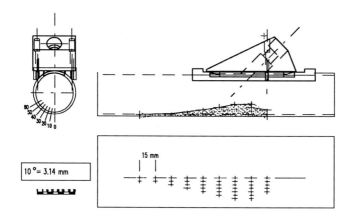

Figure 10: Test pipe with measurement locations

5.4 Results

A mixture iso-octane + 4% hexanedione is used for scanning the tube in the region where the spray impinges the wall. Figure 10 displays the measurement locations; the axial step is 15 mm, the angular step is 10 degrees. Translation and rotation stages allow displacement of the probe throughout the entire pipe.

Some curves are reported at locations lying on the intersection line between a vertical plane and the bottom of the tube (on the side opposite to the injector). We present ensemble mean cycles calculated over 100 injection cycles in Figure 11. In the upstream region, the impact of the spray is clearly shown. A delay between the onset of injection and the increase of the film thickness is visible. The injection duration is also detectable in location 3. The film thickness gradually increases downstream. The arrival of the spray on the pipe wall clearly generates a peak which is transformed into a wave in location 6. This wave is moving downstream under the influence of air flow and gravity. The film velocity can be determined from the time delay of the wave peak between two locations. The wave is not yet damped at a distance of 100 mm from the impact zone.

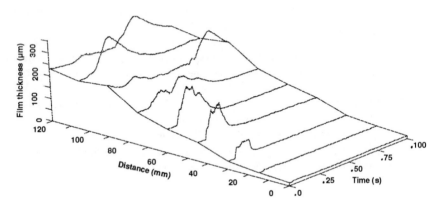

Figure 11 : Evolutions of the ensemble mean film thickness along the bottom of the tube

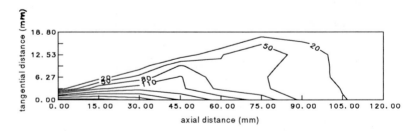

Figure 12 : Map of the total mean film thickness in the investigated area (unit : microns)

A map of the total mean film thickness is shown in Figure 12. The x-axis is a symmetry line. The film is spread on a large area in the impact zone. This area becomes narrower and the thickness increases when the liquid is flowing down. A strong stream is finally formed in the bottom of the tube, with a thickness of 250 microns.

6. CONCLUSIONS

Laser-induced fluorescence is chosen for measuring the thickness of gasoline liquid films. The main obstacle for our gasoline engine application comes from evaporation. The fluorescent tracer used has to be matched to the main compound, here iso-octane, so that its concentration in the liquid film stays as constant as possible. The best tracers found in the ketone family are 2-hexanone and 2,3 hexanedione. The latter was selected because it is excitable in the visible range and can be used with plexiglas transparent models.

An optical fibre is used for carrying the excitation light at 441 nm and collecting the fluorescence light at 485 nm in the reverse optical path. At one end a dichroic mirror separates the fluorescence from the excitation. At the other end lenses make focal point in the liquid film.

A precise liquid wedge, that makes the thickness vary from 0 to 500 microns, is used for calibration. The working tracer concentration is 4% by mass, which showed sufficient fluorescence yield and rather weak absorption by the liquid, i.e. slightly non-linear response with film thickness up to 500 microns.

An application of the technique is realised on a model intake pipe : a stationary air flow inside an inclined straight pipe, and pulsed low-pressure injection. The technique has proven to be able to show very interesting phenomena. The impact of the spray, the deposition of a large quantity of fuel and the displacement of liquid waves are typically quantified and followed with time.

Acknowledgements

The authors would like to acknowledge the technical support provided by V. Ricordeau and P. Lessart, and the calculations of liquid/vapour thermodynamic equilibriums by J. Vidal. C. Catalano from Politecnico di Milano, Italy, participated in building the optical set-up and perfecting the experimental procedure. This work was partially supported by the Commission of the European Communities within the frame of the Joule II Programme, Contract JOU2-CT92-0162, by the Swedish National Board for Industrial and Technical Development NUTEK, by the Austrian Government, and by the Joint Research Committee of European automobile manufacturers (Fiat, Peugeot SA, Renault, Rover, Volkswagen and Volvo) within the IDEA EFFECT Programme.

130

References

[1] H.P. Lenz, Mixture Formation in Spark-Ignition Engines, SAE Publications

[2] F. Vannobel, J.B. Dementhon, D. Robart, "Phase-Doppler Anemometry Measurements on a Gasoline Spray inside the Inlet Port and downstream of the Induction Valve; Steady and Unsteady Conditions", 6th International Symposium on Applications of Laser Techniques to Fluid Mechanics, Lisbon 1992, also Applications of Laser Techniques to Fluid Mechanics, edited by R.J. Adrian, D.F.G. Durao, F. Durst, M. Maeda and J.H. Whitelaw, 6, 507, Springer-Verlag

[3] M.J. Miller, C.J.E. Nightingale, P.A. Williams, "Measurement of Spark-ignition Engine Mixture Preparation and Assessment of its Effects on Engine Performance", IMechE 1991, C433/022

[4] I. Sherrington, E.H. Smith, "Experimental Methods for Measuring the Oil-film Thickness between the Piston-rings and Cylinder-wall of Internal Combustion Engines", Tribology International, Vol. 18, No. 6, Dec. 1985

[5] R.K. Sun, W.F. Kolbe, B. Leskovar and B. Turko, "Measurement of Thickness of Thin Water Film in Two-phase Flow by Capacitance Method", IEEE Transactions on Nuclear Science, Vol. NS-29, No. 1, Feb. 1982

[6] V. H. Muller, H. Bellmann, J. Himmelsbach, A. Elsasser, W. Samenfink, and M. Hallmann, "Gemischaufbereitung und Wandfilmverhalten in Saugrohren von Ottomotoren", MTZ No. 3 and No. 4, March and April 1994 (in German)

[7] A.E. Smart, R.A.J. Ford, "Measurement of Thin Liquid Films by a Fluorescence Technique", Wear No. 29, pp 41-47, 1974

[8] R.A.J. Ford, C.A. Foord, "Laser-based Fluorescence Technique for Measuring Thin Liquid Films", Wear No. 51n pp 289-297, 1978

[9] L.L. Ting, "Development of a Laser Fluorescence Technique for Measuring Piston Ring Oil Film Thickness", Journal of Lubrication Technology, Vol. 102, pp 165-171, Apr. 1980

[10] D.P. Hoult, J.P. Lux, V.W. Wong, S.A. Billian, "Calibration of Laser Fluorescence Measurements of Lubricant Film Thickness in Engines", SAE 881587

[11] B.T. Shaw II, D.P. Hoult, V.W. Wong, "Development of Engine Lubricant Film Thickness Diagnostics Using Fiber Optics and Laser Fluorescence", SAE 920651

[12] D.E. Richardson, G.L. Borman, "Using Fiber Optics and Laser Fluorescence for Measuring Thin Oil Films with Application to Engines", SAE 912388

[13] D.A. Walker, "A Fluorescence Technique for Measurement of Concentration in Mixing Liquids", J. Phys. E: Sci. Instrum. 20 (1987) pp 217-224

[14] J. Trapy, "Mesure des épaisseurs de films liquides au moyen de la fluorescence induite par laser", I.F.P. report 38687, 1991 (in French)

[15] E. Robert, personal communication, Institut Français du Pétrole

[16] A. Lozano, B. Yip, R.K. Hanson, "Acetone : a Tracer for Concentration Measurements in Gaseous Flows by Planar Laser-induced Fluorescence", Exp.in Fluids Vol. 13, pp 369-376 , 1992

[17] T.A. Baritaud, T.A. Heinze, "Gasoline Distribution Measurements with PLIF in a S.I. Engine", SAE 922355, 1992

[18] T.A. Heinze, T.A. Baritaud, "Development of Laser Induced Fluorescence Method to Visualise Gas Distribution in a S.I. Engine", 6th International Symposium on Applications of Laser Technology to Fluid Mechanics, Lisbon, July 20-23 1992

[19] W. Lawrenz, J. Kohler, F. Meier, W. Stolz, R. Wirth, W.H. Bloss, R.R. Maly, E. Wagner, M. Zahn, "Quantitative 2D LIF Measurements of Air/fuel Ratios During the Intake Stroke in a Transparent SI Engine", SAE 922320

[20] H. Neij, B. Johannsson, M. Aldén, "Development and Demonstration of 2D-LIF for Studies of Mixture Preparation in SI Engines", 25th Symp. (Int.) on Combustion, Irvine, July 31st-August 5th 1994

[21] C.N.R. Rao, Ultra-Violet and Visible Spectroscopy, Butterworths, London 1967

[22] J. Vidal, personal communication, Institut Français du Pétrole

Manifold Injection and the Origin of Droplets at the Exit of an Inlet Valve

M.Posylkin, A.M.K.P.Taylor and J.H.Whitelaw

Imperial College of Science, Technology and Medicine
Mechanical Engineering Department
Exhibition Road, London SW7 2BX, U.K.

Abstract. Distributions of droplet size, axial velocity and liquid flux have been measured downstream of a model inlet valve with a phase-Doppler velocimeter. The valve had a diameter of 40mm, a 45° bevelled edge and was arranged at the exit of a straight tube of inside diameter 34mm with the lifts of 2, 4 and 6mm. Suction was applied by a variable speed electric fan to simulate the intake stroke of a cylinder of 400cc displacement at 1200 RPM, with an air flow rate for the 4mm valve lift of 0.99m³/min, corresponding to the bulk velocity of 38m/s in the valve gap. Gasoline was supplied either by a commercial injector located 70mm from the valve or by its introduction on the surfaces of the valve and port. The injection of fuel with an open valve led to Sauter mean diameters between 80 and 100μm, lower than the 130μm appropriate to the free unconfined spray. With 2mm valve lift, the spray emerged between angles of 10° to 70° to the horizontal axis and with larger valve lifts it emerged with narrow angle close to that of the valve head surface immediately upstream of the valve bevel, that is 25°. The introduction of fuel as a liquid streams led to a Sauter mean diameter of 240μm with the maximum flux emerging from the edges of the valve and seat bevels, at angles of 50° and 40° to the horizontal axis, respectively.

Keywords. Engine, inlet valve, fuel, spray, injection, atomization, impingement, liquid films, droplets

1. Introduction

Port injection of fuel can lead to mixture control, fuel economy and low emissions particularly when combined with a three way catalyst in a spark-ignition engine. Electronic control offers an added advantage in setting of the injection timing for individual cylinders and Quader (1982) has demonstrated that axial stratification, which can be beneficial with lean mixtures, can be achieved by injecting fuel into the inlet port before the end of the intake stroke with the inlet valve open.

The droplet sizes produced by a pintle injector, see for example Hardalupas, Taylor and Whitelaw (1990) and Dementhon (1992), correspond to Sauter mean diameter of about 130μm and, for injection with an open valve, Miller and Nightingale (1990) quote 10μm as the largest droplet diameter which is able to follow the gaseous phase in the inlet port and therefore able to avoid impingement on the solid surfaces of the valve and the port. Thus, most droplets impinge on solid surfaces and the consequences depend on the orientation and location of the injector. The liquid may re-atomize, for example by impingement, or remain as liquid on the surfaces to be removed ultimately by evaporation or by the air flow.

Extensive research has been carried out to study the effect of the interaction between sprays and solid surfaces. Photographic methods were used with single droplets, for example by Stow and Hadfield (1981) and Chandra and Avedesian (1991), and have shown that the droplet impinging on a solid surface may spread, rebound or form smaller droplets according to the droplet size, velocity, surface roughness and temperature. However, the results were limited to droplets of approximately 1mm diameter and should be extrapolated with care to the substantially smaller droplets of the impinging spray produced by a gasoline injector. Hardalupas, Okamoto, Taylor and Whitelaw (1992) used phase-Doppler velocimetry to examine the consequences of a spray impinging on a inclined disk, without co-flowing air, and showed that resulting Sauter mean diameters were smaller than in the free spray upstream of the disk but up to 180μm close to the downstream edge, possibly due to impingement on the surface covered with liquid film from previous injections. Similarly, impingement on a rod led to secondary atomization and a reduction of Sauter mean diameter immediately upstream of the rod and to a liquid film with most of the liquid content of the spray in the form of a stream downstream of the rod. Measurements of the droplet sizes downstream of the inlet valve in a steady and unsteady flow simulation of the iso-thermal manifold of a spark ignition engine, for example Miller and Nightingale (1990) and Dementhon (1992), have shown diameters in excess of 200μm which are larger than those of the incident spray and probably due to the formation of liquid film. Formation of surface liquid was also conjectured from measurements of the droplet sizes in a firing engine and with injection with the inlet valve closed, Vannobel, Robart, Dementhon and Whitelaw (1994) and Posylkin, Taylor, Vannobel and Whitelaw (1994), which have shown large effects of the residence time of the fuel on the surface of the valve from which there must be evaporation, though not immediately after the start-up when the inlet valve is still cold.

Due to the complexity of induction phenomena it is difficult to distinguish between the consequences of a spray impinging on a solid surface and of the removal of liquid from a solid surface by co-flowing air. Thus, the purpose of the present work was to measure size, velocity and flux characteristics at the exit from an inlet valve of a spray produced by injection of fuel with an inlet valve open and to compare the results with these for subsequent experiments in which liquid fuel was allowed to flow down the valve stem and down the pipe wall to be subsequently removed by air flow, as with injection during a cold start with the inlet valve closed. The following section describes the experimental arrangement and instrumentation. The results are presented in the third section and the paper ends with a summary of the more important findings.

2. Experimental Arrangement

The steady flow arrangement used to investigate the nature of droplets in the valve curtain area is shown in figure 1, together with co-ordinate system, and with an injector directing a fuel spray towards a valve. Air was drawn through a flow contraction, with area ratio of 16, mounted on the top of the inlet pipe which was 400mm long, with inner and outer diameters of 34 and 40mm. The valve was located on the axis of the inlet pipe to within ±0.5mm by a spacer arrangement with a valve stem of 4mm diameter and 410mm long. The axial location was controlled and measured by a micrometer arrangement to within ±0.05mm. The bulk of the micrometer arrangement was located in the contraction and so did not disturb the flow within the inlet pipe. The valve diameter was 40mm at the bottom face and 34mm diameter at the upper face, with a 45° bevelled edge which corresponded to a similar bevelled edge on the exit of the

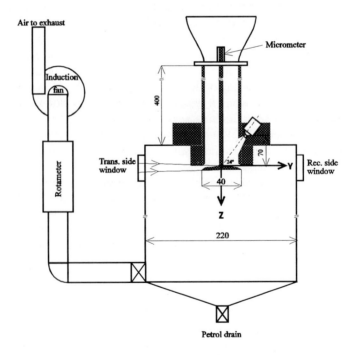

Figure 1: The experimental arrangement

inlet pipe. The valve arrangement was mounted vertically on the lid of a closed cylindrical chamber of 220mm inside diameter and a variable speed electric fan was connected to the bottom of the chamber to induce an air flow through the inlet pipe and the volumetric flow rate of the air was monitored by a rotameter.

Unleaded gasoline of density 765 kg/m^3 and viscosity 0.515 cSt at 25°C was drawn from a tank by an electric fuel pump and supplied through a diaphragm pressure regulator, which maintained a nominally constant pressure of 3 bar gauge, to either a commercial injector or, through a control valve and rotameter, to the upper valve surface or the port wall. The fuel injector was located 70mm from the valve and at 24° to the vertical axis of the pipe, so that the axis of symmetry of the injector was directed to the centre of the valve. Fuel was supplied to the surface of the valve through six radial 1mm holes located 3mm above the upper surface of the valve which had a hollow stem, and to the port wall from a pipe of 5mm diameter located 400mm above the valve from which it was allowed to spread freely over the circumference of the wall.

Two 90mm diameter glass windows provided optical access to the chamber and permitted collection of the scattered light at an angle of 30° to the axis of the transmission optics of the phase-Doppler anemometer. The optical characteristics of the anemometer are summarized in Table 1 and the electronic signal was processed by a counter triggered by a TTL pulse train of 1ms duration from the delayed gate output of an oscilloscope and synchronized with the injection pulse from the driving electronics of the injector. Size-discriminated measurements of the velocity vectors were achieved by rotation of the plane of the beams about the axis of the transmitting optics as described by Hardalupas and Liu (1992).

The accuracy of sizing was unaffected by the presence of non-spherical droplets because their velocities resulted in Weber numbers less than unity for a typical measured Sauter mean diameter of 100μm, which is far below the critical breakup value of around 12. The measurements were based on about 5000 samples at each point, resulting in statistical uncertainties of less than ±3% in the mean, ±10% in the rms of the velocity signal and ±2% in the cumulative size distribution based on the number of droplets (Tate, 1982). The liquid flux measurements are presented in arbitrary units since the uncertainty in the individual measurements is difficult to quantify and may be more than 30% in some regions of the flow.

3. Results and Discussion

The results are presented in two sub-sections which describe size, velocity and liquid flux characteristics of droplets in the valve curtain area as a result of injection with an open valve and those produced by fuel introduced entirely as a liquid stream on either the upper surface of the valve or on the port wall. The axial distributions of droplet characteristics as a result of injection with an open valve are presented for valve lifts of 2, 4 and 6mm and with air flow rates of 0.56m^3/min, 0.99m^3/min and 1.2m^3/min, respectively. Additional results are

Table 1: Characteristics of the Phase-Doppler Velocimeter[*]

	Transmitting Optics		
300 mW (nominal) Ar$^+$ laser wavelength	514.5		nm
operated at	100		mW
Beam diameter, at e^{-2} intensity	1.25		mm
imaging lens from laser to grating	80	(100)	mm
collimating lens after grating	200		mm
imaging lens to form measuring volume	600		mm
Number of lines on radial diffraction grating	16384	(8192)	
Shift frequency (nominal) due to rotation of grating	3	(1.5)	MHz
Short-term stability of shift frequency (r.m.s.)	0.3		%
Beam separation	34	(17)	mm
Measured half-angle of intersection	1.62	(0.88)	deg.
Fringe spacing	9.106	(16.69)	μm
Calculated number of fringes within e^{-2} intensity	14	(9.4)	
Frequency to velocity conversion factor	0.109	(0.0599)	MHz/ms^{-1}

	Receiving Optics	
Location of collection optics from forward scatter	30	deg.
Focal length of collimating lens in receiving optics	600	mm
dimension of rectangular apertures	45 x 6	mm
separation between apertures 1 and 2	25	mm
separation between apertures 1 and 3	50	mm
Focal length of imaging lens in receiving optics	300	mm
Width of spatial filter before the photomultipliers	100	μm
Magnification of receiving optics	2	
Effective length of measuring volume	400	μm
Phase angle-to-diameter conversion factor for channel 1 and 3	0.51 (1.02)	μm/deg.

then presented for a valve lift of 4mm which was required to avoid impingement of emerging droplets on the bevelled edge of the port, with a steady suction pressure of 2kPa gauge in the cylindrical chamber which resulted in co-flowing air flow rate of 0.99m^3/min and bulk velocities of 18.2m/s in the inlet pipe and 38m/s in the valve gap. This flow rate was about four times higher than the time averaged air flow rate into a single cylinder of 400cc displacement operating at 1200 RPM and was chosen because induction occurs during one of the four strokes. The injection duration was kept constant at 7ms, unless otherwise stated,

---

[*] Values in brackets show the modified setting used to increase the measurable range of a PDA instrument during the experiments with liquid streams.

with a corresponding charge of 25mm³/injection and an injection frequency of 10 Hz, which resulted in a fuel flow rate of 15cc/min, and also corresponded to a speed of 1200 RPM in a four stroke engine. For the experiments with introduction of steady fuel streams, the flow rate should be larger than 15cc/min to simulate the high instantaneous flow rate experienced by a solid surface during the impingement and therefore, since the 7ms injection duration corresponds to about one third of the duration of the intake stroke at 1200 RPM, the fuel flow rate was varied in the range of 12 - 42cc/min.

The results are presented as ensemble averages over the injection cycle and with time measured from the beginning of injection. The velocity results have been normalized by the bulk air velocity in the valve gap, that is $U_o = 43.7$, 38 and 31m/s for the valve lifts of 2, 4 and 6mm, respectively, and the diameter of the valve, D = 40mm, was used to normalize the radial and axial distances from the valve exit. Measurements were made at a radial distance of Y = 24mm from the valve exit which was chosen to allow at least 0.5ms between detachment of droplet from the valve and arrival at the measuring volume to ensure spherical droplets, Naber and Farrell (1993), and to remain close enough to the valve exit to preserve temporal information. The measurement locations were between Z = 1mm and 30mm below the exit, where droplets emerged as a result of injection or introduction of liquid stream on valve or port surfaces, and corresponded to $0.025 < Z/D < 0.75$ at $Y/D = 1.1$. Velocity measurements are presented for four ranges of droplet diameters, namely 24 to 30μm, 56 to 62μm, 120 to 126μm and 141 to 153μm, which are referred to as the 27, 59, 123 and 147μm droplet-size ranges. The 27μm range was representative of droplets subject to secondary atomization and able to follow the gaseous velocity most closely, the 59μm and 123μm ranges were representative of the most probable diameter and of the sizes carrying more than 50% of the liquid flux when the fuel was injected with the valve open, and the 147μm size range was typical of the most probable diameter when films were removed by air.

3.1 Spray Injected with the Valve Open

Figure 2(a-c) shows time averaged spatial distributions of arithmetic and Sauter mean diameters, mean axial velocities of most probable size class and liquid flux for valve lifts of 2, 4 and 6mm, with the velocities of the co-flowing air, measured in absence of fuel, shown by vectors. For the valve lift of 2mm, the liquid flux had a flat distribution and emerged in the range $0.05 < Z/D < 0.75$, that is in an arc between 10° to 70° to the horizontal axis, and a Sauter mean diameter of around 100μm compared with that of 130μm in the free unconfined spray. With increase in the valve lift to 4 and 6mm, the axial distribution of flux exhibited maxima at Z/D = 0.075 and 0.125 respectively, which suggests that most of the droplets emerged from the region close to that of the valve head immediately upstream of

138

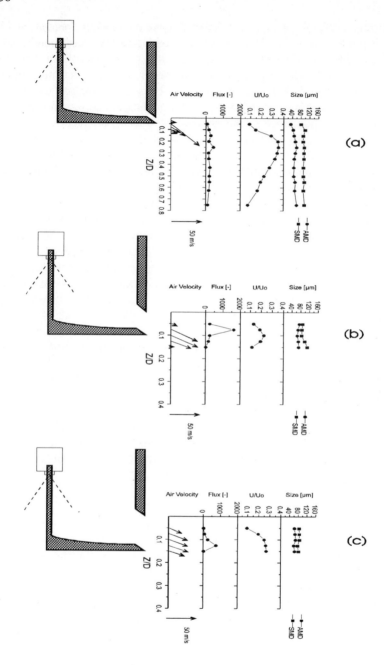

Figure 2: Spray injected with valve open: Axial distribution of mean diameters, axial velocity of 57μm size class, flux and air velocity vectors for valve lifts of (a) 2mm (b) 4mm and (c) 6mm. Y/D = 1.1

the valve bevel. Size-discriminated measurements of the axial and radial velocity components at the locations of flux maxima confirmed that the velocity vectors in the Y-Z plane were at about 25° to the horizontal axis, compared with about 35° for the co-flowing air, and did not depend on the droplet size. The Sauter mean diameters were some 20% smaller than with the 2mm valve lift, at about 80μm. The axial velocity of the most probable size class, that is 57μm, at the location of flux maxima, was about 10m/s and the magnitude of velocity vector of about 23m/s considering the trajectory of droplets. Reduction of the air flow rate, and consequently the bulk air velocity in the valve gap, by a factor of two did not affect the spatial distribution of liquid flux although there was an increase of 10% in Sauter mean diameter.

Figure 3 shows the effect of injection duration on arithmetic and Sauter mean diameters of emerging droplets, averaged over the injection cycle, with the valve lift of 4mm and at Z/D = 0.075, corresponding to the location of flux peak of figure 2(b). The Sauter mean diameter increased monotonically from 75 to 100μm as the duration of injection increased from 3 to 11ms and the larger difference between arithmetic and Sauter mean diameters with increased injection duration suggests that the relative number of large droplets increased.

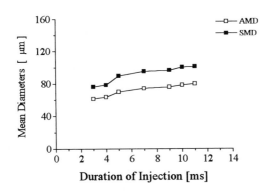

Figure 3: Spray injected with valve open: Effect of injection duration on mean diameters of droplets downstream of the inlet valve. Diameters averaged over injection cycle. Valve lift = 4mm; Y/D = 1.1; Z/D = 0.075.

Figure 4 shows the temporal development of Sauter mean diameter, axial velocities of three size ranges and fluxes at the same vertical location as in figure 3 and for injection durations of 4, 7 and 10ms. The Sauter mean diameter increased monotonically with the time elapsed from the beginning of injection, with the largest droplets emerging at the end of the injection cycle and their size

140

increasing with injection duration. The axial velocities of the large droplets were found to be lower than those of the small droplets and independent of injection duration. This temporal development of Sauter mean diameters and size-velocity relationship is opposite to that measured in the free unconfined spray by Hardalupas et al (1990) and Posylkin et al (1993). The temporal development of the liquid flux shows that the peak has been displaced towards trailing edge of the spray, as compared with 1ms delay after the spray tip arrival in the free unconfined spray of Posylkin et al (1993).

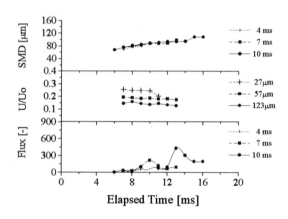

Figure 4: Spray injected with valve open: Temporal development of Sauter mean diameter, axial velocities of 27, 57 and 123μm size classes and liquid flux for injection durations of 4, 7 and 10ms. Valve lift = 4mm; Y/D = 1.1; Z/D = 0.075

The differences between characteristics of the spray at the exit of an inlet valve and these of the free unconfined spray are likely to follow from the impingement of incident spray on the solid surfaces. It is worth noting that, to avoid impingement on the valve, a droplet must acquire a radial component of velocity from the air in the valve gap and, hence, the time of flight from the injector to the valve, t_f, must be large compared to the Stokes response time, $\tau = r_l \, d_l^2 / 18\mu_a$, where r_l and d_l are the density and diameter of the liquid droplet and μ_a is the viscosity of air. Thus the Stokes number $S_t = t_f / \tau$ should be of the order of 10, Ingham et al. (1990), and a typical value of t_f is about 3ms for the present experimental arrangement, so that droplets larger than 20μm are incapable of satisfying the Stokes criterion. This suggests that more than 90% of injected into the port flux will impinge on surfaces. The increase in size of the droplets emerging from the valve with time elapsed from beginning of injection could occur due to coagulation as more droplets are accelerated towards a narrow gap, probably with some formation of a liquid film. It is unlikely, however, that a continuous and stable liquid film could cover the valve head since the maximum film thickness would be less than 20μm assuming that all injected fuel remained on the valve. Secondary atomization by splashing would result in an even thinner estimate of film thickness, as argued in Appendix A, and would be unstable. The droplets, produced as a result of impingement, would be removed by the co-flowing air to explain the temporal development of the emerging spray, figure 4, where small droplets appeared first. With increase in injection duration and time elapsed from the start of injection, impingement of droplets could occur on already wetted surfaces and to contribute to an increase of the mean diameters, as indicated by figures 3 and 4, respectively. It is not possible to deduce from the above results the relative contribution of the drag applied by the co-flowing air on films, as with injection against a closed valve, and the following sub-section considers this by introducing the fuel entirely as a liquid streams on the surface of the valve and port walls.

3.2 Fuel Introduced on the Valve and Port Surfaces

Figure 5 shows the effect of fuel flow rate when introduced as a steady liquid stream on the head of the valve on Sauter mean diameter, axial velocities and liquid flux of droplets, and the spatial distribution of these parameters will be discussed in relation to figure 6. The three volumetric flow rates of fuel were 12, 27 and 42cc/min and corresponded to those introduced by injection with durations of 5, 10 and 15ms. Increase in the volumetric fuel flow rate caused the integrated flux profiles to increase in proportion with the amount of fuel supplied but hardly

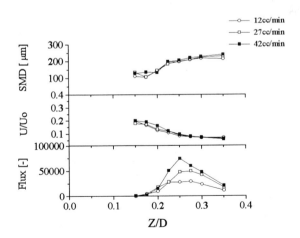

Figure 5: Steady liquid stream on the valve head: Effect of the fuel flow rate on Sauter mean diameter, axial velocity of 147μm size class and liquid flux.
Valve lift = 4mm; Y/D = 1.1

affected either the axial velocities and Sauter mean diameters of the droplets. A three-fold increase in the fuel flow rate resulted in an increase of less than 5% in the Sauter mean diameter, far below than that predicted by the relation suggested by Rizk and Lefebvre (1980) for pre-filming air blast atomizers. Variation in the fuel flow rate introduced on the port wall resulted in similar independence of the sizes of droplets downstream of the valve.

Figure 6 shows a composite of spatial distributions of droplet characteristics measured with liquid streams introduced on the valve head and the wall surface at different times. The volumetric flow rate of the fuel was 15cc/min and similar to that of injected fuel with a duration of 7ms and frequency of 10Hz, so that results can be compared with these of figure 2(b). The distributions of flux showed that most of the spray emerged from the lower edges of the valve and seat bevels, with the attached arrows indicating the flux maxima which were located at Z/D = 0.25 and Z/D = 0.05, and at angles of 50° and 40° to the horizontal axis, respectively. More than 90% of the liquid flux emerged as droplets with Sauter mean diameters larger than 200μm, increasing to about 240μm at locations of flux maxima. The axial velocities at these locations were less than 4m/s which, considering the trajectory of droplets, led to the magnitude of velocity vector of about 6m/s or 15% of that of the bulk velocity of the air in the valve gap. The remained fraction of the flux emerged in the region around Z/D = 0.15 as smaller droplets with higher velocities, possibly due to the stripping of the liquid upstream of the bevels.

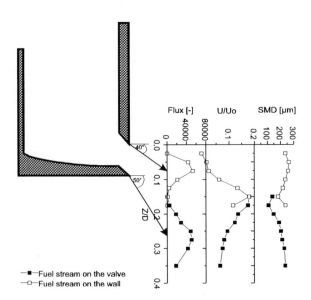

Figure 6: Steady liquid stream on the valve head and port wall: Axial distribution of Sauter mean diameter, axial velocity of 147μm size class and flux.
Valve lift = 4mm; Y/D = 1.1; Fuel flow rate = 15cc/min

The above results suggest that more than 90% of fuel was in the form of droplets with Sauter mean diameters above 200µm which were swept to the edges of the valve or seat bevels before emerging at the angles of these surfaces and with velocities which were comparatively low since the interface occurred at the intersection of liquid and air boundary layers. Thus, injection with the inlet valve closed is likely to imply that the initial characteristics of the spray are unimportant for the droplets at the exit of an inlet valve.

4. Conclusions

The following is a summary of the most important findings:

1. Injection with the inlet valve open implied reduction of the Sauter mean diameter of the spray from 130µm to around 90µm in the valve curtain. With valve lifts larger than 2mm, most of the droplets emerged from the region close to the surface of the valve head at an angle of about 25° to the horizontal axis and for the 2mm lift they emerged between angles of 10° and 70°. The magnitude of velocity vector of emerging droplets was around 20m/s and effect of the flow rate of air on their trajectory and size was small.

2. Injection with the inlet valve open led to mean diameters which increased with time from the beginning of injection, and larger droplets emerged at the end of the injection cycle. Increase in the duration of injection from 3 to 11ms led to increase in the Sauter mean diameter, averaged over the injection cycle, from 75 to 100µm.

3. The introduction of the fuel as liquid streams on the valve and port surfaces resulted in about 90% of the flux emerging from edges of the valve and seat bevels at angles of 50° and 40° to the horizontal axis, respectively. The Sauter mean diameters at locations of flux maxima were around 240µm, almost independent of the fuel flow rate, with magnitude of velocity vector of about 6m/s. These results are likely to imply that initial characteristics of the spray are unimportant for the droplets at the exit of an inlet valve with injection with the inlet valve closed.

Appendix A: Estimate of surface film thickness due to the impact of a single droplet.

An estimate of the effect of impingement by the single droplet on the valve head on the thickness of developing liquid film can be obtained from the energy balance before and after the impingement and assuming first that no splashing has occurred and that the evaporation during the period of droplet's shape transformation is negligible.

Before the impact kinetic and surface energy of the droplet are given by:

$$(1) \qquad E_{k_1} = \frac{1}{2}\rho_1 U^2 (\frac{\pi}{6}d_1^3)$$

$$(2) \qquad E_{s_1} = \pi d_1^2 \sigma$$

where σ is the surface tension of the liquid drop. After the impingement, kinetic energy was expended to transform the droplet into a disc of diameter D and with surface energy (Chandra and Avedisian, 1991)

$$(3) \qquad E_{s_2} = (\frac{1}{4}\pi D^2)\sigma (1 - \cos\Theta)$$

where Θ, defined in figure A1, is the angle between the solid surface and the tangent line to the disk front and is dependent on the solid surface temperature, with $\Theta \approx 30°$ for the surface at ambient temperature according to experimental results of Chandra and Avedisian (1991).

The work lost in deforming the liquid against the viscous forces can be estimated by assuming that the flow of expanding liquid is one dimensional, the volume of the drop is equal to the volume of the liquid once it has flattened out into the shape of disc and the time over which the expansion takes place to be equal to the time needed for the droplet of height h to go from its maximum value of d_1 to zero at velocity U. Based on the above assumptions it can be shown that the work lost, W, is:

$$(4) \qquad W = \frac{3\pi\mu U D^4}{8} \frac{1}{d_1^2}$$

From energy conservation $E_{k_1} + E_{s_1} = W + E_{s_2}$ the expression for diameter of the flattened disc D yields:

$$(5) \qquad \frac{3\mu U}{8 d_1^2}D^4 + \frac{\sigma(1-\cos\phi)}{4}D^2 - d_1^2 (\frac{1}{12}\rho U^2 d_1 + \sigma) = 0$$

and the thickness of the liquid disc t_d therefore can be obtained from:

$$(6) \qquad t_d \approx \frac{2 D^3}{3 d_1^2}$$

For the droplets produced by a pintle injector, with a Sauter mean diameter of about 100μm and impinging on the valve surface with velocities between 20m/s and 10m/s, (Hardalupas et al. 1990), the thickness of the developing disc shaped film is estimated to be 4μm and 6μm respectively, based on the analysis described above. It is unlikely that such a thickness could be sustained in practice due to the instabilities developed in the shear flow at both liquid-solid and liquid-air interfaces and the film is expected to be unstable and to break-up under the contracting action of the surface tension.

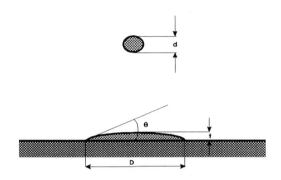

Figure A1: Impingement of a single droplet on solid surface

5. References

Chandra, S., Avedisian, C.T. (1991) On the collision of a droplet with a solid surface. Proc. R. Soc. Lond. **A432**, 13-41.

Dementhon, J.B. (1992) L'injection essence dans un moteur a etincelle: mesures granulometriques par la methode des phases Doppler et phenomnes de pulverisation. These presentee a la Faculte des Sciences de l'Universite de Rouen, Rouen, France.

Hardalupas, Y. and Liu, C.H. (1992) Size-Discriminated Velocity Cross-Correlation Measured by a Single Channel Phase-Doppler Velocimeter. In Laser Anemometry in Fluid Mechanics **6**, 145-165. (Editors: R.J.Adrian, D.F.G.Durao, F.Durst, M.Heitor, M.Maeda and J.H.Whitelaw), Springer Verlag.

Hardalupas, Y., Taylor, A.M.K.P., Whitelaw, J.H. (1990) Unsteady Sprays by a Pintle Injector. JSME International Journal, Series II, **33**, 2, 177-185.

Hardalupas, Y., Okamoto, S., Taylor, A.M.K.P. and Whitelaw, J.H. (1993) Application of phase-Doppler anemometer to a spray impinging on a disk. In Laser Anemometry in Fluid Mechanics **6**, 490-506. (Editors: R.J.Adrian, D.F.G.Durao, F.Durst, M.Heitor, M.Maeda and J.H.Whitelaw), Springer Verlag.

Ingham, D.B., Hildyard, L.T., and Hildyard, M.L (1990) On Critical Stokes' Number for Particle Transport in Potential and Viscous Flows near Bluff Bodies. J. Aerosol Sci., **21**, 935-946.

146

Kadota, T., Mizutani, S., Wu, C.Y., and Hoshino, M. (1990a) Fuel Droplet Size Measurement in the Combustion Chamber of a Motored SI Engine via Laser Mie Scattering. SAE paper 900477.

Miller, M.J. and Nightingale,C.J.E. (1990) Measurement of the charges in mixture preparation that occur during flow past the inlet valve of an SI engine. IMechE Int'l conference on Automotive Power Systems, Chester, paper C394/004

Naber, J.D., Farrell, P.V. (1993) Hydrodynamics of droplet impingement on heated surface. SAE paper 930919.

Posylkin, M., Taylor, A.M.K.P. and Whitelaw, J.H. (1993) Size and velocity characteristics of fuel droplets in the valve curtain area of a straight, steady-flow inlet manifold. Imperial College, Thermofluids Section, Mechanical Engineering Department Report no. TF/93/05.

Posylkin, M., Taylor, A.M.K.P., Vannobel, F. and Whitelaw, J.H. (1994) Fuel droplets inside firing spark ignition engine. SAE paper 941989

Quader, A.A., (1982) The Axially-Stratified-Charge Engine. SAE paper 820131

Rizk, N.K., Lefebvre, A.H. (1980) Influence of liquid film thickness on airblast atomization. Trans. ASME J. Eng. Power, **102**, 706-710.

Stow, C.D., Hadfield, M.G. (1981) An experimental investigation of fluid flow resulting from the impact of a water drop with an unyielding dry surface. Proc. R. Soc. Lond. **A373**, 419-441.

Tate, R.W. (1982) Some Problems Associated with the Accurate Representation of Droplet Size Distributions. In Proc. of 2nd Int'l Conf. on Liquid Atomisation and Spray Systems (ICLASS), Wisconsin, USA.

Vannobel, F., Robart, D., Dementhon, J.B. and Whitelaw, J.H. (1994) Velocity and drop size distributions in a two-valve production firing engine. Proc. COMODIA

PHASE-DOPPLER ANEMOMETRY (PDA)
A NEW TOOL FOR MONITORING THERMAL SPRAYING

R. Zeller, J. Domnick*, E. Schubert, H. W. Bergmann, F. Durst*

ATZ-EVUS, Rinostr.1, D-92249 Vilseck
* Lehrstuhl für Strömungsmechanik
Universität Erlangen-Nürnberg, Cauerstr.4, D-91058 Erlangen

Abstract

Thermal spraying is an expanding technology with still growing markets. Especially the variety of plasma spraying techniques offers a number of possibilities for new and advanced applications. The spraying of sophisticated materials, the demand of generating coating structures or substrate/coating combinations with novel properties and the improvement of the spraying efficiency require a more detailed understanding of the physical and chemical reactions inside the plasma, in order to achieve an appropriate process control. The present paper demonstrates that phase-Doppler anemometry (PDA) is applicable as a powerful tool for process diagnostics in plasma spraying. Yielding simultaneous measurements of particle size and velocity at discrete locations, the PDA-technique generates new possibilities to characterize and further improve the spraying process.

1. Introduction

Thermal Spraying is a multitask coating process in modern surface technology. Plasma spraying, initially developed for the aerospace industry, is one of the youngest but also most universal thermal spray processes *(Busse, 1988)*. The heat source for the processing of materials with high melting points (oxide ceramics, carbides, borides, nitrides, silicides) is a plasma jet, exceeding temperatures between 6000 K and 20 000 K. However, also low melting materials could be sprayed with this technique achieving high deposition rates *(Lugscheider et al. 1991)*. The plasma gun consists mainly of two electrodes. A rod shaped tungsten cathode is located inside a coaxial copper ring anode. A high frequency ignition forms an arc between the electrodes. By excitation, dissoziation and ionization of the fluidizing gas between the electrodes a 4-5 cm long plasma free jet is generated, exiting the nozzle at very high velocities. Applicable gases are Nitrogen, Argon, Helium and Hydrogen, but also mixtures of two or more of these gases. The gun parts are watercooled to sustain the high operation temperatures and to avoid melting of the electrodes. Particles are injected into the plasma jet, heated up, accelerated towards the substrate and finally deposited forming a functional layer *(Herman, 1988)*. Many experimental efforts were made to characterize the spraying process in order to achieve a better understanding of the governing parameters. However, due to the extreme conditions inside the plasma jet (temperature approx. 10 000 K, gas velocity between 300-600 m/s) only a few techniques are applicable. Temperature measurements with spectroscopy or enthalpy probes are common techniques *(Pfender & Spores, 1987)*, yielding an overall impression of the temperature distribution. For locally resolved measurements of the velocity field, laser-Doppler anemometry (LDA) has been used recently *(Mayr et al. 1993, Fouchais et al. 1992)*. Based on experimental results many attempts have been made to simulate the spraying process *(Pfender & Chyou 1989)*. Nevertheless, at the current state these models can not deliver a detailed description of the process because of too many simplifications necessary for modelling *(Joshi 1992)*. On the other hand great efforts were made to save expensive powders forming functional layers by an adequate process control. Up to this point all control mechanism are based on the same principle, keeping the machine setting constant and assuming that there is no change in the spray process or in the layer quality. Certainly, this is not satisfactory, but due to the complexity of the interacting parameters no successful on-line control is available at the moment. The use of the PDA-technique may be a big step towards an improvement of this situation.

2. Test Rig

The plasma spray facility used in this investigations consisted of a METCO 9MB machine mounted watercooled air plasma spray gun with power supply and control unit for the mixing of the plasma gases. The spray gun and a 4-axis ISEL traversing system were mounted in a sound proof chamber. The powder feeder was a METCO 4 MP type. Typical machine operation parameters are given in table 1.

Parameter	Argon	Hydrogen	Current
Range	30-80 l/min	1-10 l/min	300-500 A

Parameter	Voltage	Powder feed rate	Carrier gas
Range	56-80 V	5-50 g/min	1-10 l/min

Table 1: Range of variable gun settings

3. Experimental Set-up

In order to obtain a rough idea of the flow field downstream the nozzle exit thermal imaging at different spray conditions was performed. Using an infrared thermo-camera (Hörotron Infravision 500), positioned perpendicular to the spray cone at a distance of 1 m, the interesting regions of the particle stream were visualised. The position of the camera was arranged in such a way that the tip of the flame operating without powder was just at the edge of the monitor. This arrangement was necessary because the core of the plume was too bright for the selected filter combination. The camera was operated in a temperature window from 1273 K - 2273 K. The intensity reduction was achieved by using specific grey filters. This setting was appropriate for most of the powders, only a few ceramic powders have melting points exceeding a temperature of 2273 K. Preliminary measurements of the particle velocities were made using an INVENT DFLDA laser-diode based fibre optics system, operating at 6 MHz shift frequency. The backscatter transmitting probe was equipped with a 250 mm focal

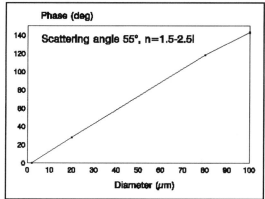

Fig.1: Calculated phase/size correlation for metal powders

length lens, yielding a probe volume with a diameter of 200 μm and a length of approx. 2 mm. For the PDA measurements, this transmitting system was completed with an INVENT PDE 2 detector receiving unit. The receiver focal length was 310 mm. Since the investigated metal particles are highly absorbing, PDA measurements could be made with scattered light based on reflection. A scattering angle of 55° was chosen, yielding a

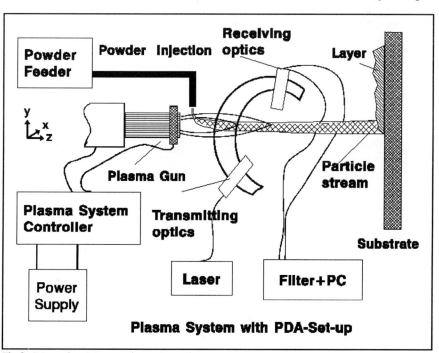

Fig.2: Schematic of the experimental set-up

reasonable scattering intensity at the particle sizes investigated. The excellent linearity of the phase/size relationship is demonstrated by fig 1, which is based on Mie-scattering calculations. The values for the complex refractive index are typical values taken from *van de Hulst* (1957). A final size range of 160 μm was achieved by the chosen arrangement. The obtained Doppler signals were filtered by an INVENT PDE filterbank and further evaluated with a QSP PDA 100 FFT processing system, yielding a frequency bandwidth of 50 MHz. Fig. 2 gives a schematic illustration of the experimental set-up in combination with the plasma spray gun. The measurement location of the LDA and the PDA was moved by traversing the gun, leaving the optical arrangement fixed. The investigated powders were a CuSn 6 bronze and a CrNi-powder. A typical SEMimage of the bronze powder with a size range of 0-63 μm in solidified state is given in fig.3. Clearly, the overall shape of the particles appears to be spherical, nevertheless some slight ellipsoids may also be recognized. The surface has a certain

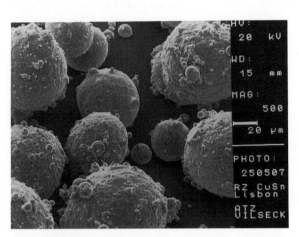

roughness, which might disappear when the particles are molten. Also, the collectives of small and large droplets could be destroyed inside the plasma flame. As mentioned above, scattered light from reflection was used for the PDA measurements without any noticable problems. The mean validation rate in the measurements was more than 90 %.

Fig.3: SEM-image of the bronze powder

4. Preliminary Investigations

Fig. 4 shows an example of images taken by the thermo camera for 4 different

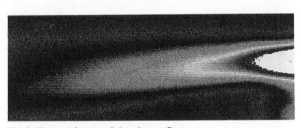

sets of spraying parameter. The picture mainly visualizes the temperature field generated by the heated and accelerated particles, thereby indicating also regions with high and low particle number density. Without

Fig.4: Thermal image of the plasma flame

going too much into details, it should be noted that the particle stream is splitting into two separate jets beginning at approximately 100 mm downstream the plasma gun. According to Pfender (1987) this is a region with highly turbulent gas flow, influencing heavily the final properties of the layer. The observation of separate jets is also confirmed by first LDA measurements. Fig. 5 exhibits the isotachs at 120 mm downstream the nozzle exit. In this diagram two distinct regions of high velocities can be detected. At these positions the mean velocity reaches maximum values around 145 m/s. Near the spray edge, a strong negative velocity gradient is existing, however, mean velocities at the outermost measurement points are still around 75 m/s. It should be noted, that the LDA measurements basically reflect the velocity of the particles, which might differ strongly from the gas velocity. Also, some size/velocity correlations were obtained in the PDA measurements. The results of the LDA measurements were used to identify interesting areas for the PDA measurements.

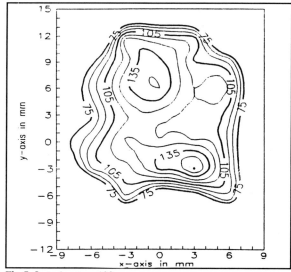

Fig.5: Isotachs at z=120 mm

5. Results

The PDA measurements were performed at the operating conditions summarized in table 2. The measurements were made at an axial distance of 150 mm from the spray gun, which corresponds to the standard working distance between plasma gun and substrate. Therefore, the results indicate the properties of the particles when reaching the target. In order to limit the powder and process gas consumption and also to reduce the heating of the

Gas flow rate	Ar 50 l/min	H₂ 7 l/min
Powder feed rate	30 g/min	
Voltage	60 V	
Current	450 A	

Table 2: Settings of the plasma gun during the measurements

152

optical system, the measurement time was limited to 10 s at each point. Therefore, the number of samples taken varied between approximately 50 and 2500 according to the local number density of the particles. In a first set of measurements, two different particle fractions of the bronze powder, i.e. size distributions from 0-63 μm and 63-160 μm, were compared. The result of this comparison, which was made at an axial distance of 150 mm and 10 mm below the nominal centerline of the spray cone, is shown in fig. 6. Clearly, two distinct

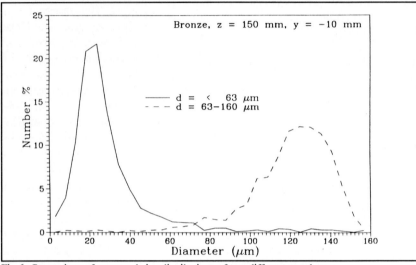

Fig.6: Comparison of measured size distributions of two different powders

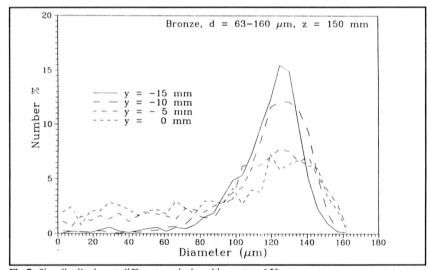

Fig.7: Size distribution at different vertical positions at z=150 mm

number weighted size distributions can be recognized with peak values around 25 and 125 μm. The obtained widths of the distributions compare very good with the nominal size ranges with only a few samples counted outside of the measuring range. From this result, it can be concluded that the PDA delivers the size of the bronze particles inside the plasma jet to a high degree of reliability. In fig. 7 the measured number distributions for the powder fraction from 63-160 μm is shown for different y-positions at an axial distance of 150 mm. These measurements were taken at x = 0.0 mm, i.e. on the vertical line of symmetry. It is evident, that the percentage of large particles at the right hand side of the size distribution is increasing with decreasing y-position, i.e. below the centerline of the spray. This can be explained by gravity effects, which influence the trajectories of the different particle sizes. The separation of the particle size classes is also indicated by the mean values given in tab.3. The velocities at these locations are relatively low,

y (mm)	d_{10} (μm)	d_{30} (μm)	u_{mean} (m/s)
0	97.7	113.0	37.3
-5	105.6	117.4	49.6
-10	120.1	123.7	74.5
-15	118.3	121.4	70.8

Table 3: Mean values at x = 0.0 mm, powder size 63-160 μm

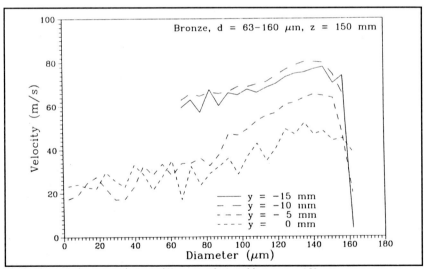

Fig.8: Size velocity correlations at different vertical positions at z= 150 mm

which is due to a non-symmetric shape of the spray cone, as will be discussed later. In fig. 8, the mean velocities are given as a function of the particle sizes, indicating a strong positive correlation between size and velocity. Obviously,

154

150 mm downstream the gun the gas flow is already decelerated with the smaller particles being able to follow this axial velocity gradient. On the other hand larger particles still keep their initial velocity. This behaviour will strongly influence the residence time of different particle sizes inside the plasma flame and, hence, the properties of the particles in terms of temperature and state when reaching the substrate. To examine a region of high mean velocities obtained in the LDA measurements, a grid of 25 measurement points was arranged at 150 mm axial distance. These points were distributed in one quarter of the spray cone below the horizontal line of symmetry, i.e. at negative y-coordinates. The measurements were made with bronze particles of a 0-63 μm size fraction. The result in terms of validated samples in 10 s is given in fig. 9. The maximum number of samples, and hence, the maximum particle number flux is obtained almost in the center of this region, strongly decreasing along x and y. The maximum measured particle arrival rate is approximately 200 1/s, which is in the order of the nominal arrival rate of 700 1/s, estimated from the powder feed rate, the volume mean diameter and the probe volume cross section, assuming a homogeneous number flux across the whole spray cone. The corresponding

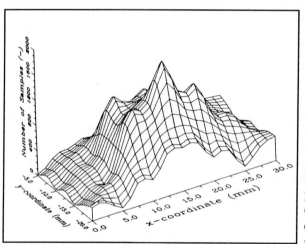

Fig. 9:   Distribution of the number of samples at z=150 mm
(d=0-63 μm)

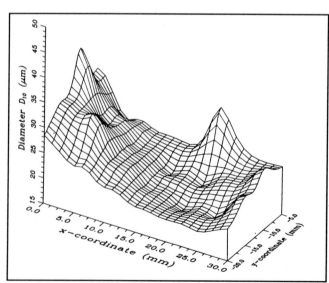

Fig.10:   Distribution of the arithmetic mean diameter at z=150 mm
(d=0-63 μm)

distribution of the arithmetic mean diameter d_{10} is shown in fig. 10. Basically, the mean diameter is increasing towards the vertical line of symmetry, reaching its maximum 15 mm below the centerline of the spray cone. Maximum arithmetic mean diameters are around 45 μm. The distribution of the volume mean diameter, which is not shown here, has a very similar shape. Surprisingly, the region of the highest number flux is not correlated with any specific particle sizes. This is also true for the axial mean velocity exhibited in fig. 11. Here, a smoothly decreasing mean velocity can be obtained moving along the x-axis. Furthermore, the mean velocity is almost con-stant along the y-axis. This is in contrast to the results shown in fig. 8; however, it should be noted that the size ranges of the particles are completely different. In general, the measurements indicate a strong asymetry of the spray. Most likely, this is the result of a nonsymmetric particle feeding system. Since the particles are injected into a region with extremely high mean velocities and velocity gradients, the distribution of the particles inside the plasma jet is heavily sensitive to any asymetry. It should be mentioned again, that the duration of the experiments and their spatial resolution had to be chosen as a compromise between the practical value of the results and the consumption of expensive plasma gases and powder.

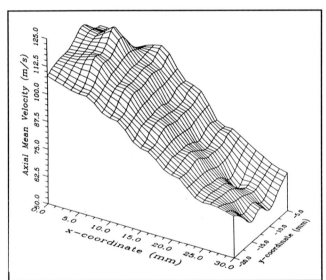

Fig.11: Distribution of the axial mean velocity at z= 150 mm (d=0-63 μm)

6. Summary and Outlook

The obtained results, although still having a preliminary character, show the possibility of using phase-Doppler anemometry as an appropriate measuring technique for the investigation of plasma sprays. Some major characteristics of the spray, i.e. a vertical separation of particles of different sizes due to gravity or the presence of positive size/velocity correlations have already been recognized. One problem that was recognized during the measurements is the strong spatial variation of the particle number density. Low particle number density regions could lead to either long measurement durations with expensive gas and powder consumption or to siginificant reductions of the number of samples taken and, hence, to reductions of the statistical reliability. Also, it should be noted that the measurements presented herein were made with initially spherical metallic powders. If non-spherical particles are used, which are not melting completely inside the plasma flame, errouneous size measurement would occur. Further problems may appear, if particles of ceramic materials are used. Here, the optical properties necessary for the layout of a phase-Doppler system, i.e. the complex refractive index, have to be determined with great care. If measurements with reflected light are not feasible, the refractive index at the process temperatures of the particles has to be measured. Nevertheless, the use of the PDA-technique in plasma sprays could yield an improved understanding of the plasma spray process through detailed investigations of the gas-particle interactions inside the plasma flame. In the future, this knowledge might be used for the development of more sophisticated and more accurate simulation models. Futhermore, small and robust PDA-systems might become a tool for an on-line control of the spraying process, rather than just keeping the spray gun settings constant during operation.

7. References

Busse, K.H.,(1988), Thermisches Spritzen Moderne Oberflächentechnik, Oberfläche and JOT, vol.9, pp 30-39

Lugscheider,E., Eschnauer, H., Müller,U., Weber, T., (1991), Quo Vadis, Thermal Spray Technology, Powder Metalurgy International, vol.25, no.1, pp18-20

Herman, H., (1988), Plasma Sprayed Coatings, Scientific American, no.9, pp.78-81

Pfender, E., Spores R., (1987), Flow Structure of a turbulent Thermal Plasma Jet, Surface and Coating technology, Vol.37, pp.251-270

Mayr, W., Landes, K., Reusch, A., (1993), Untersuchungen des Partikelverhaltensim Spritzstrahl mittels Laser-Doppler-Anemometrie, Proc. Thermal Spray Conference TS93 Aachen, pp.143-147

Fauchais, P., Coudert,J.F., Vardelle, M., Denoirjean, A., (1992), Diagnostics of Thermal Spraying Plasma Jets, Journal of Thermal Spray Technology, vol.1, pp.117-127

Chyou, Y.P., Pfender, E., (1989), Modeling of Plasma Jets with Superimposed Vortex Flow, Plasma Chemistry and Plasma Processing, vol. 9,no. 2, pp. 291-328

Joshi S.V., (1992), A Prediction Model to Assist Plasma and HVOF Spraying, Materials Letters, vol. 14, pp. 31-36

van de Hulst, H. C., (1957) Light Scattering by Small Particles, J. Wiley&Sons, New York

Part II
Applications to Combustion

Characteristics Of Counter - Gradient Heat Transfer In A Non - Premixed Swirling Flame

Y Hardalupas[1], M Tagawa[2], A M K P Taylor[1]

1   Department of Mechanical Engineering, Imperial College of Science, Technology & Medicine, London SW7 2BX

2   Department of Mechanical Engineering, Nagoya Institute of Technology, Gokiso-cho, Showa-ku, Nagoya 466, Japan

Abstract. Measurements of Reynolds shear stress and heat flux, by means of laser Doppler velocimetry and compensated fine wire thermocouples, have been made in the shear layer surrounding a swirl - induced recirculation bubble stabilising an unconfined non-premixed flame. The burner was operated at a Reynolds number of 27000, based on the air bulk velocity of 8.5 m/s and on a characteristic burner exit radius, R, of 50 mm, a swirl number of 1.07 and at a natural gas flow rate corresponding to an equivalence ratio of 0.85 and a potential heat release rate of 50 kW. The sign of the flux of the shear stress was found to be in accordance with that expected from a gradient diffusion hypothesis but the sign of the heat flux was in the counter-gradient direction for axial distances up to 0.5 R from the exit of the quarl. Arguments based on the shape of the weighted joint probability distribution of radial velocity and temperature fluctuations, $W(v'',t'') = v''t''P(v'',t'')$, show that the counter-gradient contributions are due to large amplitude, comparatively rare departures of temperature from the local mean and also suggest that the corresponding radial flux of the Favre-averaged conserved scalar, $\langle v''\xi'' \rangle$, will also be in the counter-gradient sense. Even in those regions of the flow where $\langle v''t'' \rangle$ is in the gradient sense, the magnitude of $W(v'',t'')$ in quadrants 2 and 4, which represent counter-gradient flux, remains strong in comparison with quadrants 1 and 3, which represent gradient diffusion. The quantity $\langle v''\xi'' \rangle$ is used in the prediction of non-premixed flames and these results show that predictions of this kind of flow must be based on second moment, rather than on effective viscosity, turbulence model closures so as to capture the effects of $\overline{\xi''}\partial P/\partial x_i$ on heat fluxes.

Keywords. Counter-gradient diffusion, swirling flame, laser Doppler velocimeter, compensated thermocouple.

1 Introduction

1.1 The Flow Considered

Experimental evidence of counter-gradient transport of a scalar, such as heat, or of momentum is important in that it shows that it is unreasonable to expect calculations based on effective viscosity hypotheses to be accurate and that solutions should use the Reynolds stress/flux levels of closure which, however, have the disadvantage of being computationally more expensive. In the context of combustion, the retardation of mixing, which is associated with the counter-gradient heat flux, may be a technically desirable characteristic, for example when the flame is to be formed in long radiant tube combustors. The phenomenon is to be found where the effect of mean pressure gradients, which may arise in the flow due either to mean streamline curvature or because of acceleration of gases across the flame front, preferentially accelerate packets of hot, low density gas relative to cold high density ones (e.g. Libby & Bray, 1981; Bray et al., 1985; Bray & Libby, 1991). Counter-gradient diffusion is a well-documented effect in the vicinity of bluff body *premixed* flame stabilisers (e.g. Heitor et al., 1987; Ferrão & Heitor, 1993; Duarte, Ferrão & Heitor, 1994). The abrupt streamline curvature, associated with the recirculation zone, is associated with a strong pressure gradient so that the heat flux between the incoming reactants and the hot products in the recirculation zone is in the counter gradient sense. The flow considered in this work is that of a *non-premixed* flame fired by a turbulent gaseous fuel jet surrounded by a concentric swirling (combustion) air flow which exhausts to the atmosphere through a short diffuser, sometimes known as a quarl. This configuration is typical of many burners of technical interest, where the swirl is usually sufficiently strong to generate a recirculation zone of hot products which acts to stabilise the flame.

1.2 Previous work

Takagi et al. (1984) and Takagi & Okamoto (1987) provided early evidence of the existence of counter-gradient diffusion of heat in confined non-premixed flames at swirl numbers too weak to generate an on-axis recirculation zone. They also found that the radial Reynolds stresses and the variance of temperature fluctuations were reduced in comparison with the unswirled flame. Takagi & Okamoto (1987) made a comprehensive examination of the relative magnitude of the generation terms in the transport equations for the Reynolds stresses and heat flux, based on measurements of velocity and temperature, and found that the Reynolds shear stress $\overline{\rho u'' v''}$ (where ..." denotes fluctuation from the Favre average and the overbar denotes the time average) was reduced mainly because of the large magnitude of the sink term $-\overline{u''} \partial P / \partial r$ and to a lesser extent by $\overline{\rho u'' w''} \langle W \rangle / r$ in the central core of the flame (where $\langle ... \rangle$ denotes a Favre-averaged quantity). The radial heat flux $\overline{\rho v'' t''}$ was reduced by the corresponding term $-\overline{t''} \partial P / \partial r$, to the extent that the heat flux was in the counter gradient sense: a corollary was that the production term for $\overline{\rho t''t''}$, i.e. $-2\overline{\rho v''t''}\partial\langle T\rangle/\partial r$, was negative.

The underlying mechanism for counter-gradient transport is thus due to the interaction between the radial density inhomogeneity together with the swirl-induced centripetal pressure gradient and this interaction can exist, and has been extensively studied, in inert flows. Experiments in inert flows include those reported by Ahmed et al. (1985) and Hirai & Takagi (1991) who found that the centreline mixing of an inner non-swirling Helium jet (representing the hot products of combustion, say) exhausting into a co-flowing, confined swirling air stream was strongly retarded, relative to the unswirled flow and to an air jet. In comparison, Hirai & Takagi (1991) found that the retardation of the low density Helium jet was stronger than that of a high density Carbon dioxide jet but, in a comparable experiment with sufficient swirl to generate a small reverse flow region, So & Ahmed (1987) found that the Carbon dioxide jet centreline velocity decay was more rapid than for an air jet (i.e. in the absence of density stratification), despite the larger momentum flow rate associated the CO_2 jet. Shear stress measurements were not possible because of the flow confinement in both experiments but Hirai & Takagi (1991) describe calculations, based on second moment closures, which successfully reproduce the slower mixing due to the reduced radial flux of species for the Helium jet, ultimately traceable to the $-\overline{m''}\partial P/\partial x_i$ term[1], where m is the gas mass fraction. In summary, the tendency in the *mean* flow is for high density fluid to be "centrifuged" outward while fluid of lower density tends to migrate to, or remain around, the axis of rotation. The radial *turbulent* exchange of momentum, energy and species may be *either* damped *or* augmented depending on whether the radial *density* stratification is stable or unstable.

Recently, Fernandes et al. (1994) investigated a non-premixed flame stabilised by both high swirl and a baffle which hence generated a strong recirculation zone. The high swirl, in particular, resulted in stronger radial pressure gradients than in the flow of Takagi & Okamoto (1987) and could be expected to provide strong counter gradient diffusion. In general, however, they found that the Reynolds shear stress was in the counter gradient sense only in a small region close to the baffle and, although the axial component of flux was not in accordance with expectation from gradient transport, the radial heat flux *was* in the gradient sense, in contrast to the result of Takagi & Okamoto (1987). Fernandes et al. (1994) also performed an analysis of the terms in the transport equations for the second order quantities and showed that, although $-\overline{u''}\partial P/\partial z$ was a negligible contributor to the transport of shear stress, $-\overline{t''}\partial P/\partial z$ was indeed dominant in the transport of axial flux. In the transport of radial flux, $-\overline{t''}\partial P/\partial r$ was only of the same order of magnitude as other terms and represented a sink of $\langle v''t''\rangle$.

---

[1] Somewhat surprisingly, similar calculations by Hogg & Leschziner (1989) of the flow of Ahmed et al. (1985) cast doubt on the importance of this term, and the counterpart in the Reynolds stress equations, in the representation of the flow.

There are a number of potential reasons for the difference in the direction of the radial heat flux between the two cited experiments. One is the well known result from inert flows (Bradshaw, 1969) that streamline curvature, here associated with the swirl-related shear strain (i.e. $\partial W/\partial r - W/r$), can modify the *turbulent* field and lead to *either* a decrease in the production of shear stresses $\overline{v'w'}$ and $\overline{u'v'}$, where the single prime denotes fluctuation from an unweighted mean, for mean rotation resembling a forced vortex, *or* an amplification of these stresses for free-vortex-like motion. In swirling flames, both types of vortex profiles are found. Thus, for the simple example of flow through a pipe rotating about its axis, it is found experimentally that the mean axial velocity profile progressively "laminarises" with increasing rotational speed and that the swirl mean velocity profile is non-linear. Similar phenomena are observed in the analogous swirling heated pipe flow, where the heat transfer rate from the wall decreases with increasing swirl of the pipe. It is useful to note that both trends can be reproduced through the use of second moment closures. For the velocity characteristics, Hirai *et al.* (1988) have found by such calculation that the Reynolds shear stress $\rho\,\overline{u'v'}$ is reduced as swirl increases because of smaller production of $\rho\,\overline{u'v'}$ caused, primarily, by the shear stress $\rho\,\overline{u'w'}$ becoming negative so that the generation term in the transport equation for $\rho\,\overline{u'v'}$, namely $\overline{u'w'}\,W/r$, becomes a sink although the confined nature of the flow prevented direct measurement of the shear stress levels. For the heat transfer characteristics, Hirai and Takagi (1988) suggest that the magnitude of $\overline{v't'}$ is reduced because of the term $\overline{w't'}\,W/r$, where t' represents the fluctuation of temperature from the time-mean value, acting as a sink in the transport equation for $\overline{v't'}$ and these explanations can be expected to hold in non-premixed flames also.

1.3 The current contribution

The preceding section shows that the existence of counter-gradient radial heat flux has been demonstrated only in a flame with swirl which was too low to generate an on-axis recirculation zone. Indeed, it is possible to argue that it is unlikely for extensive counter-gradient flux, which implies a reduction in mixing, to exist in a flow with sufficiently strong swirl to generate on-axis recirculation. This is because it is well-known that an increase in swirl is accompanied by a shortening of the flame which is due to *increased* mixing between the fuel and the air because the flow is more turbulent. The observed disappearance of a region of counter-gradient axial mass transport in inert coaxial jets when the outer jet is swirled (Johnson & Bennett, 1981; Roback & Johnson, 1983; reported by Ahmed *et al.*, 1985) seems to support this view. The work of Fernandes *et al.* (1994) also partially supports this view. The counter argument is that the increase in swirl is accompanied by an increase in the magnitude of the radial adverse pressure gradient and in the steepening of the radial (negative) gradients of temperature, and hence increasing the radial (positive) gradients of density, which could be expected to promote the term $-\overline{t''}\,\partial P/\partial r$ and, hence also, counter gradient diffusion. Given that many industrial burners operate with on-axis recirculation zones, it is of interest to determine which argument is correct.

There are three purposes in this work. First, to demonstrate that counter-gradient diffusion of heat flux, $\langle v''t'' \rangle$, *does* exist in a swirling non-premixed burner when the swirl number of the combustion air is sufficiently large to generate a large internal recirculation zone. The other distinguishing features of the present flow from the investigation of Takagi *et al.* (1984) and Takagi and Okamoto (1987) is that the flame is unconfined, as is the flow of Fernandes *et al.* (1994) although here we have no central bluff body, so that difficulties which arise from the subcritical nature of the confined swirling recirculating flow (as discussed, e.g., by Escudier & Keller 1985) are absent. Second, to present evidence which suggests that counter-gradient diffusion is also likely to exist in the flux of the conserved scalar, $\langle v''\xi'' \rangle$. It is important to demonstrate this because the predominant method of calculating turbulent non-premixed flames is through the conserved scalar formalism and *not* through the solution of the energy and turbulent heat flux equations. Third, to show that even in those regions of the flow where there is *no* evidence counter-gradient diffusion, the underlying driving mechanism for counter-gradient diffusion remains strong and that therefore the so-called second moment closures *must* be used to calculate the flow. The method of data analysis is also different from Takagi & Okamoto (1987), Heitor *et al.* (1987) and Fernandes *et al.* (1994), in that here so-called quadrant analysis is applied to the heat fluxes, as used in non-reacting boundary layer analysis (e.g. Perry & Hoffman, 1976; Nagano and Tagawa, 1988). Its application in the analysis of flames is, by comparison, rare presumably because of the difficulty in making the required measurements, although Cheng and Ng (1985) have used it, in the analysis of the sublayer of hot products of a turbulent boundary layer with premixed combustion. They show that the observed increase in turbulent kinetic energy *and* simultaneous reduction in the kinematic Reynolds shear stress, as compared to the non-reacting counterpart, is due overwhelmingly to the reduction in the second quadrant contribution, the so-called burst motion.

The rest of the paper is organised as follows. The next section, §2, describes the experimental method and the results are described and discussed in §3. The paper closes with a summary of the more important conclusions.

2 Experimental Method

2.1 Experimental Configuration

The burner geometry is shown in figure 1, which provides essential dimensions and details of the fuelling arrangement, and is identical to that used by Milosavljevic et al. (1990). The gaseous fuel (94% methane by volume; net calorific value 35 MJ m^{-3}) was drawn from a compressor, metered by a rotameter and delivered to the throat of the quarl through a pipe with 15 mm internal and 18 mm external diameter which was centred within the surrounding annular duct with 50.8 mm internal diameter. The gaseous fuel was injected radially by fitting a bluff body at the exit of the central pipe, as shown in figure 1, which is known to result in a predominantly symmetrical and attached flame for a wide range of

Reynolds and swirl numbers (Milosavljevic et al. 1990; Milosavljevic, 1993). The swirl number is defined as follows :

$$S = \frac{2G_\theta}{G_z \, D_e}$$

where $G_\theta = 2\pi\rho \int_{r=r_i}^{R_e} Wr \, Ur \, dr$ is the axial flux of angular momentum and G_z

$= 2\pi\rho \int_{r=r_i}^{R_e} U \, Ur \, dr$ is the axial flux of axial momentum, R_e and D_e are the radius and the diameter of the entrance of the quarl and r_i is the radius of the gas fuel injection tube. Swirled and unswirled air were metered by rotameters separately and supplied to the annulus and the swirl number was continuously variable from zero to a value beyond unity by varying the swirled and unswirled air flow rates while the total air flow rate was kept constant. Details of the air feed arrangement can be found in Milosavljevic et al. (1990). Measurements were obtained for three swirl numbers and two equivalence ratios, but results are presented here only for the highest of the swirl numbers and equivalence ratios. The burner was operated at a total air flow rate of 900 l/min, resulting in a Reynolds number of 27000, based on the kinematic viscosity and bulk velocity of the air at room temperature of 8.5 m/s in the annulus and on the diameter at the entrance of the quarl, swirl number of 1.07 and at a gas flow rate of 85 l/min corresponding to an equivalence ratio of 0.85 and a potential heat release rate of 50 kW.

2.2 Instrumentation

Velocities were measured using a single channel laser Doppler system with optical characteristics as given in table 1. As argued elsewhere (Heitor *et al.*, 1985; Goss, 1993) the laser Doppler velocimeter measures density- (i.e. Favre-) averaged velocities because the instantaneous concentration of seeding particles (and hence the probability of occurrence of Doppler signals) is proportional to the instantaneous gas phase density. The cross correlation $\langle u'' v'' \rangle$ was measured by taking readings of the mean and variance of the Doppler frequency after positioning the plane of the laser beams in directions $0°$ and $\pm 45°$ to the burner axis at each point, using the approach outlined by Melling and Whitelaw (1975). The transmitting optics (DISA model 55X) were based on a beam splitter arrangement and a single Bragg cell crystal operating at 40 MHz for frequency shifting. The scattered light was collected by a 300 mm focal length lens and focussed on a photodetector arrangement (DISA model 9055X). Seeding particles were Al_2O_3, nominally 1 μm, and were dispersed in the air flow by passing a part of the total air flow rate supply to the burner through a reverse cyclone feeder arrangement similar to that of Glass and Kennedy (1977). The Doppler signals were fed in the frequency shift mixer unit (DISA model 55N12), where the initial 40 MHz frequency shift was removed, and were amplified and filtered.

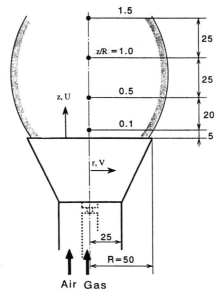

Fig. 1 The burner geometry

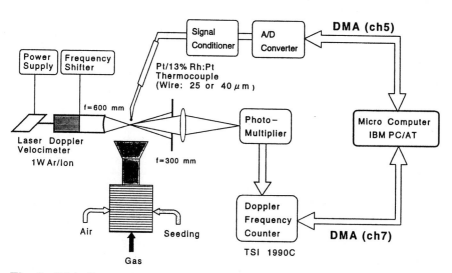

Fig. 2 Digitally compensated thermocouple and a single channel laser Doppler anemometer.

The Doppler frequency was measured by a zero crossing counter (TSI model 1990C) which was interfaced to an IBM compatible 286 computer using a purpose built 16 bit DMA card. The signals were validated if their frequency over the first 10 of 16 cycles was the same to within 3%.

Table 1. Principal Characteristics Of The Laser Doppler Anemometer

Ar+ laser wavelength	514.5 nm
power	500 mW
laser beam diameter, at e^{-2} intensity	1.25 mm
focal length of imaging lens	553 mm
shift frequency (nominal)	3 MHz
beam separation	58.5 mm
measured half-angle of intersection	3.028 deg
calculated dimensions of beam	0.315 mm
intersection volume at e^{-2} intensity	0.315 mm
	5.95 mm
fringe spacing	4.87 μm
calculated number of fringes	64
frequency to velocity conversion factor	4.87 (ms^{-1})/MHz

Mean and rms values of temperature were measured by a digitally compensated Pt / 13%Pt-Rh thermocouple with 25 and 40 μm wires. The thermocouple signal was conditioned and amplified by a factor of 373 before being digitised by a 16 bit A/D converter (Data Translation model DT2827) with maximum sampling rate of 100 kHz. The gain of the amplifier was adjusted to ensure that the amplitude of the thermocouple signal was amplified substantially to approach the 10 V limit of the A/D converter and improve in this way the resolution of the temperature measurements.

The turbulent heat fluxes $\langle u''t'' \rangle$ and $\langle v''t'' \rangle$ were measured by means of a digitally compensated thermocouple and a single channel laser Doppler anemometer, which was arranged to measure the corresponding velocity component, as shown schematically in figure 2. The instantaneous temperatures measured by a fine-wire thermocouple are unweighted (Heitor and Moreira, 1993) and hence the mean and rms temperatures presented below are unweighted time-averages: where the temperature measurement is combined with the velocity signal from the anemometer to provide the heat flux, the latter measurement becomes density weighted. The distance between the thermocouple wire and the laser Doppler probe volume was about 1 mm and the axis of the thermocouple wire was always arranged to be normal to the temperature gradient in the flame to avoid errors due to spatial averaging occurring due to the finite length of the wire. The heat fluxes were modified by less than 10%, if the distance between the wire and the probe volume increased by 0.5 mm. Tests indicated that placing the thermocouple wire at a position 1 mm upstream of the laser Doppler probe volume resulted in smaller interference with the gaseous flow.

The measurement cycle was controlled by a C language computer program. Channels 5 and 7 of the DMA controller were programmed to allow data transfers of temperature and velocity measurements respectively. By enabling both DMA channels, the A/D converter started digitising the temperature signals and stored them in the computer memory buffer for DMA channel 5. The DMA channel 7 enabled the TSI counter to obtain a single velocity measurement. Once a Doppler signal was successfully processed by the counter and transferred to the computer memory buffer of DMA channel 7, an interrupt was generated by the terminal count flag of the DMA channel 7. The interrupt service routine allowed six more temperature measurements to be digitised by the A/D converter on DMA channel 5 after the time of the velocity measurement and then disabled the DMA channel 5. The single velocity measurement and twelve temperature measurements, six before and six after the time of the velocity measurement, were read from the memory buffers of the two DMA channels and stored in memory and on a datafile on the hard disk. Then the interrupt routine enabled the two DMA channels again and the procedure repeated till another successful velocity measurement. In this way, 2048 measurements were obtained at each point, limited by the available memory on the computer. The sampling rate was varied between 10 and 20 kHz during the heat flux measurements depending on the measurement location, with the higher sampling rate required at the shear layer. After the measurement was complete, the temperature signals were reprocessed to compensate for the finite temporal response of the thermocouple according to :

$$T = T_0 + \tau_c \ dT_0 / dt$$

The time constant, τ_c, was calculated from the Collis and Williams (1959) law together with the mean velocity and the mean temperature at each point. The temporal derivative of temperature, dT_0 / dt, was estimated by fitting the digitised points using first to third order polynomials. The time constant, τ_c, varied between 9 and 20 ms for most of the measurement points. The basic features of compensation have been described by Heitor (1985) and recent analysis for non-premixed flames has been provided by Mastorakos (1993).

Table 2 provides estimates of the errors in positioning the anemometer and the thermocouple; in the measurement of the air and gas flow rates by rotameters; of the velocity and temperature characteristics; and of the heat fluxes. The estimation of the tolerance on the measurement of the velocity characteristics is based on the specification of the anemometer, of the electronics for processing the signal (e.g. the resolution of the crystal frequency in the counter) and on the size of the sample used to estimate the moments of the velocity distributions at each point. The mean and rms of the velocity fluctuations were formulated at each point from around 3000 measurements, which result in errors of 1% and 3% respectively (Yanta & Smith 1978).

Table 2. Estimates Of Systematic And Random Errors

Quantity	Systematic	Random
position	1 mm	±1 mm
air and gas flow rates	-	5%
mean axial velocity, $\langle U \rangle$	+3%	1%
rms of axial velocity, $\sqrt{\langle u''^2 \rangle}$	- 3%	3%
mean radial velocity, $\langle V \rangle$	+3%	2%
rms of radial velocity, $\sqrt{\langle v''^2 \rangle}$	- 5%	5%
correlation coefficient, $\langle u''v'' \rangle / \sqrt{\langle u''^2 \rangle \langle v''^2 \rangle}$	7%	6%
mean tangential velocity, $\langle W \rangle$	+3%	2%
rms tangential velocity, $\sqrt{\langle w''^2 \rangle}$	- 5%	5%
mean temperature, $\overline{T}$	- 3%	3%
rms of temperature, $\sqrt{\overline{t''^2}}$	- 5%	15%
axial heat flux, $\langle u''t'' \rangle$	-10%	10%
radial and tangential heat flux, $\langle v''t'' \rangle$ & $\langle w''t'' \rangle$	-10%	15%

The estimation of the tolerance on the measurement of the temperature characteristics and heat fluxes is based on the errors in the instantaneous compensated value of temperature, the spatial separation between the measurement locations of the velocity and the temperature and on the limited population size used to estimate the moments of the temperature distributions at each point. The mean temperature is systematically underestimated by around 5% due to radiation losses. According to Heitor *et al.* (1985), a 40 μm thermocouple systematically underestimates the energy of the spectrum of the temperature fluctuations resulting in systematic errors of the order of 5% in the value of $\sqrt{\overline{t''^2}}$. This was confirmed by testing the temperature fluctuations measured after compensating the signals from a 25 and 40 μm wire. Uncertainties of 10% in the estimate of the time constant of the thermocouple cause random errors of the order of 15% in the value of $\sqrt{\overline{t''^2}}$. The largest systematic error in the measurement of heat fluxes is an underestimation by 10% due to a 10% underestimation of the time constant of the thermocouple. The largest random error in the value of heat fluxes is due to the radial displacement of the measurement locations of the velocity and temperature and was found to be around 10% and 15% of the maximum value of the axial and radial heat fluxes respectively.

169

Fig. 3 Radial profiles of mean axial, radial and swirl velocity with downstream distance from the exit of the quarl as parameter for a swirl number of 1.07.

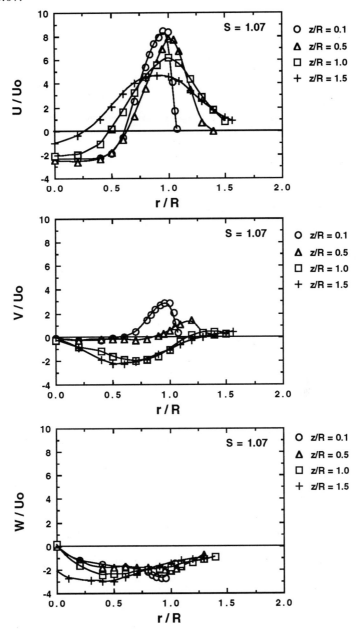

170

3 Results and Discussion

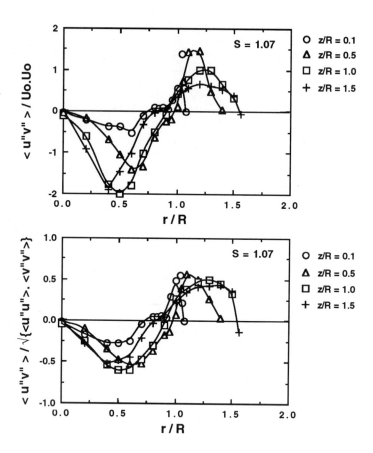

Fig. 4 Radial profiles of Reynolds shear stress and associated correlation coefficient with downstream distance from the exit of the quarl as parameter for a swirl number of 1.07.

Milosavljevic (1994) has comprehensively analyzed the flow in terms of the field distributions of velocity characteristics, mean species, and hence mean mixture fraction, and temperature (with swirl number, quarl geometry and fuel nozzle geometry as parameters) and those results should be considered in conjunction with those presented here. Thus, we concentrate on the results of direct interest to this paper only. Figures 3 and 4 show, for a swirl number of 1.07, radial profiles of mean axial, radial and swirl velocity, Reynolds shear stress and associated correlation coefficient with downstream distance from the exit of the quarl as parameter. Note that R is the radius of the *exit* of the quarl, namely 50 mm. The origin of the cylindrical co-ordinate system is at the centre of the exit of the quarl.

The mean flow has high rate of strain, $\partial\langle U\rangle/\partial r$, leaves the quarl at a large angle to the vertical and approximately equal to half the included angle of the quarl and the relative magnitude of $\langle V\rangle$ and $\langle W\rangle$ confirms that the swirl number is approximately unity. The main feature of interest here is that the sign of the shear stress is that to be expected from a gradient diffusion hypothesis, that the maximum and minimum values of the correlation coefficient are of the order expected in a shear flow and therefore the counter-gradient flux observed by Fernandes *et al.* (1993) is absent, even in the region closest to the quarl exit.

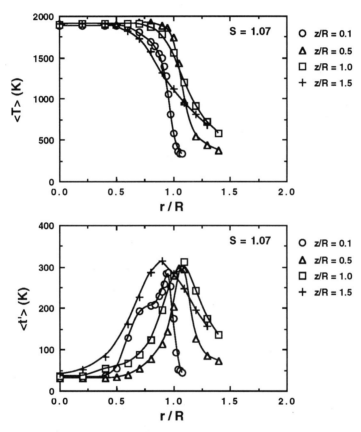

Fig. 5 Radial profiles of time-mean and -rms temperature with downstream distance from the exit of the quarl as parameter for a swirl number of 1.07.

Figures 5 and 6 show, for the same swirl number, the radial profiles of time-mean and -rms temperature and the axial, radial and azimuthal components of heat flux. The mean temperature profile has a negative gradient and, as expected,

172

Fig. 6 Radial profiles of axial, radial and azimuthal components of heat flux with downstream distance from the exit of the quarl as parameter for a swirl number of 1.07.

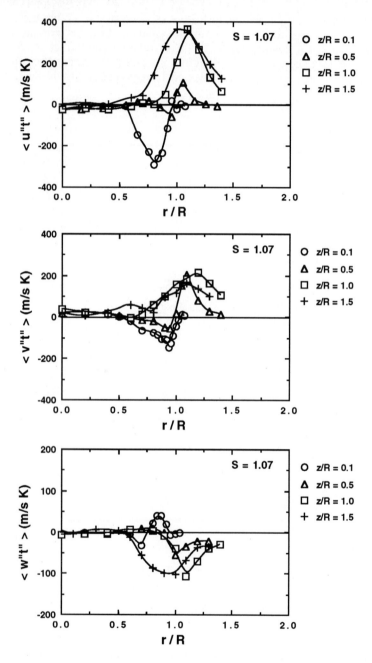

the rms temperature profile is narrowest closest to the quarl and diffuses radially as the flow expands downstream. The gradients of the axial heat flux are large partly because the flame brush is at an angle to the coordinate system used to analyze the flow corresponding approximately to the angle of the quarl. The main interest is provided by the profiles of the radial component of the heat flux, where the expectation, based on the gradient transport hypothesis, is that the sign of the flux will be positive everywhere. The profiles at z/R = 0.1 and 0.5, however, show clear regions of negative heat flux, which is in the counter-gradient direction, and the results of Takagi and Okamoto (1987) strongly suggest that the term responsible will be the strength of $-\overline{t''}\partial P/\partial r$ in the transport equations for radial heat flux.

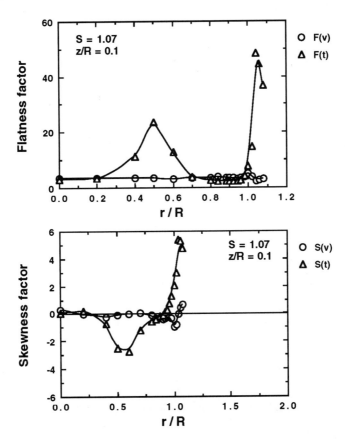

Fig. 7 Radial profiles of the corresponding skewness and flatness factors of the velocity and temperature signals at z/R = 0.1.

Figure 7 shows the corresponding skewness and flatness factors of the velocity and temperature signals at z/R = 0.1. It is clear that the probability

Fig. 8 Normalised joint probability distributions of radial velocity and temperature, $P(v'',t'')$, at location $r/R = 1.06$, $z/R = 0.5$, corresponding to strong gradient flux.

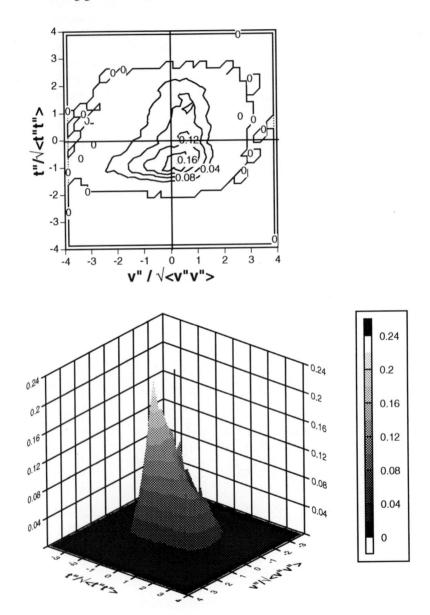

Fig. 9 Normalised joint probability distributions of radial velocity and temperature, P(v", t"), at location r/R = 0.94, z/R = 0.1, corresponding to strong counter-gradient flux.

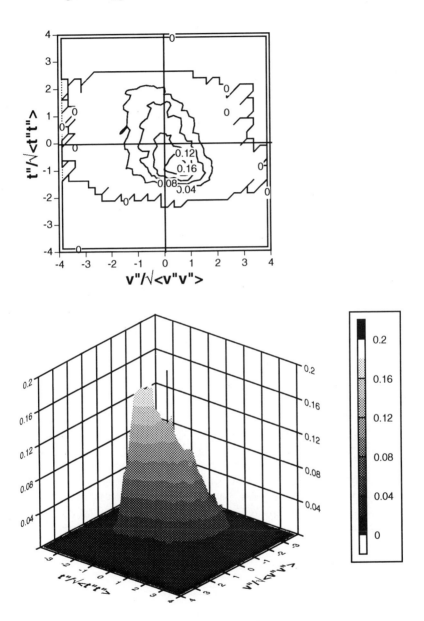

distribution of the velocity signals is close to Gaussian by both measures and presents no remarkable feature. In contrast, the temperature probability distribution departs strongly from Gaussian: in particular, the high skewness factors are indicative of a strongly intermittent flow (see Nagano and Tagawa, 1988). As is commonly observed, the flatness is smallest where the scalar variance is maximum.

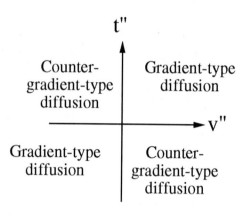

Fig. 10 Sketch showing the physical significance of the quadrants.

Figures 8 and 9 show the joint probability distributions of radial velocity and temperature at two locations, $P(v'', t'')$, the one corresponding to strong gradient flux ($r/R = 1.06$, $z/R = 0.5$) and the other to strong counter-gradient flux ($r/R = 0.94$, $z/R = 0.1$). The axes have been normalised by the respective rms values. In examining these figures, it is necessary to recall that the physical significance of the quadrants is as shown in figure 10 but it is not self-evident from inspection of figures 8 and 9 that the joint probability distributions at these two locations result in different signs for the correlation $\langle v'' t'' \rangle$. This does, however, become more apparent by examination of the weighted probability distributions:

$$W(v'', t'') = v'' t'' P(v'', t'')$$

so that

$$\langle v'' t'' \rangle = \iint W(v'', t'') dv'' dt''$$

Fig. 11 Normalised, weighted joint probability distribution of radial velocity and temperature, $W(v", t")$, at location $r/R = 1.06$, $z/R = 0.5$ corresponding to strong gradient flux, indicating the relative importance of the entries in quadrants 1 and 3.

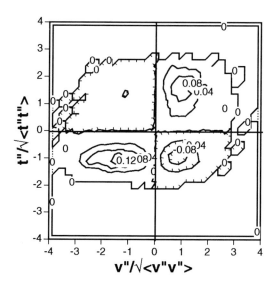

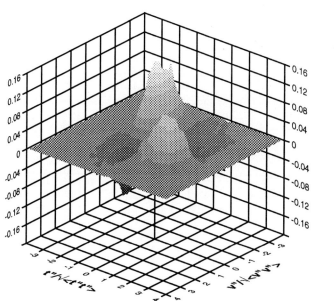

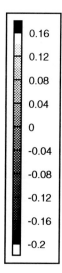

Fig. 12 Normalised, weighted joint probability distribution of radial velocity and temperature, $W(v'', t'')$, at location $r/R = 0.94$, $z/R = 0.1$ corresponding to strong counter-gradient flux, indicating the relative importance of the entries in quadrants 2 and 4.

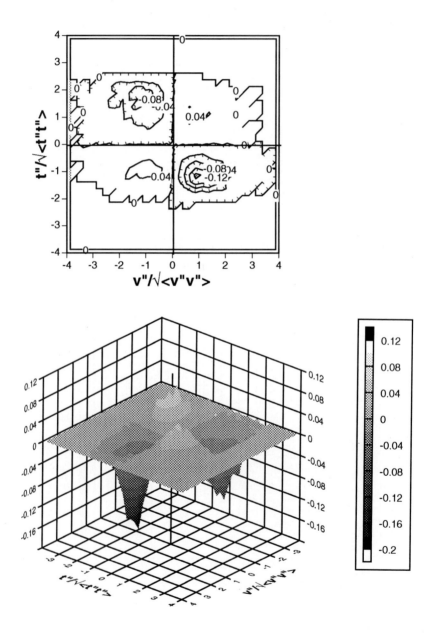

The corresponding plots of $W(v", t")$ are shown in figures 11 and 12 and the relative importance of the entries in the four quadrants is clear with quadrants 1 and 3 dominating in the case of gradient diffusion and quadrants 2 and 4 dominating for counter-gradient diffusion. The entries in quadrants 2 / 4 arise when the radial velocity $v"$ which is smaller / greater than the local mean are associated with values of temperature $t"$ which are greater / smaller than the mean. The magnitude of the local mean radial velocity, $\langle V \rangle$, is large relative to its rms value and hence it should be carefully noted that $v" < 0$ *does not* correspond to an "insweep" motion, which *is* the correct physical interpretation *for boundary layers only*, because there the mean cross-stream velocity, $\overline{V} \approx 0$.

The preceding text has shown that there is counter-gradient diffusion of heat flux in the region of $z/R < 0.5$. However, the calculation of non-premixed flames usually proceeds through the Favre-averaged transport equations for the mean and the variance or the turbulent flux of the *conserved scalar*, $\langle \xi \rangle$, $\langle \xi"^2 \rangle$, together with the use of a presumed-shape pdf method (for a good recent overview see Jones & Khaki, 1995). Note that the transport equations for mean energy and turbulent heat flux are *not* used[2]. The question of interest here is thus whether the observed counter gradient diffusion of $\langle v"t" \rangle$ *also* implies counter gradient diffusion of $\langle v"\xi" \rangle$ because it would make necessary the use of the modelled transport equations for the scalar flux, $\langle u_i"\xi" \rangle$, rather than an eddy viscosity hypothesis[3], for the solution of the transport equation of $\langle \xi \rangle$. To answer this question, it is necessary to refer again to the distribution of $W(v", t")$.

Figures 11 and 12 show that the strongest contributors to the counter gradient heat flux are from the regions of

$$t" / \sqrt{\langle t"^2 \rangle} > 2$$

$$t" / \sqrt{\langle t"^2 \rangle} \approx \text{-1.0 to -1.5.}$$

These values are sufficiently far from the mean to imply that these are not only large fluctuations but also comparatively rare events (which give rise to the large values of the skewness factor remarked on above). Temperature is a stepwise-linear, but multi-valued, function of mixture fraction and hence it is not possible,

---

[2] A transported pdf approach (see Khaki & Jones, 1995) might be able to provide both heat flux and conserved scalar flux information. Thus, some experiments have reported the flux of conserved scalar, measured either directly using marker nephelometry (Heitor *et al.*, 1993), or indirectly by measuring sufficient major species to reconstruct the conserved scalar itself, using a variety of expensive optical techniques.

[3] It should be noted that there is no generally applicable, straightforward relationship between $\langle v"t" \rangle$ and $\langle v"\xi" \rangle$.

in general, to relate t" to ξ". However, the location of strongest counter-gradient diffusion corresponds to a mean temperature of about 1000 K and an rms temperature of the order of 300 K and the limits above show that t" ranges from just above ambient temperature to below the adiabatic flame temperature. It is thus plausible to believe that the temperature fluctuations contributing strongly to counter-gradient heat flux are caused by excursions $0 < \xi < \xi_S$ (the latter symbol denoting the stoichiometric value of the mixture fraction, a range over which temperature is a linear *and* single valued function of mixture fraction. Hence, in this region, the counter-gradient diffusion of $\langle v"t"\rangle$ is also associated with counter-gradient diffusion of $\langle v"\xi"\rangle$ [4].

Finally, it is noted that counter-gradient diffusion was observed up to z/R ≈ 0.5, *i.e.* a relatively short distance compared to the axial length of the flame. Thus, it is pertinent to ask whether the relatively isolated occurrence of counter-gradient diffusion is an important feature of this flow. An answer can, once again, be deduced from examination of the distribution of W(v",t"), on this occasion that corresponding to the location at which *gradient* diffusion of $\langle v"t"\rangle$ occurs. The contributions from the entries in the "counter-gradient" quadrants, 2 and 4, are of course smaller than those of the "gradient" quadrants, 1 and 3. However, those from the former are far from negligible and this shows that the transport of $\langle v"t"\rangle$, and hence $\langle v"\xi"\rangle$, is still influenced by the source terms which drive counter gradient diffusion and that therefore the solution of the transport equations for $\langle u_i"\xi"\rangle$ will remain necessary. Thus, the sign of the flux being in accordance with that expected from a gradient diffusion hypothesis is *not* a reliable test for deciding on the relative importance of terms such as $-\overline{u_i"}\partial P/\partial x_i$ and $-\overline{\xi"}\partial P/\partial x_i$ in the transport equations for the velocity or scalar flux.

4. Conclusions

1. Measurements of Reynolds shear stress and heat flux have been obtained in a swirling flow at a swirl number strong enough to result in a recirculation zone which stabilises an unconfined flame.

2. In contrast to other measurements in the literature, we have found evidence for the existence of counter-gradient heat flux for axial distances up to

---

[4] The relative importance of the "additional" terms in the transport equations for $-\overline{u_i"}\partial P/\partial x_i$ and $-\overline{\xi"}\partial P/\partial x_i$ is usually found by examining each term in the appropriate transport equation which is amenable to experimental measurement. This requires much effort and, particularly when the scalar under consideration is temperature rather than the conserved scalar, the effort is directed at an equation which is not directly relevant to the calculation of the flame. In contrast, the use of the weighting functions in v"t" space is easily implemented and provides direct physical insight that can be extended, here, to v"ξ" space.

0.5 R from the exit of the quarl but none for Reynolds shear stress. The primary driving mechanism for this result is expected to be $-\overline{\xi}''\partial P/\partial x_i$.

3.   The quantity of primary interest in the calculation of non-premixed flames is the conserved scalar and in particular $\langle u_i''\xi''\rangle$. Arguments based on the shape of the weighted joint probability distribution of radial velocity and temperature fluctuations, $W(v'',t'') = v''t''P(v'',t'')$, show that the counter-gradient contributions are due to large amplitude, comparatively rare departures of temperature from the local mean and also suggest that the corresponding radial flux of the conserved scalar, $\langle v''\xi''\rangle$, will also be in the counter-gradient sense.

4.   Even in those regions of the flow where $\langle v''t''\rangle$ is in the gradient sense, the magnitude of $W(v'',t'')$ in quadrants 2 and 4, which represent counter-gradient flux, remains strong in comparison with quadrants 1 and 3, which represent gradient diffusion.

5.   These results show that predictions of this kind of flow must be based on second moment, rather than on effective viscosity, turbulence model closures so as to capture the effects of $-\overline{\xi}''\partial P/\partial x_i$ on heat fluxes.

6.   Analysis of the quadrants of has been used, rather than a term-by-term analysis of the transport equations for heat flux, with the advantage that this permits insight into the scale of the fluctuations of temperature and velocity which give rise to the counter-gradient contribution *and* allows discussion of the probable value of the conserved scalar flux .

Acknowledgements

Mr. J. R. Laker constructed the amplifiers for the thermocouple signals and the interface card for the TSI counter.

References

Ahmed, S A, So, R M C, Mongia, H C, 1985. Density effects on jet characteristics in confined swirling flow. Expts. Fluids 3, 231 - 238.

Bradshaw, P, 1969. The analogy between streamline curvature and buoyancy in turbulent shear flow. J Fluid Mechanics, 36, 177 - 191.

Bray, K N C, Libby, P A, 1991. Premixed turbulent combustion. In Fluid Dynamical Aspects of Combustion Theory (M Onofri and A Tesei, eds.), pp. 113 - 140, Longman.

Bray, K N C, Libby, P A and Moss, 1985. Combust. Flame, 61, 87 - 102.

Cheng, R K and Ng, T T, 1985. Conditional Reynolds stress in a strongly heated turbulent boundary layer with premixed combustion. Phys. Fluids, 28, 473 - 488.

Collis, D C, Williams, M J, 1959. Two dimensional convection from heated wires at low Reynolds number. J. Fluid Mech., vol. 6, 357-384.

Duarte, D, Ferrão, P, Heitor, M V, 1994. Flame structure characterisation based on Rayleigh thermometry and two-point laser-Doppler measurements. Seventh International Symposium on Applications of laser techniques to fluid mechanics. Paper 2.3.

Escudier, M P and Keller, J J, 1985. Recirculation in swirling flow: a manifestation of vortex breakdown. AIAA J, 32, 111 - 116.

Fernandes, E C, Ferrão, P, Heitor, M V, Moreira, A L N, 1994. Velocity-temperature correlations in recirculating flames with and without swirl. Experimental Thermal and Fluid Science, 9, 241 - 249.

Ferrão, P, Heitor, M V, 1993. Turbulent mixing and non-gradient diffusion in baffle-stabilised flames. *Paper 28-2-1* in Ninth Symposium on Turbulent Shear Flows, Kyoto, Japan.

Glass, M and Kennedy, I M, 1977. An improved seeding method for high temperature laser Doppler velocimetry. Combust. Flame, 29, 333 - 335.

Goss, L P, 1993. CARS instrumentation for combustion applications, in Instrumentation for flows with combustion (Ed. Taylor, A M K P), pp 251 - 322. Academic Press, London.

Heitor, M V, 1985. Experiments in turbulent reacting flows. Ph D Thesis, University of London.

Heitor M V and Moreira, A L N, 1993. Thermocouples and sampling probes for combustion studies. Prog. Energy Combust Sci, 19, 259 - 278.

Heitor, M V, Stårner, S H, Taylor, A M K P, Whitelaw, J H, 1993. Velocity, size and turbulent flux measurements by laser Doppler velocimetry, in Instrumentation for flows with combustion (Ed. Taylor, A M K P), pp 113 - 250. Academic Press, London.

Heitor, M V, Taylor, A M K P, Whitelaw, J H, 1985. Simultaneous velocity and temperature measurements in a premixed flame. Experiments in Fluids, vol. 3, 323-339.

Heitor, M V, Taylor, A M K P, Whitelaw, J H, 1987. The interaction of turbulent pressure gradients in a baffle-stabilised premixed flame. J Fluid Mech., vol. 181, 387 - 413.

Hirai, S and Takagi, T, 1988. Prediction of heat transfer deterioration in turbulent swirling pipe flow. JSME Int J Series II, 31, 694 - 700.

Hirai, S, Takagi, T and Matsumoto, M, 1988. Predictions of the laminarization phenomena in an axially rotating pipe flow. Trans. ASME J Fluids Eng., 110, 424 - 430.

Hirai, S and Takagi, T, 1991. Numerical prediction of turbulent mixing in a variable-density swirling pipe flow. Int. J. Heat Mass Transfer, 34, 3143 - 3150.

Hogg, S and Leschziner, M A, 1989. Second-moment-closure calculation of strongly swirling flow with large density gradients. Int. J Heat and Fluid Flow, 10, 16 - 27.

Johnson, B V and Bennett, J C, 1981. Mass and momentum turbulent transport experiments with confined coaxial jets. NASA CR-165574.

Jones, W. P. and Kakhi, M., 1995. "Mathematical Modelling of Turbulent Flames" in Unsteady Combustion (Eds F. Cullick, M. V. Heitor and J. H. Whitelaw), Kluwer Academic Pub., Dordrecht. In the press.

Libby, P, Bray, K N C 1981. Counter-gradient diffusion in premixed turbulent flames. AIAA J., vol. 19, 205-213.

Mastorakos, E, 1993. Turbulent combustion in opposed jet flows. Ph D Thesis, University of London.

Melling A and Whitelaw J H, 1975. Turbulent flow in a rectangular duct. J Fluid Mech., 78, 289 - 315.

Milosavljevic, V D, Taylor, A M K P, Whitelaw, J H, 1990. The influence of burner geometry and flow rates on the stability and symmetry of swirl stabilized non-premixed flames. Comb Flame, vol. 80, 196 - 208.

Milosavljevic, V D, 1993. Natural gas, kerosene and pulverised fuel fired swirl burners. Ph D Thesis, University of London.

Nagano, Y and Tagawa, M, 1988. Statistical characteristics of wall turbulence with a passive scalar. J Fluid Mech., vol. 196, 157 - 185.

Perry, A E and Hoffman, P H, 1976. An experimental study of turbulent convective heat transfer from a flat plate. J Fluid Mech., 77, 355 - 368.

Roback, R and Johnson, B V, 1983. Mass and momentum turbulent transport experiments with confined swirling coaxial jets. NASA CR-168252.

So, R M C and Ahmed, S, 1987. Behaviour of carbon dioxide jets in a confined swirling flow. Heat and Fluid Flow, 8, 171 - 176.

Takagi, T, Okamoto, T, Taji, M and Nakasuji, Y, 1984. Retardation of mixing and counter-gradient diffusion in a swirling flame. Twentieth Symposium (International) on Combustion, pp. 251 - 258. The Combustion Institute, Pittsburgh.

Takagi, T & Okamoto, T, 1987. "Direct measurements of the turbulent transport of momentum and heat in the swirling flame", in Laser diagnostics and modelling of combustion (Iinuma, K et al., Eds.), pp 273 - 280. Springer Verlag, Berlin.

Yanta, W J, Smith R A 1978. Measurements of turbulence-transport properties with a laser Doppler velocimeter. AIAA paper 73-169, 11th Aerospace Science Meeting, Washington.

Flame Structure Characterization Based On Rayleigh Thermometry And Two-Point Laser-Doppler Measurements

D. Duarte, P. Ferrão, and M.V. Heitor

Technical University of Lisbon
Instituto Superior Técnico
Dept. of Mechanical Engineering
Av. Rovisco Pais
1096 Lisboa Codex, PORTUGAL

Abstract. Turbulent premixed flames stabilized in a bluff-body are experimentally investigated in order to discuss flame structure and morphology. Integral scales were measured within the reacting flow, based on the integral of the lateral velocity correlations obtained by using two laser-Doppler velocimetry systems. The quantitative results obtained are compared with flame shadowgraphs and complemented with simultaneous measurements of time-resolved velocity and temperature obtained by combining laser-Doppler velocimetry and laser-Rayleigh scattering. The results are used to quantify the relative magnitudes of the terms involving the balance equations of momentum, turbulent kinetic energy and heat flux. In addition, the interaction of pressure gradients and temperature fluctuations is shown to affect the heat release in the flame studied, revealing the existence of large zones characterized by non-gradient scalar fluxes.

1 Introduction

In the past many attempts have been performed to extend the knowledge on laminar flames and non-reacting fluid mechanics, to turbulent combustion, but turbulent mixing in flames is altered by the accompanying heat release and can, as consequence, be qualitatively different from that occurring in non-reacting flows. Examples include counter-gradient diffusion, e.g. Libby and Bray (1981), Bray et al. (1985), in either confined non-premixed swirling flames, Takagi et al. (1985), or in unconfined premixed flames, Heitor et al. (1987), Ferrão and Heitor (1993).

Although turbulent flames are characteristic of most of the industrial burning devices, their physical structure cannot be regarded as completely understood at the present. As pointed out by Borghi (1985), the detailed study of the fine-scale fluctuating structure of turbulent flames is very difficult to investigate experimentally and, consequently, theoretical studies based on physical assumptions suffer from lack of validation. This is the main motivation of the work reported in this paper.

The discussion on turbulent flames structure and morphology emerged from the fifties when turbulent flames were divided in two different categories, namely "wrinkled laminar flames" and "distributed reaction zones". Work performed during the sixties in Russia, e.g. Schelkin (1968) and Talantov et al. (1969), suggested that turbulent flames could be represented by either regimes, depending on the magnitude of turbulent intensity and the turbulent scales compared to laminar flame velocity and thickness. The need for an intermediate combustion regime was emphasized by Zimont et al. (1979), who justified the introduction of the "thickened -flame regime".

It is now generally accepted that the definition of turbulent combustion regimes according to the flame structure and morphology is based on the magnitude of turbulent intensity and the turbulent scales compared to laminar flame velocity and thickness. The resulting diagrams representing the combustion regimes, e.g. Borghi (1985), are frequently used, although they should be analysed with care and require the quantification of characteristic scales which are not generally easy to measure. It should be noted that the boundaries between combustion regimes are not as sharp as derived from physical analysis as discussed, for example, by Poinsot et al. (1991), through the direct numerical simulation of a flame front under strong curvature effects. The experimental work reported in this paper intends to contribute with quantitative data on these aspects, and includes the characterization of a flame in the "distributed reaction zone" regime. The main characteristics theoretically associated with this regime are analyzed and compared to the experimental data obtained.

The work follows that of Ferrão and Heitor (1993), which has assessed the extent to which a laser-Doppler velocimeter, LDV, may be used for the analysis of turbulent heat transfer in strongly sheared disc-stabilized propane-air flames through its combination with laser-Rayleigh scattering. The present paper complements this information and reports measurements of lateral integral scales within the reacting flow.

The remainder of this paper includes three sections. The experimental techniques are briefly discussed in the next section. The results obtained are presented and discussed in the third section and the last section summarizes the main conclusions.

2 Flow Configuration And Experimental Method

The analysis considered in this paper refers to a premixed propane-air flame stabilized on a disc of D = 0.056m diameter located at the exit section of a contraction with a diameter of 0.080m. The flame is open to the atmosphere to facilitate optical access. The annular bulk velocity at the trailing edge of the disc was equal to 19.5 m/s, resulting in a Reynolds number based on the disc diameter of 71,000. For the experiments reported here, the equivalence ratio was 0.60 and the corresponding adiabatic flame temperature is 1717 K. The rear stagnation point of the flame is located at an axial position of 110 mm, or x/D= 2.0.

The instrumentation used in this work is based on the combined LDV/LRS system schematically represented in Fig. 1a), which uses a single laser light source (5W Argon-ion laser), the details of which can be found in Ferrão (1993).

The velocimeter was based on the green light (514.5 nm) of the laser and was operated in the dual-beam, forward scatter mode with sensitivity to the flow direction provided by a rotating diffraction grating. The calculated dimensions of the measuring volume at e^{-2} were 606 μm and 44 μm. The Rayleigh scattering system was operated from the blue line (488 nm) of the same laser source, which was vertically polarized and made to pass through a 5:1 beam expander. The light converged in a beam waist of 50 μm diameter, and was collected at 90° from the laser beam direction, through a slit of 1μm. The collected light was filtered by a 1 nm interference filter and passed through a polarizer in order to increase the signal-to-noise ratio. A calibration procedure was implemented in order to compensate for number density dependence on the chemical composition. The uncertainty on the average temperature was quantified as 4%.

For the measurements of spatial velocity correlations, the laser light source was operated in multiline and the LRS optical system was replaced by a second dual-beam laser velocimeter mounted on a positioning system which allows the displacement of the control volume of this system relative to that described before, figure 1b). This second LDV was based on the green light (514.5 nm) of the laser and operated in backscatter mode, with sensitivity to the flow direction provided by a bragg cell. Two interference filters of 1nm bandwith were used in each optical collection systems, in order to avoid optical interference between the two systems.

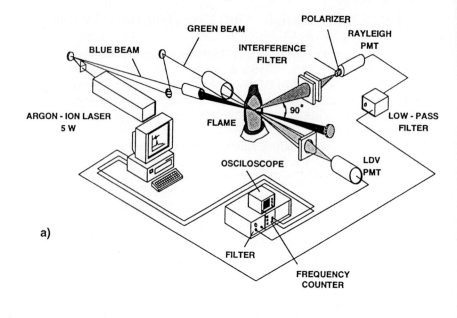

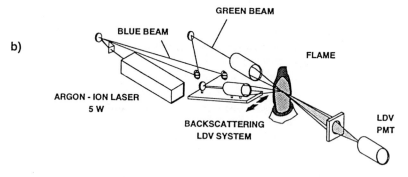

Fig.1 Schematic diagram of the instrumentation used:
  a) combined LDV/Rayleigh scattering system
  b) velocity correlation measuring system

Two frequency counters (DANTEC, model 55L96) were used to process the Doppler signals and a maximum coincidence time of 1μs was used to decide upon the simultaneity of the measurements. Each measurement was based in a population of N=6144 valid simultaneous velocity time series.

Correlation measurements at small separations can be expected to be influenced by the spatial resolution of the system. However, this does not noticeably affect the determination of the integral length scales, which are the purpose of the present work. The consequence of the limited spatial resolution is that the velocity correlation measured for nominally zero separation is never equal to unity. In practice, this is mainly because the measuring values are bigger than the smaller scales of the flow.

The experimental procedure included the measurement of the axial velocity fluctuations with the two LDV systems, respectively u_1 and u_2, at each measuring location, ζ, which was followed by successive displacement, ζ, of the backscatter system up to a maximum distance of 17 mm. This procedure allowed the measurement of the lateral velocity correlation coefficient, $g(r,\zeta,)$ for a location r, defined as :

$$g(r, \zeta) = \frac{1}{(N-1)} \frac{\overline{u_1'' (r) \, u_2'' (r + \zeta)}}{\sqrt{\overline{u_1''^2 (\zeta)}} \sqrt{\overline{u_2''^2 (r + \zeta)}}} \qquad [1]$$

were $\widetilde{u}_1''$ represents the turbulent velocity fluctuations of u_1 .

For the results reported here, the velocity correlations obtained were divided by the value measured at zero separation, which was generally about 0.7.

3   Experimental Results And Discussion

The analysis and discussion of the experimental results presented here is divided in three parts. The first discusses the measurements of the lateral velocity correlation coefficient and the evaluation of the integral scales along the flame. The values obtained are used to define the combustion regime representative of the flame, which is compared to theoretical considerations. The last section summarizes the main mechanisms inherent to the turbulent transport of kinetic energy and heat fluxes in the flames considered.

3.1   Integral Scales Of Turbulence

The integral length scale of turbulence is, to a certain extent, a measure of the longest connection, or correlation distance, between the velocities at two points of the flow field. It is reasonable to expect, as pointed out by Hinze (1975), that the degree of correlation will decrease as the distance between the two measuring points, ζ, is increased and that, beyond some finite distance, L, this correlation will be practically zero. Thus, the integral length scale of turbulence, l, is defined by:

$$l = \int_0^\infty g(\zeta)\, d\zeta = \int_0^L g(\zeta)\, d\zeta \tag{2}$$

The measurements of the lateral velocity correlation coefficient were performed in several radial positions located at two characteristic axial locations of the flame studied, namely at the rear stagnation point, x/D=2.0, and at the location of maximum width of the recirculation zone, x/D=1.2. Some results obtained for g(ζ) at different points for each location are represented in Figs. 2 and 3.

The shape of g(ζ) is qualitatively similar for all the conditions studied, but it can be clearly concluded that there are considerable differences in the values of the integral length scale of turbulence. The lateral velocity correlation coefficient rapidly decreases to zero in the points located at the reacting shear layer, as represented in Fig.3, while within the recirculation zone the correlation do not reach zero within the measuring distance.

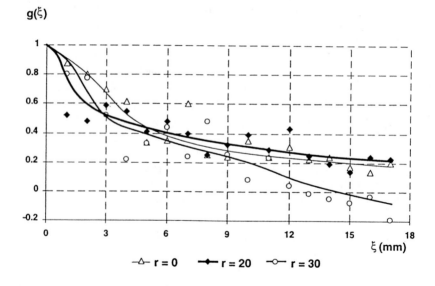

Fig.2 Lateral velocity correlation coefficient obtained at the rear stagnation point, x/D=2.0

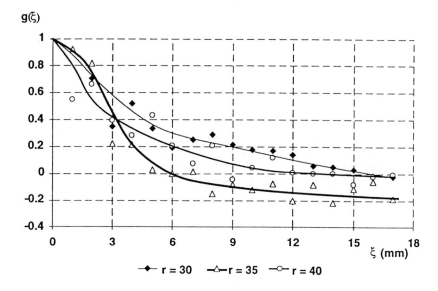

Fig. 3 Lateral velocity correlation coefficient for the axial location of, x/D=1.2

The results can be used to obtain of the length scales and the turbulence intensity characteristic of each location measured, as represented in Table 1, where l_f and s_L represent the laminar flame thickness and velocity, respectively. It is shown that the length scales inside the recirculation zone are of about 10-12 mm and decrease to about 3-4 mm within the reacting shear layer. These units may be compared with the qualitative information provided by the shadowgraphs of Fig. 4, although this technique integrates over the line of sight and the information along that line is not unique and must be considered with some precaution.

Table 1. Integral length scales and characteristic parameters.

X/D	r (mm)	l (mm)	u´ (mm)	u´/s_L	l/l_F	Ka	Da
1.2	0	12	4	27	30	25	1
1.2	15	9	4	27	23	29	1
1.2	30	5	4	27	13	39	1
1.2	35	3	2	13	8	18	1
1.2	40	4	2	13	10	15	1
2.0	0	8	4	27	20	31	1
2.0	15	7	4	27	18	33	1
2.0	30	4	3	20	10	28	1

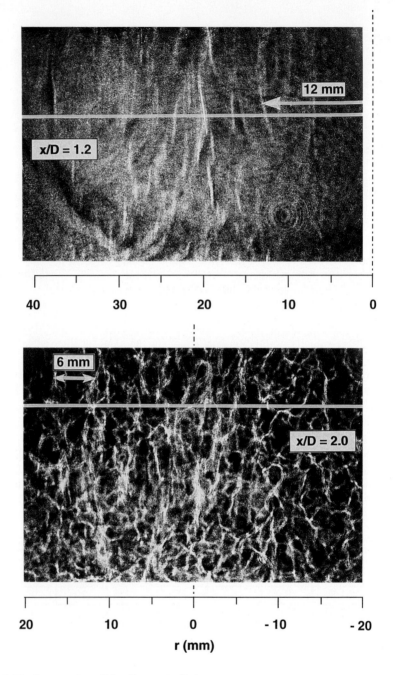

Fig. 4 Shadowgraphs of the flame studied

Nevertheless, qualitative information on integral scales derived from Fig.4 agrees with the results reported in Table 1 Once the technique is sensitive to the second derivative of the index of refraction, i.e. the density, the results suggest the existence of regions of strong curvature of the flame front.

3.2 Combustion Regime Characterization

The non-dimensional Damköler (Da) and Karlovitz (Ka) numbers reported for the flame studied are presented in Table 1, and show that the flame is in the transition of the "distributed reaction zone" regime, because Da≈1 and Ka>1. At this combustion regime, thin reaction zones may be generated locally and quenched by flame stretching, thus characterizing a turbulent flame as a thickened-wrinkled flame with possible extinctions, Borghi (1985). As a consequence, the probability of finding gases in an intermediate state is not negligible, which means that a probability density function of temperature in any point should have a near bimodal shape with a significant population between combustion reactants and products.

This analysis was experimentally validated from the simultaneous velocity-temperature measurements performed in the reacting shear layer and shown in Fig.5, which represents the joint probability density function of velocity and

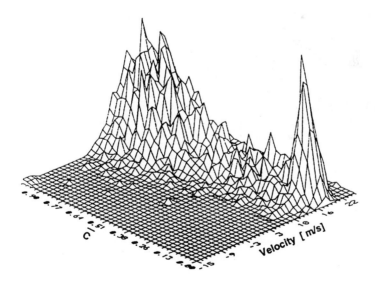

Figure 5 Typical joint probability density function of velocity and progress reaction variable for the reacting shear layer

the progress reaction variable, $c=(T-T_0)/(T_{ad}-T_0)$. The results confirm that the reacting shear layer is characterized by large temperature and velocity fluctuations were local extinction in the inner reaction layer is expected to occur due to straining caused by the small eddies. The results validate the theoretical analysis available for the combustion regime identified before.

3.3 Balance Equations For Turbulent Kinetic Energy And Heat Fluxes

This section gives details of the characteristics of the flame at X/D=1.2, and summarizes the main mechanisms inherent to the transport of turbulent kinetic energy and heat fluxes. Fig. 6 represents the mean velocity, progress reaction variable and turbulent velocity fluctuations across this axial section of the flame. It can be concluded that the recirculation zone is characterized by homogeneous temperatures and that its limit is located at r/D≈0.4. The maximum turbulence intensity occurs at the location of the maximum gradients of mean velocity as predicted by gradient models, and the turbulence is not isotropic, as the axial velocity fluctuations in the shear layer is about twice the intensity of the radial and the tangential velocity fluctuations.

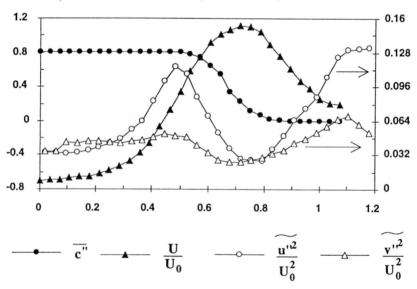

Figure 6 Radial profiles of mean velocity, progress reaction variable and turbulent velocity fluctuations for the axial location of x/D=1.2

The terms in the transport of axial and radial momentum have been calculated from the measurements and are shown in Figs. 7 and 8.

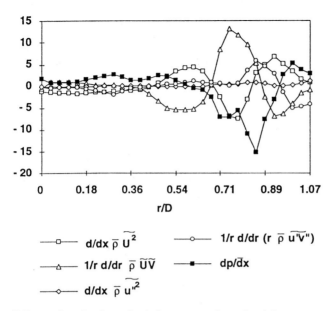

Fig. 7 Terms involved on the balance equation of axial momentum

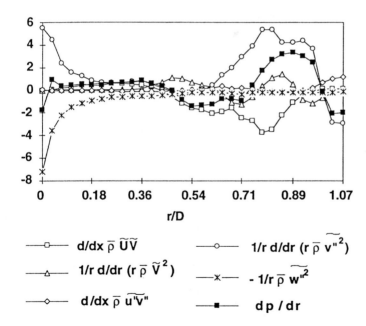

Fig. 8 Terms involved on the balance equation of radial momentum

The mean pressure gradients have been found by addition and are dominated by the difference between two large terms, namely the convection terms (also the diffusion term in the balance of radial momentum). The pressure gradients represent an important source or sink of radial momentum, which is to be expected because of the large streamline curvature.

The mean pressure gradient also appears in the transport equations for turbulent kinetic energy, whose most significant terms are represented in Fig.9.

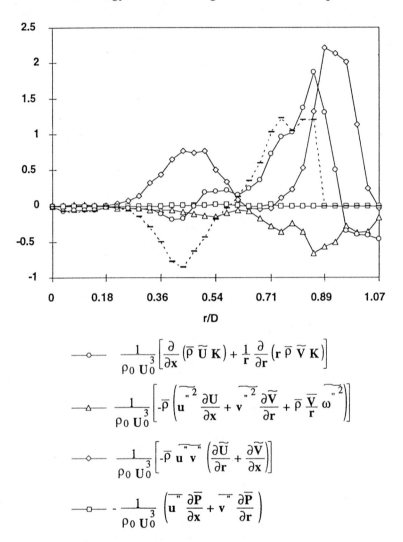

Fig. 9 Terms involved on the balance equation of turbulent kinetic energy

In the core of the annular jet, turbulence production by normal stresses is negative and convection is the largest term and represents a loss. In the shear layer, the major contribution to the production of turbulent kinetic energy is due to the interaction between shear stresses and shear strains. The production by the interaction of pressure gradients and velocity fluctuations is negligible all over the flow field.

The turbulent axial and radial heat fluxes measured with the LDV/LRS system across this axial location are represented in Fig.10 and are restricted to the thin zone along the shear layer where the radial gradients of mean temperature are large. These quantities represent the exchange rate of reactants responsible for the phenomena of flame stabilization and the results show that the axial fluxes are higher than the radial values. This reveals non-gradient transport of heat, as the isotherms in this location are axially aligned. It should be noted that radial fluxes are positive, as expected in a recirculating flame (Heitor et al., 1987).

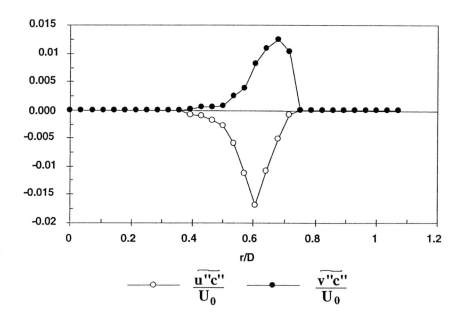

Fig. 10 Axial and radial velocity correlation components

The flame is then stabilized by the heat transfer between the hot products and the cold reactants with the sign of the radial heat flux opposite to that derived from gradient-transport models. Similar behaviour has been observed in the turbulent premixed flames of Ferrão and Heitor (1993) Heitor et al. (1987) and has also been predicted analytically by Bray et al. (1985).

The non-gradient heat transport has been associated to the preferential deceleration of the light products of combustion relative to the cold reactants, due to the interaction between the mean adverse pressure gradients, typical of the present recirculating flow, and the large density fluctuations characteristic of reacting flows. This can be confirmed by the analysis of the terms involved in the transport equations for the axial and radial heat flux that are represented in Figs. 11 and 12, respectively. The evidence is that the interaction between the mean adverse pressure gradients and the large temperature fluctuations that occur in the present flame is important and influence the turbulent heat transport in recirculating flames.

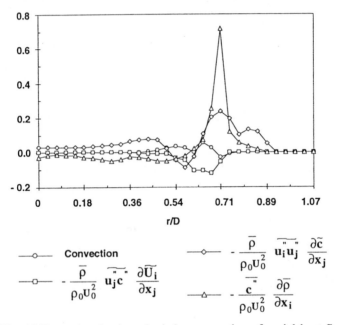

Fig. 11 Terms involved on the balance equation of axial heat flux

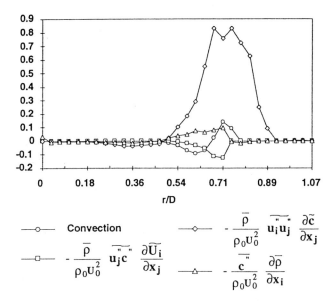

Fig. 12 Terms involved on the balance equation of radial heat flux

4   CONCLUSIONS

Time-resolved simultaneous measurements of velocity and temperature, together with those of the spatial correlations of velocity, have been obtained by combining laser-Doppler velocimetry and laser-Raleigh scattering and using two laser velocimeters, respectively, in a turbulent recirculating premixed propane-air flame.

The spatial velocity correlations within the flow were used to analyze the extent to which turbulent mixing interacts with combustion, changing the flame structure and morphology. The experimental results were used to validated currently accepted theoretical descriptions of turbulent combustion regimes.

The velocity-scalar correlations measured were used to quantify the relative magnitudes of the terms involving the balance equations of momentum, turbulent kinetic energy and heat flux. It was concluded that the interaction of pressure gradients and temperature fluctuations is shown to affect the heat release in the flame studied, revealing the existence of large zones characterized by non-gradient scalar fluxes.

Acknowledgments

The assistance of Messrs. Carlos Carvalho and Jorge Coelho in the preparation of this paper is gratefully appreciated. Financial support has been provided by the SCIENCE programme of the DGXII of the European Commission, under the contract SC1-0459.

References

Borghi, R. (1985). "On the Structure and Morphology of Turbulent Premixed Flames". Recent Advances in the Aerospace Sciences, Ed.Corrado Casci, Plenum Publishing Corporation : 117-137.

Bray, K.N.C., Libby, P. and Moss, J.H. (1985). "Unified Modelling Approach for Premixed Turbulent Combustion - Part I: General Formulation". Comb. & Flame, 61: 87-102.

Ferrão P. and Heitor M.V. (1992). "Simultaneous Measurements of Velocity and Scalar Heat Transfer in Recirculating Flames" 6th Symposium on Applications of Laser Techniques to Fluid Mechanics, 20-23 July, 1992.

Ferrão, P. (1993). "Análise Experimental de Chamas Turbulentas com Recirculação". Ph.D Thesis, Technical University of Lisbon (In Portuguese).

Ferrão, P. and Heitor, M.V. (1993). "Turbulent Mixing and Non-Gradient Diffusion in Baffle-Stabilized Flames". "Turbulent Shear Flows-9", ed. Springer Verlag, Japan: 28-2.

Heitor, M.V., Taylor, A.M.K.P. and Whitelaw, J.H. (1987). "The Interaction of Turbulence and Pressure Gradients in a Baffle-Stabilized Premixed Flame". J. Fluid Mechanics, 181: 387-413

Hinze, J.O. (1975). Turbulence, 2nd Edition. McGraw-Hill.

Libby, P. and Bray, K.N.C. (1981). "Counter-Gradient Diffusion in Premixed Turbulent Flames". AIAAJ., 19 : 205-213

Poinsot, T., Veynante, D. and Candel, S. (1991). "Quenching Process and Premixed Turbulent Combustion Diagrams". Journal of Fluid Mech., 228:561-606.

Schelkin, K.I. (1968). "Combustion Hydrodynamics". Fiz. Goreniya Vzryva, 4 : 455-468.

Takagi, T., Okamoto, T., Taji, H. and Nakasuji, Y. (1985). "Retardation of Mixing and Counter-Gradient Diffusion in a Swirling Flame". Twentieth Symposium (Int'l) on Combustion, The Combustion Institute: 251-258

Talantov, A., Ermolac, V.H., Zotin, V.K. and Petrov, E.A. (1969). "Laws of Combustion of a Homogeneous Mixture in a Turbulent Flow". Comb. Explos. Schock Waves, 5 : 73-75.

Zimont, V.L. (1979). "Theory of Turbulent Combustion of a Homogeneous Fuel Mixture at High Reynolds Numbers", Fiz. Goremiya Vzryva, 15 : 23-32.

Simultaneous Measurement of Temperature and OH Concentration Fields in Turbulent Combustion Using One Single Laser Source and One Single CCD Camera

S. Kampmann and A. Leipertz

Lehrstuhl für Technische Thermodynamik (LTT),
Universität Erlangen-Nürnberg, Am Weichselgarten 8, D-91058 Erlangen,
Germany

Abstract. A new method is described for the simultaneous detection of the two-dimensional (2D) temperature and the 2D OH concentration distributions in a high turbulent premixed flame inside a contained combustion chamber. The temperature information is obtained by Rayleigh scattering and the OH concentration from laser-induced predissociated fluorescence (LIPF). A newly designed detection optics for the separation of the synchronously induced Rayleigh and fluorescence signals allows the use of only one single laser source in combination with only one single 2D detector.

1 Introduction

In technical combustion systems high turbulent premixed flames are of increasing importance due to their low NO_x emissions. The investigation of the interaction between turbulence and combustion kinetics in premixed flames requires the knowledge of the instantaneous temperature field and of the OH concentration field as well. From accurately measured and spatially resolved temperature distributions the flame thickness can be evaluated (Kampmann 1994). This can be used to explain the influence of turbulence on the combustion process in the context of a diagram which has been introduced by Borghi (1988) in order to characterize different areas of turbulent premixed combustion (Hofmann et al. 1994). However, strong turbulence can also result in flame quenching, which may cause wrong interpretation of the temperature data if not taken into account appropriately.

The detected concentration of the OH radical can be used as an indicator for the intensity of the reaction taking place and thus for the reaction zone. Therefore the simultaneous measured OH distribution is an important and sometimes necessary information for the interpretation of flame thickness data considering flame quenching.

Investigations have to be performed at similar operating conditions as used in technical flames, e.g., highly turbulent flames in combustion chambers, if the transferability of the results should be guaranteed. Therefore the appropriate diagnostic technique must be applicable also to contained combustion chambers which allows operation under those similar conditions.

2 Two-dimensional Rayleigh Thermometry

The applicability of 2D Rayleigh scattering for gas thermography in laboratory flames has been shown nearly one decade ago by Long et al. (1985). Recently Kampmann et al. (1993) presented for the first time its use in a contained technical flame for the investigation of temperature structures in an industrial double cone swirl combustor. Design and construction of the burner have been done by Asea Brown Boveri (ABB) and is described in detail by Sattelmayer et al. (1990). The Rayleigh signal was generated by a pulsed frequency-doubled Nd:YAG-laser and detected with an ICCD slow-scan camera. Interferences of the Rayleigh scattered light with glare from burner surface and Mie scattered light from dust particles could be reduced by using high efficient filter elements in the gas supplies and by a sophisticated design of the optical access to the combustion chamber. Figure 1 shows schematically the experimental situation and the orientation of the light sheet and the observation direction in combination with the combustion chamber.

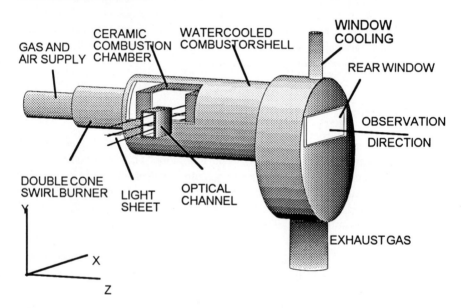

Fig. 1 Scheme of the combustion chamber with optical access

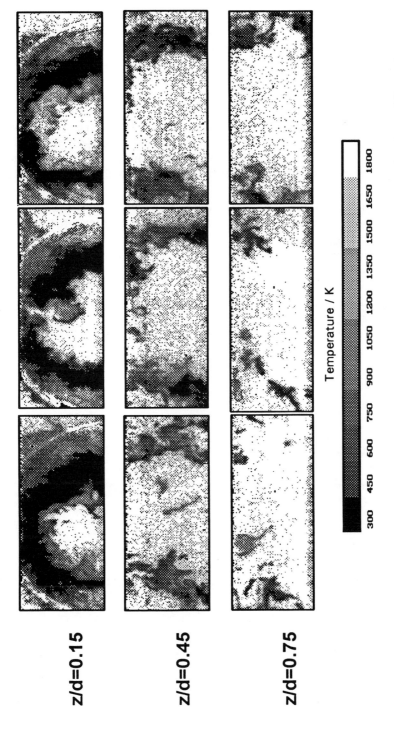

Fig. 2 Instantaneous two-dimensional temperature fields detected at three different downstream positions for three different instants.

It is a rather complicated procedure to obtain real quantitative temperature results from Rayleigh scattering. This is described in detail elsewhere (Kampmann et al., 1993). Figure 2 exhibits the development of the structure of the investigeated turbulent combustion field by instantaneous single laser shot 2d temperature distributions taken at three different downstreams positions. The ring of premixed unburnt gas disappears further downstream leading to extended areas of high gas temperature. In the outer regions of the unburnt gas ring high temperature areas can be recognized which seem to indicate an additional flame front in these outer regions. But this also could be areas of hot burnt gas coming back to these regions due to reverse flows in the outer parts. From the temperature information alone it is not possible to find the correct answer.

The extension of the technique from the visible to the UV for a more efficient generation of the Rayleigh signal due to the $1/\lambda^4$ relation of the Rayleigh cross section was recently done by Kampmann et al. (1995) using a KrF excimer laser source (λ=248nm). Figure 3 shows schematically the used experimental arrangemant and gives details on typical setup data, on the detected 2d field and on the spatial resolution obtained. The absorption prism in front of the telescope was needed for the simmultaneous detection of 2d temperature and OH concentration fields which is explained in more detail in chapter 4. For gas temperature measurements alone, no prism was used.

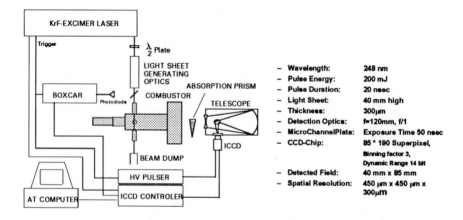

Fig. 3 Schematic of the electro-optical part of the used experimental setup

The inaccuracy of the temperature data being evaluated from the shot-noise limited Rayleigh signals was smaller than ±5 % for a typical combustion temperature of 1550 K. As an example, Figure 4 shows one typical result when crossing the flame front within an instantaneous temperature distribution at one particular position in the turbulent flame.

Temperature [K]

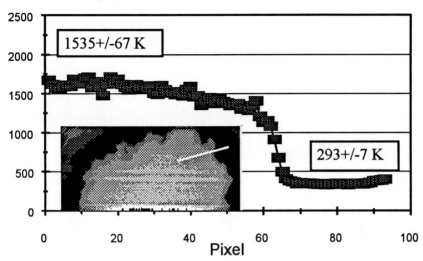

Fig. 4 Cross section through the flame front indicating the accuracy of the excimer laser Rayleigh probe for instantaneous temperature field measurements

3   Two-dimensional Planar LIPF OH Measurements

Planar laser-induced fluorescence (LIF) measurements of the OH radical by the excitation with a KrF excimer laser has been used successfully for the determination of flame structures and reaction zones in laboratory flames as well as in technical applications by Versluis et al. (1992) and Koch et al. (1993) making use of the predissociated fluorescence technique (LIPF) introduced by Andresen et al. (1988). For the OH measurements the absorbtion prism has been used in front of the telescope, see Fig. 3. It absorbs the Rayleighs contributions in the signal (see chapter 4).

Figure 5 represents a few examples of OH measurement for two different operation points of the burner at a downstream position of $z/d = 0.1$. Figure part a shows the average of 50 single-shot measurements of which three different are displayed in the figure parts b-d. From the detected OH distributions one can realize, that the burning process is much closer to the burner's surface for operation point 1 and that for both conditions no burning process occurs in the outer combustor regions where we had identified higher gas temperatures (Fig. 2). Thus, the high gas temperatures we found at these positions must be brought there by the recirculation of already burnt gas from regions further downstream.

operation point 1   operation point 2

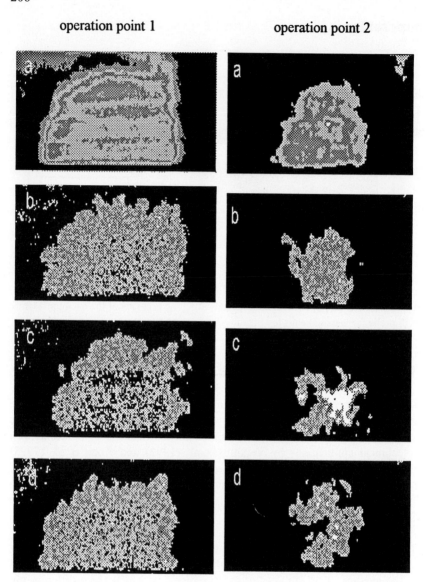

Fig. 5 OH concentration fields at z/d=0.1 d at three different instants (b-d) for two different operation points of the burner and the average (a) for 50 single-shot measurements

4 Simultaneous 2D Rayleigh and Planar OH Measurements

For collecting good quality Rayleigh signals an UV-telescope with high aperture (f/1.0) had to be used for the combined Rayleigh/LIPF measurements. A refractive filter optics has been developed in order to separate the Rayleig signal from the OH fluorescence signal in front of the telescope. Figure 6 shows a scheme of the choosen optical configuration. A prism was formed with two parallel plates of fused silica. The volume between the plates was filled with butylacetate, which acts as a longpass filter at a cutoff wavelength at 250 nm (see Fig. 7). This prism is placed in front of the telescope in such a way that it covers only a part of the aperture. The other part of the signal which passes the uncovered area of the aperture will unaffectedly be imaged on the CCD detector placed behind the telescope and gives thus a combined Rayleigh/LIF signal distribution $I_{Ray,LIF}(x_1, y_1)$. From that part of the signal which is detected through the prism, the Rayleigh signal is absorbed, and the transmitted LIF signal is deviated, which results in a LIF signal distribution at a new location on the detector $I_{LIF}(x_2, y_2)$. However, for temperature evaluation, the pure Rayleigh signal distribution $I_{Ray}(x_1, y_1)$ has to be known. If the ratio of the LIF signal at location 1 to the LIF signal at location 2 can be measured

$$R = \frac{I_{LIF}(x_1, y_1)}{I_{LIF}(x_2, y_2)} \tag{1}$$

the pure Rayleigh signal can be calculated from the simultaneously collected data by

$$I_{Ray}(x_1, y_1) = I_{Ray,LIF}(x_1, y_1) - R \cdot I_{LIF}(x_2, y_2) \tag{2}$$

The required temperature fields can be calculated from these data sets following the method described elsewhere (Kampmann et al., 1993). In order to measure the ratio R an UG11 filter was used in front of the array detector. This filter blocked the Rayleigh contribution of the combined Rayleigh/LIF signal (see Fig. 7) providing the same transmission at both signal paths. R can be determined from the resulting LIF signals at both locations. Variations of R could easily be achieved by adjusting the position of the prism in front of the telescope. As best done we regard a LIF signal as low as necessary and a combined Rayleigh/LIF signal as high as possible for reliable temperature information.

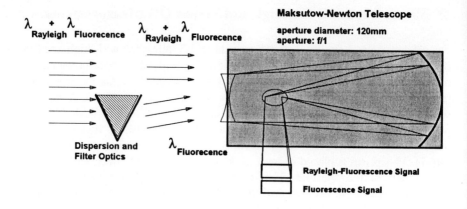

Fig. 6 Scheme of the detection arrangement

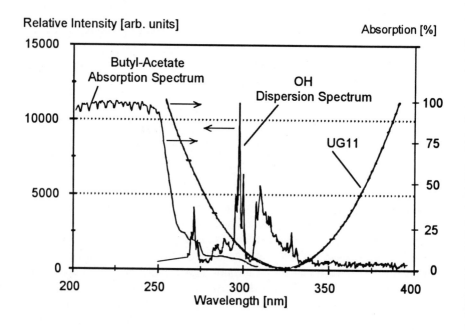

Fig. 7 Absorption spectra for butylacetat and the used colour glas filter UG 11 relative to the OH dispersion spectrum

The Rayleigh signal, however, can also be interfered by resonant fluorescence signals. There are two possible ways to overcome this problem:
a) Using a weak excitation transition, so that the interference is negligible. This however works only with detector systems with high dynamic range, e.g., slow-scan CCDs, as the wavelength-shifted fluorescence signal becomes also smaller and has to be detected together with the Rayleigh signal within the dynamic range of the CCD detector.
b) Estimation of the contribution of the resonant fluorescence signal by two measurements of the scattered light intensities at laser wavelength in a laminar flame, for one tuning the laser on an OH excitation wavelength and for one tuning it off. The difference of the signal intensities gives the intensity of the resonant fluorescence signal. This can be taken into account appropriately by the knowledge of the wavelength shifted fluorescence distribution.

5 First Results

First measurements have been performed in the turbulent combustion field of the Asea Brown Boveri (ABB) burner. Figure 8 shows the first results of our simultaneous Rayleigh/LIF experiments. Picture part (a) exhibits the combined Rayleigh/LIF signal distribution, picture part (b) the LIF distribution alone. From these data the pure Rayleigh signal distribution was reconstructed, from which the temperature distribution could be calculated, which is shown in picture part (c).

A comparison of the temperature distribution (c) and the OH concentration field (b) shows very good agreement in the observed structures. At locations where ignition temperature is reached also OH starts to appear and increasing concentration indicates the higher combustion intensity.

6 Acknowledgements

The authors gratefully acknowledge financial support for parts of the work by the Deutsche Forschungsgemeinschaft, by the Bundesministerium für Forschung und Technologie and by the Max-Buchner Forschungsstiftung. They also thank for the very fruitful cooperation they have with Drs. K. Döbbeling, W. Polifke, J. Haumann, and Th. Sattelmayer of the Aerodynamic Group of Asea Brown Boveri Research Center, Baden/Dättwil (Switzerland), who also provided the technical burner and the combustion chamber for this investigation.

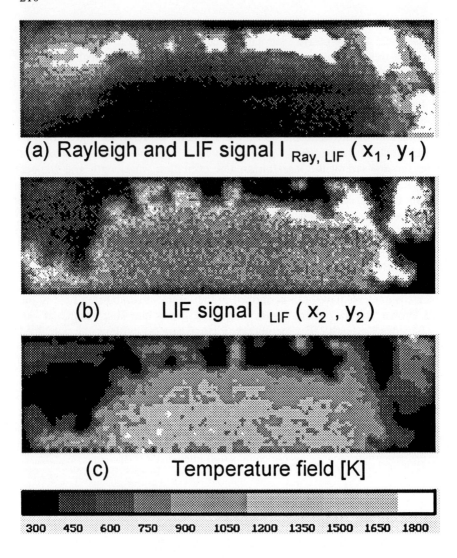

(a) Rayleigh and LIF signal I $_{Ray, LIF}$ (x_1 , y_1)

(b)   LIF signal I $_{LIF}$ (x_2 , y_2)

(c)   Temperature field [K]

| 300 | 450 | 600 | 750 | 900 | 1050 | 1200 | 1350 | 1500 | 1650 | 1800 |

Fig 8 Results of the simultaneous measurements

7 References

Andresen, P., Bath, A., Gröger, W., Lülf, H.W., Meijer, G. & ter Meulen, J. J. 1988, Laser Induced Fluorescence with Turnable Excimer Lasers as a Possible Method for Instantaneous Temperature Field Measurements at High Pressures: Checks with an Atmospheric Flame, Applied. Optics, Vol. 27, p. 365

Borghi, R. 1988, Turbulent Combustion Modelling, Progress in Energy and Combustion Science, Vol. 14, p. 245

Hofmann,D., Kampmann, S., Leipertz, A., Polifke, W., Döbbeling, K. & Sattelmayer., Th. 1994, Experimental investigation of the spatial structure of lean-premixed highly turbulent flames, submitted November 1994

Kampmann, S. 1994, Entwicklung einer zweidimensionalen Laser-Rayleigh-Temperaturmeßsonde zur Untersuchung hochturbulenter Verbrennungsvorgänge in einem industriellen Drallbrenner, Berichte zur Energie- und Verfahrenstechnik (BEV), Heft 94.2, Esytec, Erlangen

Kampmann, S., Leipertz, A., Döbbeling, K., Haumann, J. & Sattelmayer, Th. 1993, Two-Dimensional Temperature Measurements in a Technical Combustor with Laser Rayleigh Scattering, Applied Optics, Vol. 32, p. 6167

Kampmann, S., Seeger, T. & Leipertz, A. 1995, Simultaneous CARS and 2D Laser Rayleigh Thermometry in a Contained Technical Swirl Combustor, Applied Optics, Vol. 34 (in press)

Koch, A., Chryssostomou, A., Andresen, P. & Bornscheuer, W. 1993, Multi-Species Detection in Spray Flames with Turnable Excimer Lasers, Applied Physics B, Vol. 56, p. 156

Long, M.B.; Levin, P.S. & Fourguette, D.C. 1985, Two-dimensional mapping species concentration and temperature in turbulent flame, Optics Letter, Vol. 10, p. 267

Versluis, M., Boogaarts, M., Klein-Douwel, R., Thus, B., de Jongh, W., Braam, A., ter Meulen, J.J., Meerts, W.L. & Meijer, G. 1992, Laser-Induced Fluorescence Imaging in a 100kW Natural Gas Flame, Applied Physics B, Vol. 55, p. 167

Structure of Spark-Ignited Spherical Flames Propagating in a Droplet Cloud

Fumiteru AKAMATSU, Kazuyoshi NAKABE, Masashi KATSUKI, Yukio MIZUTANI and Toshikatsu TABATA

Department of Mechanical Engineering, Osaka University
2-1 Yamada-oka, Suita, Osaka 565, Japan

ABSTRACT. A droplet cloud of liquid fuel produced by an ultrasonic atomizer was ignited by a spark, and the flame ball propagating outward was observed in order to elucidate the mechanism of flame propagation and complicated group combustion behaviors of spray flames. For that purpose, the instantaneous images of droplet clusters, OH-radical chemiluminescence and C_2-band flame luminosity were taken simultaneously. Furthermore, the light emission signals in OH- and CH-bands, Mie-scattering signal from droplets, and the size and velocity of droplets were monitored simultaneously in time series. It was found that a nonluminous flame propagated ahead of a luminous flame, and that droplets disappeared in the luminous flame zone due to rapid evaporation, where a number of small-scaled droplet clusters were burning in diffusion combustion mode associated with solid-body emissions.

Keywords. Spray Combustion, Flame Propagation, Flame Structure, PDA, Light Emission, Mie Scattering

1   INTRODUCTION

The combustion of a single droplet or a droplet array has been investigated so far by numerous researchers (for example, Williams, 1973; Law, 1982). However, detailed structure and propagation mechanism of a flame propagating through a droplet cloud has not been discussed extensively yet. A spray flame is thought to be composed of a number of flamelets where droplets and fuel vapor burn in a complicated manner interacting with each other (Faeth, 1977; Chigier, 1981; Nakabe et al., 1991). Apparently, random and discontinuous flame propagation occurs in a liquid fuel spray as far as one observes luminous spray flames by conventional high-speed direct photography (Mizutani and Nishimoto, 1972; Nakabe et al., 1988). This fact implies that the spray burns in a complicated group combustion mode (Suzuki and Chiu, 1971; Chiu and Liu, 1977; Chiu et al., 1982) being divided into many clusters and subclusters corresponding to the number density distribution of droplets. Since the conventional high-speed photography cannot visualize nonluminous flames which are thought to play an important role in the flame propagation process, the planar laser-induced fluorescence (PLIF) (Allen and Hanson, 1986) or chemiluminescence of OH-radicals should be observed to elucidate the actual flame propagation mechanism in a spray. The simultaneous visualization of luminous and nonluminous flames along with unburned droplets is also very useful for understanding of the detailed spray flame behavior.

In the previous study (Mizutani et al., 1993), we observed a spark-ignited flame ball propagating through a droplet suspension, freely falling and entraining surrounding air, in order to observe the flame propagation mechanism and the detailed flame structure of droplet clusters burning in sprays under the minimal influences of atomization process and fluid motion. Firstly, the characteristics of the noncombusting droplet suspension were studied by the simultaneous measurement of the velocity and diameter of droplets with three laser beams, two He-Ne laser beams and one Ar^+ laser beam crossing with each other at one point. Secondly, the behavior of a growing flame ball was examined by comparing the local light emission signals of OH- and C_2-radical bands with the processed images of the flame ball obtained by a high-sensitivity CCD camera and a high-speed camera.

In the present study, a pair of short-exposure images of OH-radical chemiluminescence and either of flame luminosity in C_2-radical emission band or of Mie-scattering from droplet clusters were taken simultaneously to clarify the spacial relation between the nonluminous and luminous flames and unburned droplet clusters. Furthermore, this observation was compared with local continuous measurements, where the size and velocity of droplets, signals of light emissions in OH- and CH-radical bands and Mie-scattering signal from droplets were monitored simultaneously. It was found that a nonluminous flame first propagated continuously through a coexisting region of small droplets and gas-phase mixture and that a number of small-scaled droplet clusters burned randomly associated with discontinuous luminous flames behind the nonluminous flame front.

2   EXPERIMENTAL APPARATUS

Figure 1 shows an illustration of the experimental apparatus. An ultrasonic atomizer of ferrite magnetostriction type (resonance frequency = 18.5 kHz) was installed at the top center of a vertical square duct (280 mm X 280 mm X 1325 mm) which shields the droplet suspension from surrounding disturbances. A spark gap, 4.0 mm wide, was located 400 mm below the atomizer tip. A liquid fuel (kerosene) was fed to the atomizer by a microsyringe pump. The freely falling droplet suspension was ignited by intermittent electric sparks. A pulse-delay-generator (Stanford Research Systems,

214

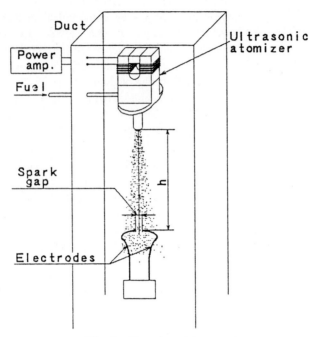

Fig. 1 Experimental apparatus

WC Model DG535) was used to control the triggering timing of each instrument and device. Preliminary observations with a high-speed video system (Nac, HSV-1000) demonstrated an ignition delay of several milliseconds, almost spherically propagating flame ball for several tens of millisecond and the final deformation of flame ball due to buoyancy. Therefore, the observations were concentrated in the period of spherical flame propagation. Throughout the present experiment, the fuel injection rate was kept 4.44 cm³/min and the ignition discharge duration time was 20 ms.

The optical and image processing system used to obtain various simultaneous flame images is shown in Figure 2. The planar beam of an Ar^+ laser tailored by a cylindrical lens illuminated a vertical plane involving the central axis of the apparatus. The thickness of the laser sheet was approximately 1.5 millimeters. The light, perpendicularly scattered by the droplets in the rectangular area (indicated in Fig.2) and passed through a dichroic mirror (DM) and an interference filter (F) (transmission peak wavelength of 514.5 nm), was focused by a UV Nikkor lens onto the photocathode of an image intensifier (I.I.). On the other hand, OH- chemiluminescence, emitted in the same direction, reflected by the dichroic mirror and filtered by another interference filter (transmission peak wavelength of 308.5 nm), was focused on the photocathode of another image intensifier. In order to obtain the OH-chemiluminescence image and the spectral luminosity image of a flame in the C_2-radical band simultaneously, the laser was turned off and the optical interference filter in the droplet image path was replaced

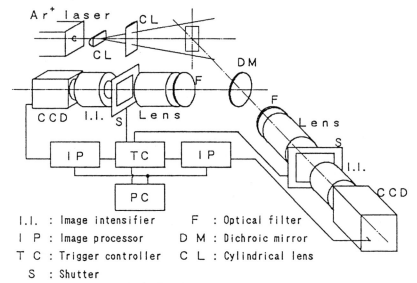

I.I. : Image intensifier F : Optical filter
I P : Image processor   D M : Dichroic mirror
T C : Trigger controller C L : Cylindrical lens
S : Shutter

Fig. 2 Optical image processing system

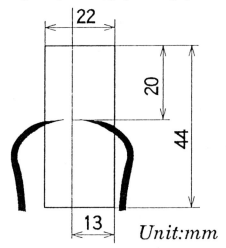

Unit:mm

Fig. 3 Imaging region

with the one for C_2-band. This system consisted of a pair of image processors (IP) and mechanical focal plane shutters (S), both of which were controlled by the trigger controller (TC) connected to a personal computer (PC). Image data were transferred to the computer and stored in floppy disks. The ordinary exposure were set at 1/250 (4 ms). A rectangular area for visualization are indicated in Fig.3.

Figure 4 shows the continuous monitoring optical system for local velocity and diameter of droplets, Mie-scattering signal from droplets and light emission signals in

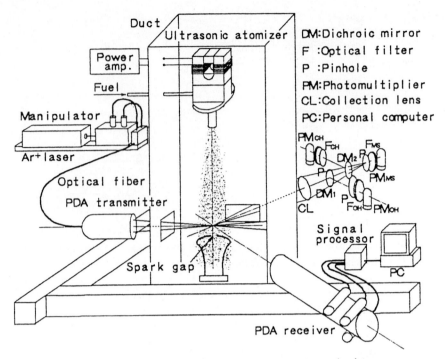

Fig. 4 Optical system for local continuous monitoring

OH- and CH-radical bands. The monitoring location was chosen at 25 mm directly above the spark gap, where the flame propagated upward steadily. The light source for Mie-scattering and PDA transmitter was an Ar$^+$ laser (Spectra Physics, Stable 2016,

Table 1 Specification of PDA system

Probe:	Focal length	500	mm
	Spot diameter	244.8	μm
	Fringe spacing	5.0	μm
	Fringe number	49	
Range:	Maximum diameter	190	μm
	Maximum velocity	1.5	m/s
	Minimum velocity	-0.5	m/s
Focal length of receiving optics		310	mm
Wavelength of laser beam		514.5	nm
Bandwidth of signal processor		0.4	MHz
Frequency shift		40	MHz

wave length = 514.5 nm). The light emissions in OH- and CH-radical bands and Mie-scattering from droplets were detected in the direction perpendicular to the axis of the PDA transmitter using dichroic mirrors (DM), optical interference filters (F) and photo-multipliers (PM).

The size and velocity of droplets were also measured simultaneously using a PDA receiver (Dantec, Model 57X10) and a signal processor (Dantec, Model 58N10). Scattering angle was set at 70.3 degrees. This detection angle was selected so as for the first order refracted light to be dominant and for the measured values by PDA not to be sensitive to the change in refractive index of droplets (Pitcher et al., 1990). The specification of the PDA system is shown in Table 1.

The Mie-scattered light corresponds to the number density of droplets or the passage of droplet clusters. The OH-band luminosity indicates the existence of combustion reaction in case of spray flames because relatively low temperatures (less than 2000 K) prevail over the whole flame. On the other hand, the light emission in CH- and C_2-band are biased by the continuous spectrum of solid-body emission, so that they correspond to each radical emission superposed by the solid-body emission from soot particles (Mizutani et al., 1989). For luminous spray flames, the sensitivity of the detectors for CH- and C_2-band emissions should be lowered by an order of magnitude as compared with the OH-band detector.

3 RESULTS AND DISCUSSION

3.1 Image Processing

Figures 5 (a) and 5 (b) show typical simultaneous images in OH- and C_2-band, respectively, at 35 ms after the electrical discharge. Only a part of the upward propagating flame ball is visualized in the pictures. As mentioned previously, in the case of luminous flame like a spray flame, the light emission in C_2-band is biased by the continuous solid-body light emission. Because its intensity is much stronger than that of OH-band, the sensitivity of the I.I. in C_2-band must be lowered by an order of magnitude than that in OH-band. Therefore, the region where an intense OH-band emission and a weak C_2-band emission coexist is thought to be the nonluminous flame region. On the other hand, the region where both OH- and C_2-band images are strong is thought to be the luminous flame region associated with intense solid-body light emissions from soot. Furthermore, the region of intense C_2-band emission not synchronized with OH chemiluminescence is regarded as the high temperature burnt gas region. It is noticed from these images that the flame first propagates spherically, although the flame eventually drifts upward due to buoyant effect as mentioned above.

The comparison between the images in OH- and C_2-band reveals that the region of intense OH-band emissions associated with weak C_2-band emissions is located in leading part of the propagating flame, that is, in the upper portion of the images in Fig.5, whereas the intense C_2-band emission region is confined in the inner (lower) portion of OH image; i.e. the nonluminous flame region is followed by the luminous

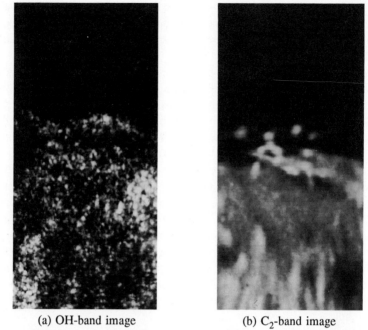

(a) OH-band image (b) C_2-band image

Fig. 5 Short-exposure images of OH-chemiluminescence and flame luminosity in C_2-band

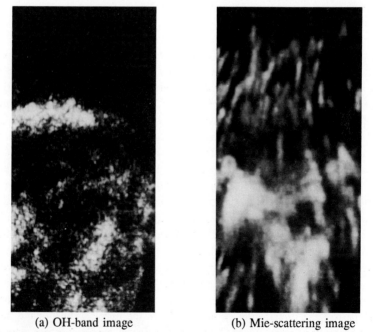

(a) OH-band image (b) Mie-scattering image

Fig. 6 Short-exposure images of OH-chemiluminescence and droplet distribution

flame region. The bright dots of C_2-band emission are regarded as small-scaled droplet sub-clusters buried in a nonluminous flame.

Figures 6 (a) and 6 (b) show typical images of OH-band emission and Mie-scattering from droplets, respectively, simultaneously taken at 35 ms after the electrical discharge. The thin white streaks in Fig.6 (b) are droplet clusters illuminated by the Argon-ion laser sheet. Since an intense flame luminescence can pass through the optical interference filter for Mie-scattering (F_{MS}) (transmission peak wavelength of 514.5 nm), both Mie-scattering image and flame luminescence image are superimposed in Fig.6 (b) but they are distinguishable from each other since the luminous flame images are blotted ones different from streaks.

By comparing these two images, it is found that the intense OH emission appears in the outer layer of the flame ball and that the luminous flame is located inside the flame ball. Furthermore, in the outer region of the flame ball, there are clear streaks of droplet clusters regardless of the existence of OH emission. Consequently, it is considered that a nonluminous flame is propagating continuously through the region of small droplets and gas-phase mixture and the remaining droplet clusters penetrate into the nonluminous flame. A number of droplet clusters are ignited randomly behind the nonluminous flame and burn associated with luminous flames.

3.2 Local Continuous Monitoring

Although two-dimensional information can be derived from flame images, temporal transition is ambiguous yet. So, here we compare the above results with the signals continuously monitored at a point. The monitoring location was chosen at 20 mm above the center of the spark gap where the flame propagated upward steadily as mentioned previously. The reason why we adopted light emission in CH-band instead of that in C_2-band during the local continuous monitoring was that it was difficult to detect both C_2-band light emission and Mie-scattering simultaneously because their bands are so close to each other. The C_2-band is located on the longer wave length side of the CH-band, so the C_2-band is more biased by continuous solid-body light emission than CH-band. Nevertheless, we have confirmed in the preliminary experiment that flame luminosity both in CH- and C_2-bands show almost the same tendency.

Figure 7 shows the five simultaneous signals at the local point; that is, OH emission (I_{OH}), CH-band emission (I_{CH}), Mie-scattering from droplets (I_{MS}), droplet velocity (V_h) and droplet diameter (D). The abscissa is the elapsed time after the spark initiation in milliseconds. Since we know that the flame propagates upward steadily crossing this point, these time series signals represent the structure of the upper portion of the flame ball. The noncombusting droplet number density was estimated about 15 particles/mm^3 by the calibration method by Saffman (1987).

The comparison between the signals in OH- and CH-bands reveals that both of them appears at the same time. (Note that the scale of coordinate for I_{CH} signal is fifty times larger than that for I_{OH}.) The CH-band emission synchronizes with OH chemiluminescence, so that it is thought to be undisturbed CH radical chemiluminescence by the solid-body emission. We have checked this by use of the common scale

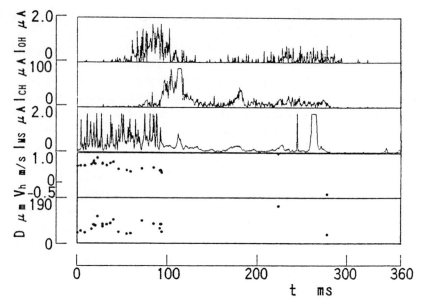

Fig. 7 5 simultaneous signals at 20 mm above spark gap

for the two signals. Thus, nonluminous flames (I_{OH} signal) begin to appear at around 50 ms after the electrical discharge, followed by luminous diffusion flames accompanied with solid-body light emission from soot (I_{CH} signal) with a delay of several tens of milliseconds. The fact well corresponds to the images in Figs. 5 and 6.

The Mie-scattering signal I_{MS} gives us the information on the behavior of droplet clusters during the flame passage through the optical control volume. Droplets randomly fall through the control volume until the luminous flame front arrives there. Only a few droplets, on the other hand, are detected after the luminous flame region appears, since the population density of droplets rapidly decreases due to the thermal expansion, evaporation and combustion reaction.

Thus, we reach the same conclusions as already derived from the image processing that a nonluminous flame is propagating continuously through the region of small droplets and gas-phase mixture and that remaining droplet clusters burn randomly and discontinuously associated with solid-body emissions behind the nonluminous flame front.

3.3 Conditional Sampling of PDA data

Figures 8, 9 and 10 show the size distribution of droplets, the correlation between diameter D and axial velocity of droplets V_h and the mean axial velocity and its fluctuation of the droplets in diameter range of every 5 μm, respectively. A solid circle in Fig. 10 indicates the mean velocity, and the line segments correspond to ±

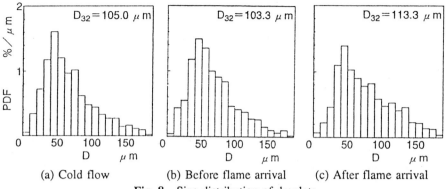

(a) Cold flow   (b) Before flame arrival   (c) After flame arrival

Fig. 8   Size distribution of droplets

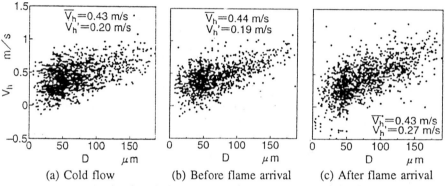

(a) Cold flow   (b) Before flame arrival   (c) After flame arrival

Fig. 9   Correlation between size and velocity of droplets

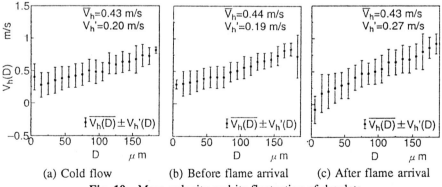

(a) Cold flow   (b) Before flame arrival   (c) After flame arrival

Fig. 10   Mean velocity and its fluctuation of droplets

root-mean-square (rms) values of velocity fluctuations. Sub-indices (a), (b) and (c) in Figs. 8, 9 and 10 denote the cold flow, before the flame arrival and after the flame arrival, respectively. The data in (b) and (c) were obtained in the way where the droplet sizes before and after the abrupt rise of OH signal were accumulated separately over several hundreds runs.

The patterns in (a) and (b) considerably resemble each other, which confirmed straightforward that the behavior of droplets is not affected by the flame until its arrival. Comparing the patterns in (a) and (b) with that in (c) to examine the droplet behavior behind the nonluminous flame front, the size distribution patterns in Fig. 8 does not vary noticeably except for the slight increase in the fraction of droplets over 80 μm. The mean velocity in Figs. 9 and 10 does not show significant differences, but the width of fluctuation is exaggerated after the flame arrival in every size class. This fact suggests that droplets are affected by the thermal expansion after the flame passage.

4   CONCLUSIONS

A spark-ignited freely falling droplet suspension was observed to elucidate the flame propagation mechanism and the detailed flame structure. A pair of short-exposure images of OH-radical chemiluminescence and either droplet clouds or flame luminosity pattern in the C_2-radical emission band were taken simultaneously to observe the spacial relation between nonluminous and luminous flames and droplet clusters distribution. These data were compared with local continuous measurements, where droplet velocity and diameter, signals of the light emissions in OH- and CH-radical bands and Mie-scattering signal from droplets were monitored simultaneously. The results were summarized as follows;

(1)   A nonluminous flame first propagates continuously through a region of small droplets and gas-phase mixture.

(2)   Remaining droplet clusters are swallowed into the nonluminous flame and then a number of droplet clusters burn randomly and discontinuously associated with luminous flames behind the nonluminous flame front.

ACKNOWLEDGEMENT

The authors wish to express their gratitude to Professors K. Nakajima and Y. Ikeda and Mr. N. Kawahara of Kobe University for their instruction and help in designing the PDA system, and Mr. T. Osaki of Osaka University for his cooperation in experiments. The authors are also obliged to Nac Co, Ltd. for the high-speed video system.

REFERENCES

Allen, M.G. and Hanson, R.K. 1986, Digital Imaging of Species Concentration Fields in Spray Flames, Proc. 21st Symp. (Int.) on Combustion, pp.1755-1762.

Allen, M.G. and Hanson, R.K. 1986, Planar Laser-induced-fluorescence Monitoring of OH in spray flame, Optical Engineering, vol.25, pp.1309-1311.

Chigier, N. 1981, Energy, Combustion and Environment, pp.248, McGraw-Hill, New York.

Chiu, H.H., Kim, H.Y. and Croke, E.J. 1982, Internal Group Combustion of Liquid Droplets, Proc. 19th Symp. (Int.) on Combustion, pp.971-980.

Chiu, H.H. and Liu, T.M. 1977, Group Combustion of Liquid Droplets, Combust. Sci. Technol., vol.17, pp.127-142.

Faeth, G.M. 1977, Current Status of Droplet and Liquid Combustion, Prog. Energy Combust. Sci., vol.3, no.4, pp.191-224.

Law, C.K. 1982, Recent Advances in Droplet Vaporization and Combustion, Prog. Energy Combust. Sci., vol.8, no.3, pp.171-201.

Mizutani, Y., Nakabe, K., Fuchihata, M., Akamatsu, F., Zaizen, M. and El-Emam, S.H. 1993, Spark-Ignited Spherical Flames Propagating in a Suspended Droplet Cloud, Atomization and Sprays, vol.3, pp.125-135.

Mizutani, Y., Nakabe, K., Matsumoto, Y., Saeki, T. and Matsui, T. 1989, Processing of Luminescent Radical Images for Flame Diagnostics, JSME International Journal, Series II, vol.32, pp.455-463.

Mizutani, Y. and Nishimoto, T. 1972, Turbulent Flame Velocities in Premixed Sprays: Part I. Experimental Study, Combust. Sci. and Technol., vol.6, pp.1-10.

Nakabe, K., Mizutani, Y., Hirao, T. and Fujioka, H. 1991, An Experimental Study on Detailed Flame Structure of Liquid Fuel Sprays With and Without Gaseous Fuel, Combust. Flame, vol.84, no.1, pp.3-14.

Nakabe, K., Mizutani, Y., Hirao, T. and Tanimura, S. 1988, Burning Characteristics of Premixed Sprays and Gas-Liquid Coburning Mixtures, Combustion and Flame, vol.74, p.39-51.

Pitcher, G., Wigley, G. and Saffman, M. 1990, Sensitivity of Dropsize Measurement by Phase Doppler Anemometry to Refractive Index Changes in Combusting Fuel Sprays, Applications of Laser Techniques to Fluid Mechanics 5th Int. Symp., Lisbon, pp.227-247, Springer-Verlag.

Saffman, M. 1987, Automatic Calibration of LDA Measurement Volume Size, Applied Optics, vol.26, pp.2592-2597.

Suzuki, T. and Chiu, H.H. 1971, Multi-Droplet Combustion of Liquid Propellants, Proc. 9th Int. Symp. on Space Technology and Science, pp.145-154.

Williams, A. 1973, Combustion of Droplets of Liquid Fuels: A Review, Combust. Flame, vol.21, pp.1-31.

Part III
Instrumentation for Velocity
and Size Measurements

A Phase Screen Approach to Non-particle Laser Anemometry

Lars Lading[1], Robert V. Edwards,[2] and Mark Saffman[1]

[1] Risø National Laboratory - Association Euratom
Optics and Fluid Dynamics Department
DK-4000 Roskilde, Denmark

[2] Chemical Engineering Department
Case Western Reserve University
Cleveland, OH 44106, USA

Abstract Velocity measurements based on the transport of turbulence induced refractive index structures are investigated. Particle scattering is assumed to be negligible. The scattering from collective structures is analyzed with a phase screen model. A system based on reference beam detection and combining Doppler and time-of-flight concepts may provide the best spatial resolution and utilize a broad band region of the high frequency turbulence for light scattering. In the optical region it is so that a long wavelength is preferable and reference beam detection is necessary. An potential hybrid system for plasma diagnostics is presented. It is shown that time resolved measurements may be possible.

Keywords. Laser anemometry, non-particle anemometry, phase screen, turbulence.

1 Introduction

Laser anemometry based on particle scattering is well established. However, in certain fluids particles cannot be present at all, or the concentration is very low. In a fusion plasma there are no large scattering particles; in high speed wind tunnels the concentration of particles may be extremely low. In such systems it may be possible to measure transport properties of large scale fluctuations on the basis of light scattering/diffraction from small scale structures convected by the larger scales.

Light scattering from collective structures has been utilized for velocity measurements. Radar scattering from refractive index fluctuations in the atmosphere has been used for the measurement of velocity on very large scales.

Strauch (1985). Simple configurations based on visible light have been devised for measuring *cross wind* velocities, Clifford, Ochs, and Wang (1975). For diagnostics in connection with plasma fusion configurations similar to the original reference beam laser Doppler anemomenter have been applied, Slusher and Surko (1978), Truck et al (1991). The method has recently been applied to the measurement of fluid velocity in high speed wind tunnels, Grésillon et al (1992).

We shall briefly discuss scattering from refractive index fluctuations. The essential types of fluctuations encountered are identified. The nonpropagating types of fluctuations may be used for measuring the fluid velocity.

The trade offs in terms of optical wavelength, spatial resolution, measurement configuration, and turbulence parameters are discussed.

The reference beam LDA has so far been the configuration that has been applied. Direct detection (as commonly used in LDA based on particle scattering) can only be applied in cases with very strong fluctuations. Surprisingly it is so that the longer the optical wavelength is the better is the spatial resolution. However, even at the wavelength of the CO_2 laser (10.6 µm) the resolution along the optical axis is inadequate. We propose a system that combines the time-of-flight configuration with the Doppler system. Such a system provides for a significantly better spatial resolution and also utilize a larger part of the spectral content of the refractive index fluctuations.

Time resolved measurements have so far to our knowledge not been performed with this type of scattering. We outline briefly the considerations on which the feasibility of time resolved measurements can be decided. It is shown that even in a fusion plasma time resolved measurements may be possible.

2 Basic Concepts, Definitions, and Assumptions

Light scattering in fluids may be classified as either inelastic or elastic. Fluorescence and Raman scattering are examples of inelastic scattering. These types of scattering are well suited to identify the chemical composition of the medium. Fluorescence may also be applied to velocity measurements in high speed flows, Hanson (1994) or even in particle LDA. The intrinsic spectral broadening of this type of scattered light is in general so large that only supersonic velocities can be accurately measured.

Quasi-elastic scattering is scattering where a possible shift in frequency only is caused by the movement (velocity) of the scatterer and not by a change in the quantum-mechanical state of the particle. This type of scattering is the basis for laser anemometry but also for measuring diffusion and molecular dynamics. However, incoherent molecular scattering will in general give a spectral broadening proportional to the thermal velocity of the molecules. In most flows this type of broadening is much larger than the fluid velocity making it unsuited for anemometry.

Despite the large thermal velocity of molecules *collective* light scattering may provide a signal with a spectral width much smaller than the thermal broadening. Collective scattering is scattering from a large number of scatterers that in some way support a large scale (large compared to the mean free path) scattering structure. A sound wave may be an example of such a wave structure. It is necessary that the spatial scale of the structure is larger than the mean free path.

In laser anemometry as well as in other types of measurements based on quasi-elastic scattering the light collection and detection mode may be either coherent or incoherent. Let us - to establish definitions - assume a large number of particles in the measuring volume. *Coherent detection* implies that the photocurrent is proportional to the squared sum of the fields from the particles, i.e. that

$$ i \propto \left| \sum_j U_j \right|^2 . \tag{1} $$

We note that the photocurrent contains a sum of contributions from the individual particles plus a sum of cross-beat terms. Coherent detection requires that the phase fronts are parallel over the detector area implying a small collector aperture. *Incoherent detection* implies that

$$ i \propto \sum_j \left| U_j \right|^2 . \tag{2} $$

In this case the collector aperture is very large and the cross particle contributions vanish. Incoherent detection is in general not feasible if collective scattering is to be applied.

In the case of a very high particle concentration it may be preferable to apply a continuum model. We shall do that and consider the random fluctuations of the refractive index.

We summarize this section by the following definitions:

Collective light scattering is scattering/diffraction from random fluctuations of the refractive index of a medium through which light is propagating.

Dynamic light scattering: Scattering where the frequency shift (or broadening) exclusively is caused by the motion of the scatterer(s).

Applying this type of scattering and detection we may measure transport of refractive index fluctuations, which may represent:

- materials transport,
- propagating waves or,
- convection of small scale turbulence by larger scales.

Applications are found in several quite different areas, e.g. velocimetry (anemometry), waves in bulk media and on surfaces (visco-elastic), diffusion, and particle sizing.

3 A Continuum Model Based on Phase Screens

We shall here outline a model for calculating the photocurrent and its correlation function. A so-called *phase screen* model is applied (originally introduced by Lee and Harp (1969)) in order to analyze wave propagation through random media; the phase screen model is also used in the analysis of thick holograms).

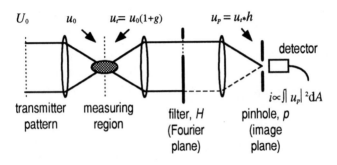

Figure 1. Fourier optical model. For the reference beam LDA the transmitter pattern consists of two parallel beams of which the top beam is much stronger than the bottom beam.

3.1 The Photocurrent

The analysis is based on a Fourier optical model as shown in Fig. 1. (Lading, Mann, and Edwards). The field in the measuring plane is given by the Fourier transform of the transmitted field pattern. The interaction region is divided into a number of screens perpendicular to the optical axis (Fig.2). Screen j is assigned a transmission function given by

$$t_j = \exp\{ik_\lambda \Delta z \delta n_j(x, y)\} \cong \exp\{ik_\lambda \Delta z\}(1 + i\Delta z \delta n_j(x, y)) \tag{3}$$

where $k_\lambda \equiv 2\pi/\lambda$, $\Delta z =$ is the thickness of the screen and δn the deviation of the refractive index from unity. The field out of screen j is

$$u_{out,j} = (u_{in,j} t_j) * h_{\Delta z}, \tag{4}$$

where

$$h_{\Delta z} = \frac{\exp\{ik\Delta z\}}{2\pi i(\Delta z / k)} \exp\left\{-\frac{1}{2}\frac{x^2 + y^2}{i\Delta z / k}\right\}. \qquad (5)$$

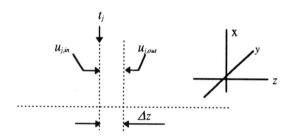

Figure 2. Phase screen.

Summing the contributions from all the phase screens letting Δz go to zero and the number of screens go to infinity yields an expression for the total field. The field is referred to the measuring plane. Convoluting with the filter function h and multiplying with the pinhole function p gives the field on the detector. Squaring and integrating over the detector area gives the photocurrent.

The photocurrent contains in general three terms (as in normal particle laser anemometry):

1. A dc current with no information about the dynamics of the measuring object.
2. A 'homodyne' term caused by interference between purely scattered/diffracted beams.
3. A 'heterodyne' term caused by mixing between scattered/diffracted light and reference beam(s).

The first term is unavoidable since the photocurrent cannot be negative. The second term is the dominating dynamic part of the photocurrent in most laser anemometry systems based on particle scattering and here the mixing does only occur between beams scattered by the same particle. Non-particle laser anemometry may be based on either the second or the third term, but the third term is often preferred because a parametric amplification of the signal is obtained since the amplitude of the dynamic part of the photocurrent is given by the product of the scattered field and the reference field. This may be necessary if a wavelength is applied where no photomultipliers or avalanche diodes are available (e.g. at the wavelength of the CO_2 laser). It is also so that the statistics of the photocurrent is simpler if the third term is the dominating dynamic component.

3.2 The Correlationfunction

The correlationfunction is evaluated using the same procedure as in Lading, Mann, and Edwards (1989). For the reference beam mode (heterodyne) we get that the correlation function is

$$R(\tau) \propto \int_{x-y\,\text{plane}} S(k,\tau)|F(k)|^2 \, dk \,, \tag{6}$$

where $S(k,\tau)$ is the 2D spatial Fourier transform of the 2D space-time correlation for the phase perturbations $(k = (k_x, k_y))$. $|F(k)|^2$ is an instrument function given by the modulated part of the field distribution in the measuring volume.

For the homodyne mode we get

$$R(\tau) = \left| \int_{x-y\,\text{plane}} S(k,0)|U(k)|^2 \, dk \right|^2 + \left| \int_{x-y\,\text{plane}} S(k,\tau)|U(k)|^2 \, dk \right|^2 \tag{7}$$

Note that in Eq. (6) $R(\tau)$ is given directly by the first order space-time correlation, whereas in Eq. (7) it is given by a more complicated expression of the first-order correlation.

Now, the expressions are given by the 2D correlation function for the refractive index fluctuations, but the measurements are here anticipated to be performed in a 3D medium. In the analysis of beam propagation through turbulent media it is customary to assume δ-correlation along the optical axis (i.e. no width of the correlation function in the z-direction). This may be legitimate if the interaction region is much larger than the beam diameters, and the spatial wavelength of the turbulence that is considered. However, this is not the case here. Other investigators, Grésillon (1992), have used the 3D correlation in a way where it is difficult to assess the impact of the finite correlation length of the turbulence in the direction of the optical axis. Being short of a rigorous procedure we shall do the following: phase screens with a spacing smaller than the selected wavenumber are added coherently, whereas screens with a spacing larger than the selected coherence length are added incoherently.

The spectrum of the refractive index fluctuations (from which the spectrum for the phase perturbations is obtained) will in general contain two terms: (1) a term that does not propagate relative to the fluid with a coherence time given by the thermal diffusivity, and (2) a term given by propagating sound waves in general with a very small damping.

4 Specific configurations

Let us investigate specific configurations and compare them in relation to spatial resolution and signal strength.

4.1 The Reference Beam Doppler Configuration

The principles of the configuration is illustrated in Fig. 3. The transmitter is essentially as for a homodyne mode LDA except for the fact that one of the transmitted beams is much weaker than the other. The detection is done in the direction of the weakest of the transmitted beams. The dynamic part of the signal is then given by scattered light from the strong beam that is heterodyned with directly transmitted light of the weak beam. (The configuration is similar to the very first Laser Doppler Velocimeter). The few practical implementations that so far have appeared are all using a CO_2 laser ($\lambda = 10.6\ \mu m$). The fringe spacing has to be selected so that it is larger than the Kolmogorov length scale; otherwise the system will not be able to detect any (weak) turbulence induced refractive index fluctuations.

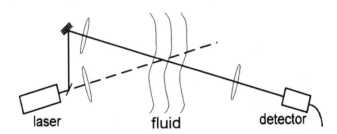

Figure 3. Basic scheme of a reference beam Doppler configuration. The actual implementation may be quite different. Frequency shift is generally applied and proper spatial mode matching should also be ensured.

Let the fringe spacing be Λ and the optical wavelength λ. The angle θ between the two transmitted beams is then given by

$$\theta = 2\sin^{-1}\left(\frac{\lambda}{2\Lambda}\right). \tag{8}$$

The spatial resolution along the optical axis, $\pm\ \Delta z$, is

$$\Delta z = 2r_0 \frac{\Lambda}{\lambda}, \tag{9}$$

where $2r_0$ is the beam diameter in the measuring plane. We note that a large optical wavelength implies a good spatial resolution!

The system shown in Fig. 3. is sensitive to refractive index fluctuations in a spatial frequency region with a center frequency given by $2\pi/\Lambda$ and a relative bandwidth of Λ/r_0.

4.2 The Time-of-flight Configuration

A so-called time-of-flight configuration (also called a two-spot system) has been applied as an alternative to the LDA in some cases where a good spatial selectivity along the optical axis is mandatory. Since it is a major problem with non particle laser anemometry to obtain a good spatial resolution along the optical axis it could be worth while to investigate the feasibility of the laser time-of-flight anemometer (LTA). The principles of an LTA for nonparticle laser anemometry could be envisaged to be as illustrated in Fig. 4. Scattered/diffracted light from the two focal regions should heterodyne with the directly transmitted light on the detectors. However, for scattering/diffraction caused by a weak phase object this will not happen. This is essentially the same problem as the classical phase contrast problem. Unfortunately, the remedy used in phase contrast imaging cannot be applied here to cure the problem. Evans et al. (1982) have 'solved' the problem by truncating the pinhole in front of the detector. This implies that a signal will appear; however, at the expense of spatial resolution and power efficiency.

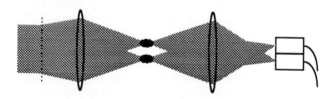

Figure 4. An envisaged LTA for non-particle laser anemometry. However due to the fact that the signal caused by weak phase perturbations is in phase quadrature with the directly transmitted beams the system cannot operate in a reference beam mode.

Let us briefly illustrate the problem: the transmission function of a weak phase object can be approximated as follows

$$e^{i\phi(x,y)} \cong 1 + i\phi(x, y). \tag{10}$$

The problem is that because of the i in front of $\phi(x, y)$ the signal is in phase quadrature with the reference. The phase contrast principle applied in microscopy to solve this problem cannot be applied here (Lading 1983). Introducing a spatial carrier frequency by tilting the reference beam will work - but then it becomes an LDA.

Away from the fundamental phase quadrature problem it is also problematic to have the full power of the directly transmitted beams impinge on the detectors; they may be damaged by the power.

4.3 A Hybrid Configuration

Let us summarize the features of the LDA and the LTA respectively. This is done in the table 1. Each of the two spots of the LTA have a diameter equal to the fringe spacing of the LDA. This implies that the spatial bandwidth of the LTA is much larger than the bandwidth of the LDA. The focal depth of a focused single mode beam is

$$\Delta z = 2\pi\, r_0^2 \, / \, \lambda \tag{11}$$

giving a length for the LTA much smaller than the intersection region of the two intersecting beams of an LDA. The directional sensitivity is small for a two spot system because a structure contributing to the signal has to pass both beams. Elliptic spots may enhance the acceptance angle, but it cannot be $\pm 180°$ as for an LDA with frequency shift.

The problem with the LDA in the present context is that a large number of fringes is needed in order to allow for a good frequency estimate. This implies a spatial resolution that is unacceptable in relation to many applications. Note here that the fringe spacing has to be larger than the smallest turbulence scale. The LTA could provide an order of magnitude better spatial resolution and it would utilize a larger part of the turbulence spectrum; however, under most conditions it is insensitive to the type of disturbances that here provides the signal.

Table 1. Comparing an LDA with an LTA.

	LDA	LTA
bandwidth	small	large
spatial resolution	poor	good
directional sensitivity	small	high
reference beam	controllable	fixed

In order to obtain some of the advantages of the two systems we propose a hybrid configuration: A dual LDA system where each fringe pattern only has -

236

say - 3-4 oscillations. The velocity is not determined from the frequency, but from the time of flight between the two fringe patterns. The system is illustrated in Fig. 5.

For a laser anemometer we can define a *code* that is a function that defines the way in which the velocity information is encoded into the system. For an LDA the code is a wave packet; for an LTA it consists of two displaced peaks. The code of the hybrid system is shown in Fig. 6. It consists of two displaced wave packets each with only a few oscillations.

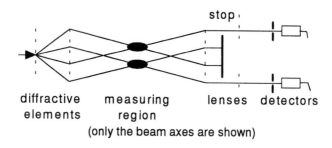

Figure 5. A hybrid LTA/LDA configuration. A diffractive element (to the left) generates four beams. A second element refracts and focuses these beams into two spatially separated fringe patterns, each with only a few fringes. The beams that impinge on the detectors will heterodyne with the scattered light from the two other beams. The reference beam is much weaker than the beams from which scattered light is utilized.

In Fig. 7 is shown a spatial turbulence spectrum . Also shown are the 'filters' of the LDA and the hybrid system, respectively. The filters select the spectral regions from which refractive index perturbations can contribute to the signal. It is seen that the LDA selects a much narrower region than the hybrid system. For a given spatial resolution this implies that the hybrid will give a much stronger signal. However, if the purpose of a given investigation is to investigate the propagation of different spatial wavenumbers a narrow spatial frequency range is mandatory. For 'frozen' turbulence a large spectral region must be preferable.

In discussing the spectral regions of the turbulence spectrum we must remember that the purpose may be to get information about the turbulence - possibly about its temporal evolution. With the present scheme it is so that we can only 'track' turbulence in a spectral region below that which gives rise to the signal: the high frequency turbulence serves the same purpose as particles do in normal laser anemometry.

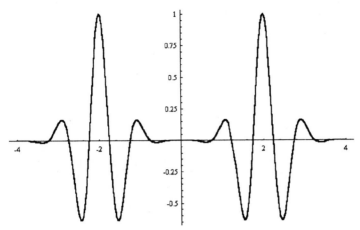

Figure 6. The intensity distribution (deviation from the mean) as seen by the detectors of a hybrid laser anemometer, which defines the *code* of the system.

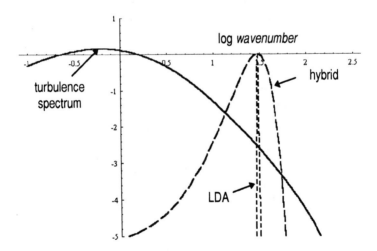

Figure 7. A turbulence spectrum (solid line) with the filter functions of an LDA (the narrow filter) and a hybrid LA, respectively. Notice that the hybrid utilizes a much larger part of the turbulence spectrum. (Log-log plot).

5 Signal Processing

The signals encountered here have similarities with signals of a laser anemometer with many particles in the measuring volume. It is a continuos signal that exhibits fluctuations caused by the randomness of the scattering object.

The expected correlation function of the signals is essentially given by the displaced cross-correlation of one of the bursts shown in Fig. 6, if the macroscopic velocity is constant. However, imperfections of the system may cause a difference in the phase of the oscillatory parts of the two wavepackets. This could give an erroneous peak of the cross-correlation. In order to avoid this the envelopes of the signals are cross-correlated as shown in Fig. 8. This implies that for a constant velocity the peak of the correlation corresponds to the time-of-flight between the two centres of the spots.

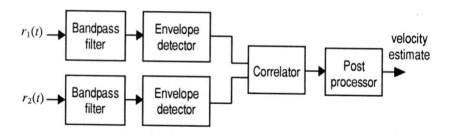

Fig. 8. Signal processing scheme for the hybrid configuration. The envelope detectors are used in order to make the system insensitive to small optical phase errors. The post processor is in general assumed to do a curve fitting from which relevant parameters are taken.

The perturbations in the flow that cause the signal will in general be a combination of convected (non-propagating) and propagating refractive index fluctuations. The propagating fluctuations (sound waves) have a very low damping, where the coherence time of the nonpropagating fluctuations is given by the thermal diffusivity. Figure 9. shows an example of a correlation function assuming that the velocity of sound is three times the convection velocity and that the thermal decay time is four times the time-of-flight between the two illuminated volumes. The peak is in general shifted towards a smaller delay relative to the true time-of-flight because of the thermal decay. The effect is hardly noticeable if the decay time is much larger than the time-of-flight. A broader peak is encountered if the convection velocity fluctuates within the averaging time.

Time resolved measurements may be possible. The situation is similar to normal laser anemometers operating with many particles in the measuring volume. The statistics of especially LDA systems have been investigated in great detail over the years. Some of the results are adaptable to the systems considered here. We shall use the results of Lading and Edwards (1993).

In the type of systems considered here there are two fundamental stochastic processes that may limit the performance: the photon noise and the random process of the signal itself. In addition to these processes other types of fluctuations may hamper the performance. For reference beam systems it is often so that laser noise (caused by power supply ripples, mode competition, intermode beats, mechanical instability etc.) is the limiting factor. However, we shall here assume that an adequately quite laser is available.

The relative measurement uncertainties can be expressed as follows:

fluctuations of the
scattering/diffracting object $\quad$ *relative variance* $\approx 1/N$
itself;

photon noise $\qquad\qquad\qquad$ *relative variance* $\approx \dfrac{\kappa^2}{\langle n_s \rangle} \dfrac{1}{N}$
(additive here)

The relative variance is the variance of the velocity estimate normalised by the square of the true velocity, i.e. $\mathrm{var}\{\hat{v}\}/v^2$. N is the number of degrees of freedom over which the averaging takes place, defined by the product of signal bandwidth times averaging time (assuming that the averaging time is larger than the reciprocal signal bandwidth). n_s is the number of detected scattered photons within the coherence time of the signal (= 1/signal bandwidth, also equal to the transit time through the illuminated volume(s)). κ is the reciprocal number of fringes in the measuring volume (LDA), or the ratio between beam radius and beam spacing (LTA).

Now, in order to perform time resolved measurements the averaging time must of course be smaller than the time scale of the fluctuations to be resolved. In order to get the best temporal resolution the photon noise should be negligible relative to the intrinsic signal fluctuations, i.e.

$$\kappa^2 / \langle n_s \rangle \ll 1. \qquad\qquad (12)$$

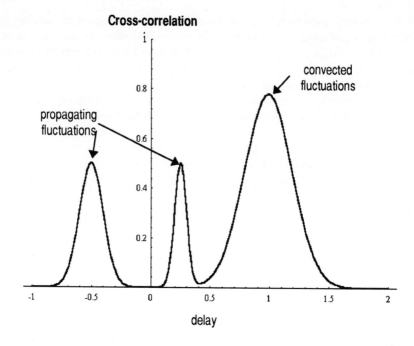

Fig. 9. Calculated cross-correlation assuming that $v_{sound} = 3\ v_{fluid}$ and $\tau_{decay} = 4\tau_0$. Note that the peak is in general shifted towards a smaller delay relative to the true time-of-flight because of the thermal decay. It is assumed that the convection velocity is constant. The cross-correlation is normalised with the square root of the product of the maximum values of the autocorrelations; the time scale is arbitrary.

6 A Specific Example for Plasma Turbulence

Turbulence in fusion plasmas is of major concern because it has an essential impact on the confinement and stability of the plasma. It also represents a type of fluid where no particles are allowed to appear. We believe that some of the considerations relevant for plasma measurements may also be relevant for measurements in other types of fluids. Going through the calculations in detail is beyond the limitations of this presentation. However, we shall summarize the numbers we arrived at in table 2. In the calculation of the signal strength (givn as # photons/coherence time of signal) we have assumed the following: (1) isotropic turbulence, (2) a coherent superposition of the refractive index fluctuations in the direction of the optical axis over a length scale of 0.5 mm (the diameter of each spot in the measuring direction) and, (3) an incoherent superposition over scales from 0.5 mm to 30 mm (the length of the measuring volume).

An overview of the conditions in a tokamak plasma is given by Callen, Carreras, and Stambaugh (1992)).

Table 2. Potential parameters for a plasma diagnostic set-up

Relative refractive index fluctuations $\Delta n/n$	10^{-5}
Fraction of refractive index fluctuations utilised	10^{-3}
Velocity	10^4 m/s
Overall efficiency	10 %
Laser Power	5 Watts
# photons/coherence time of signal	20
focal (intersection) regions	0.5 mm × 30 mm
wavelength of spatial filter	0.25 mm
spacing between focal regions	30 mm
transmitter/receiver spacing	3 m
transmitter aperture	300 mm
receiver aperture	2×100 mm
laser power	2 W

Since $\kappa < 1$ and the number of estimated scattered photons within the coherence time of the signal is larger than unity, time resolved measurements appear feasible (Eq. 12).

7 Conclusion

Principles of non-particle laser anemometry have been discussed. A novel hybrid laser anemometer configuration is identified in order to enhance the spatial resolution and overcome the phase detection problem. A potential application to diagnostics in a fusion plasma is discussed and it is shown that time resolved measurements may be possible.

It should be mentioned that a number of important problems in non-particle laser anemometry have either only been mentioned briefly or not mentioned at all. Most important is the problem of separating propagating and non-propagating refractive index fluctuations.

References

Callen, D.C., Carreras, B.J., and Stambaugh, R.D. (1992), "Stability and Transport Processes in Tokamak Plasmas", Physics Today, January, pp. 34-42.

Clifford, S.F., Ochs, G.R., and Wang, T.I. (1975), "Optical wind sensing by observing the scintillations of a random scene", Appl. Opt., **14**, pp. 2844-2850.

Evans, D.E., von Hellermann, M., and Holzhauer, E. (1982), "Fourier Optics Approach to Far Forward Scattering and Related Refractive Index Phenomena in Laboratory Plasmas", Plasma Physics **24**, No. 7, pp. 819-834.

Grésillon, D., Bonnet, J.P., Cabrit, B., and Frolov, V. (1992), "Collective Light Scattering and the Diagnostics of Velocity Fluctuations in Transonic Flows", Sixth Int. Symp. Appl. Laser Tech. Fluid Mech. pp 2.6.1. - 2.6.6.

Lading, L. (1983), "Estimating time and time-lag in time-of-flight velocimetry", Appl. Opt., **22**, pp. 3637-3643.

Lading, L., Mann, J.A., and Edwards, R.V. (1989), "Analysis of a surface-scattering spectrometer", J. Opt. Soc. Am. A. **6.**, No. 11.

Lading, L. and Edwards, R.V. (1993), "Laser velocimeters: lower limits to uncertainty", App. Opt., **32**, No. 21.

Lee, R.W. and Harp, J.C. (1969), Proc. IEEE **57**, 375.

Patrie, B.J., Seitzman, J.M., and Hanson, R.K. (1994), "Instantaneous 3-Dimensional Flow Visualization by Rapid Acquisition of Multiple Planar Flow Images", Opt. Eng., **33**, No. 3, pp. 975-980.

Slusher, R.E. and Surkom C.M. (1978), Phys. Rev. Lett. 40, p. 400.

Strauch, R.G. (1985), "Radar wind Profilers", Proc. NASA Symp. on Global Wind Measurements, July 29 - August 1, Columbia, MD, pp. 133-137.

Truck, A. et al, "ALTAIR" (1992): "An infrared laser diagnostic on the TORE SUPRA tokamak", Rev. Sci. Instrum. **63**, pp. 3716-3724..

GAUSSIAN BEAM ERRORS IN PHASE-DOPPLER ANEMOMETRY AND THEIR ELIMINATION.

G. Gouesbet, G. Gréhan.

Laboratoire d'Energétique des Systèmes et Procédés,
INSA de ROUEN, URA.CNRS.230, CORIA.
BP08, 76131, Mont-Saint-Aignan Cédex, France.
Tel: (33) 35 52 83 87, Fax: (33) 35 52 83 90

abstract. GLMT (generalized Lorenz-Mie theory) and its applications to the design of phase-Doppler instruments are discussed. A particular emphasis is stressed on Gaussian beam errors (trajectory ambiguity effects) and their elimination. Very recent concepts, even if not yet fully developed and/or implemented, are discussed too.

Keywords. Generalized Lorenz-Mie theory, Light scattering, Particle sizing

1 Introduction

A basic rule of science is to permanently take care of the necessary dialogue between theory and experiments. As a special case, this rule must obviously be satisfied in the field of opticle particle sizing (and characterization). However, most of the modern intruments used in this field rely on the interaction between laser beams (Gaussian beams, laser sheets, top-hat beams, to name a few) and particles (such as spheres or cylinders). Until recently, the required dialogue was therefore essentially forbidden by the fact that available theories assumed that the illumination beam was a plane wave, for instance in LMT (Lorenz-Mie theory) describing the interaction between a plane wave and a sphere. In the eighties neverteless, LMT has been generalized to GLMT describing the interaction between an arbitrary illuminating beam and a sphere. Thanks to GLMT, the required dialogue between theory and experiments has been restored. The aim of this paper is to review applications of GLMT to the analysis of phase-Doppler instruments, and in particular to the study of Gaussian beam defects and their elimination. Some connected topics will also be considered, such as the study of laser sheets or the development of the theory of interaction between an arbitrary incident beam and an infinite cylinder.

Even if GLMT is now essentially completed, many avenues remain to be explored. An increasing number of researchers is now following these avenues on a fundamental point of view. A still larger number of researchers is working on the applications of GLMT to concrete problems. In this paper, we shall try, beside the main objective presented by the title, to inform the reader about the more recent news. This would be impossible if we decided to go into details. Therefore, it has been decided that this paper would be written essentially in a qualitative way, dismissing in particular any mathematical expression (and figures). But, on the other hand, the reader is directed on the literature from which he could accede, hopefully, to any desired information. In other words, this paper is a guide to a sightseeing tour, not the tour by itself.

The paper is organized as follows. Section II is devoted to the GLMT itself. Section III is devoted to phase-Doppler instruments, Gaussian beam effects and their elimination. Section IV is devoted to connected topics and contains possibly some prospective ideas, trying to somehow forecast what we could expect from the near future. Section V is a conclusion.

2   Generalized Lorenz-Mie theory

GLMT describes the interaction between an arbitrary incident beam and a sphere, using partial wave expansions. The first archival paper is Gouesbet and Gréhan, 1982, although the existence of this theory and some early results may be found in Gréhan et al, 1980. The development of the theory took about ten years, and may essentially be considered as completed after Gouesbet et al, 1988 and Maheu et al, 1988. A rather recent exhaustive review of GLMT and applications is available from Gouesbet, 1994-a.

A major issue to which much effort is still devoted is the speeding-up of numerical computations. For a long time, the most difficult task from this point of view has been the numerical evaluation of beam shape coefficients which may be considered as generalized Fourier co-efficients expressing the partial wave expansions of the incident beam. A breakthrough has been the introduction of the so-called localized approximation in Gréhan et al., 1986-a, for the so-called beam waist center case, generalized in Gouesbet et al.,1990, to the case of an arbitrary location of the particle, i.e.for the so-called off-axis case. An efficient algorithm to evaluate the beam shape coefficients in the framework of the localized approximation is discussed in Ren et al., 1992-a.

Although the accuracy of the localized approximation was very good (typically one part per 10^5), a drawback has been a lack of rigorous mathematical justification. A first step toward such a justification has been achieved by Lock in the on-axis case by using a stationary phase method (Lock, 1993). Unfortunately, the same method could not be extended to the general so-called off-axis case, even leading there to conflicting results, likely meaning that the success of the proof was in part incidental. By using another technique relying on Taylor expansions, a final proof of the validity of the localized approximation has afterward been obtained for both the on-axis and the off-axis cases (Lock and Gouesbet, 1994 and Gouesbet and Lock, 1994), with, as a by-product, a very refined understanding of the nature of laser beams (Gouesbet et al, 1994). The use of the localized approximation to handle GLMT allows fast and systematic study of phase-Doppler instruments, with a reasonable amount of CPU time. Furthermore, algorithm improvements are likely not exhausted. Let us for instance mention Lock, 1994, which implements a substantial amount of such improvements, in particular within the framework of the localized approximation. A pertinent example is given in which the Gaussian beam program runs about 70 times slower than the corresponding plane wave program. It may be forecasted that the combination of further improvements in software, and hardware, should make the difference between GLMT and LMT computational times fairly unessential within a few years. The same might become true if we compare GLMT and geometrical optics, with the advantage that

GLMT is a rigorous theory.

3   Gaussian beam defects in phase-Doppler instruments

3.1   Generalities.

Phase-Doppler anemometry (PDA), an extension of laser Doppler velocimetry (LDV), is nowadays the most popular technique for simultaneously measuring the size and the velocity of individual particles. Examples of pioneering and significant papers are Durst and Zaré, 1975, Bachalo and Houser, 1984, Saffman et al., 1984, Al-Chalabi et al, 1988, among many others. The principle of the technique may possibly be summarized as follows. Let us consider two interfering laser beams B1 and B2, and a spherical particle crossing the fringe control volume. At a point P in space, waves W1 and W2 scattered by the particle from the incident beams B1 and B2 respectively, interfere together. The resulting wave at point P is obtained by summing the amplitudes of the waves W1 and W2. After evaluating intensities and integrating on the surface of a detector D, one finds that the detected signal is characterized by a phase P. Let us now consider two detectors D1 and D2 leading to phases P1 and P2, respectively. Then, under some ideal circumstances, including the plane wave character of the incident beams, it has been predicted that the phase difference (P1-P2) is related to the particle diameter through a linear relationship.

The physical process aforementioned is amenable to a full theoretical simulation when a light scattering theory is used. Such scattering theories may be LMT, geometrical optics, or GLMT. LMT is a rigorous theory but, being limited to the plane wave case, it cannot predict the influence of the Gaussian character of the incident beams. This Gaussian character may be implemented in geometrical optics. But geometrical optics is not a rigorous theory with the consequence that the induced loss of accuracy may possibly be dramatic in some cases, although, in other cases, it can be quite insignificant. Finally, GLMT is a rigorous theory which correctly incorporates the Gaussian character of the incident beams. The price to pay is more extensive computational times but, as discussed previously, this drawback is likely to be somewhat provisional. Geometrical optics and LMT have been applied to phase-Doppler instruments in earlier stages. The first applications of GLMT to PDA dates back to Gréhan et al., 1990.

3.2 Gaussian beam defects

In many cases, the size of the scatter centers is not small enough with respect to the beam size, i.e. the incident beams cannot be safely considered as being plane waves. Then departures from the linear phase/diameter relationship may be observed, possibly leading to considerable errors in diameter measurements. Nowadays this phenomenon is referred to as a trajectory ambiguity effect (TAE) or, to better stress the origin of the problem, as a Gaussian beam defect (GBD).

TAE has been theoretically predicted by Bachalo and Sankar , by using geometrical optics (Bachalo and Sankar, 1988, Sankar and Bachalo, 1991, Sankar et al, 1992), and then confirmed by using GLMT (Corbin et al, 1991). A simple qualitative explanation may be provided in terms of the competition between refracted and reflected rays induced by the Gaussian profile (Gouesbet, 1994). Following these preliminary observations, a considerable systematic effort has been devoted to the analysis of TAE and to their elimination as testified by Gréhan et al., 1991, 1992-a, 1992-b, 1993-a, 1994-b, 1994-c, Gouesbet et al., 1993, Naqwi et al., 1993, Aizu et al., 1993, 1994-a, 1994-b and Onofri et al., 1994-a. . Remedies are discussed in the next sub-section.

3.3 Elimination of Gaussian beam defects

A rather exhaustive list of remedies to Gaussian beam defects is given below. Details may be found in References cited in the previous section. Basically, remedies may be splitted in two individual classes, one corresponding to the optimization of the optical design, and the second to the refinement of the data processing. Of course, there exists also an hybrid class in which both optical design optimization and signal processing refinement may cooperate.

(i)It is possible to manipulate the incident beam polarizations in order to modify the balance between reflected and refracted contributions as suggested by Saffman et al, 1984 and Saffman, 1986. This strategy relies on the basic idea that interferences between reflected and refracted components spoil the desired linear relationship between phases and diameters. This concept has been investigated by using GLMT (Gréhan et al., 1992-b). The conclusion is that doing so it is possible to improve the quality of the measurements but not necessarily in an effective enough way.

(ii)It has been suggested to use a discrimination criterion relying on the measurement of two phase differences, P12 from detectors D1 and D2, and P13 from detectors D1 and D3. Then the discrimination to retain or reject measurements is based on the value of the ratio P12/P13. A GLMT-analysis of this concept is for instance given in (Gréhan et al., 1992-a, 1992-b). Again, improvements are observed but not effective enough.

(iii)Let P be the plane defined by the incident laser beams. The analysis of Gaussian beam defects then demonstrates that good measurements are obtained when the detectors and the particle are on the same side with respect to P. Conversely, departures from the ideal relationship arise when the detectors and the particle are located on opposite sides. Then, we may use an optical design in which two detector assemblies are used, symmetrically located with respect to P. The use of two symmetric assemblies ensures that there will always be one assembly for which relationships are satisfactory. Furthermore, the comparison between the two responses allows one to determine which response is correct. This technique is effective, although it increases the complexity of the instrument. Details are available from Gréhan et al., 1991, 1992-b, 1993-a.

(iv)Among many possibilities, Doppler signals may be analysed by using a discrete Fast Fourier Transform (DFFT), leading to the evaluation of the Doppler frequency, i.e. to the measurement of one velocity component of the scatter center. In the phase-Doppler technique, one may rely on the use of an extended version generating a cross-spectral density (CSD) function, allowing the evaluation of the frequencies from detectors D1 and D2 and of the phase difference between both signals (Domnick et al. 1989). Because CSD does not analyse phase evolution in the case of Gaussian beam defects, it may lead to significant errors in diameter measurements by averaging phase differences along the duration of the signals. Therefore, the use of a wavelet transform which has temporal localization properties may allow the detection of phase evolution in a signal, and therefore provide a discrimination procedure (Onofri et al. 1992, Gouesbet 1994). It must be however noted that a systematic investigation of this concept has not yet been carried out.

(v)In the standard design, the detectors are located outside of the plane defined by the incident laser beams, and particles pass through the control volume, essentially perpendicularly to the non moving fringes. As discussed above, such a design is subject to Gaussian beam defects. This is due to the fact that trajectories are parallel to isophase lines. Therefore, the phase is constant but depends on the particle location.

In the so-called planar system, conversely, the detectors are located in the same plane than the incident beams. In this case, the particles cross the isophase lines. Therefore, there is a phase evolution along the burst but, limiting data processing to the part of the signal around the maximum, it is possible to show that the measured phase corresponds to the plane wave case predictions (Aizu et al,. 1993, 1994-a, 1994-b). In this case, TAE elimination is obtained by an appropriate modification of the optical design with respect to the standard system. A similar effect may also be obtained by an appropriate design of a two-colour PDA standard system (Xu and Tropea, 1994), leading to a design which may be called the modified standard system. It is also possible to use a two-colour dual-mode system, combining a planar system based on one colour and a modified standard system based on another colour (Tropea et al. 1994). An advantage of this dual-mode technique is that it allows a measurement of the particle refractive index, of sufficient quality for material recognition. Furthermore, each colour component of the system determines a diameter, i.e. actually a curvature radius, therefore leading possibly to non sphericity identification.

All the above techniques based on modifications of the optical design and/or on data processing refinements are devoted to the elimination of TAE considered as defects. On the other hand, TAE also contain relevant information and, therefore, instead of working towards their elimination, it might be very advisable to recover, if possible , the information that they contain. This different attitude leads to the so-called dual-burst system.

3.4 Taking advantage of TAE

In the dual-burst system, the control volume is choosen very small, therefore enhancing TAE to better take advantage of the information that they carry. Apart of that, the optical design may be either a planar system or a modified standard system, but data processing is carried out by recording the phase evolution along the duration of the signal. Then, the theoretical analysis, confirmed by experiments, shows that one part of the signal is free of TAE while another part, easily identified, contains extra information associated with TAE. Therefore, we may measure velocities and diameters by analyzing the TAE free part of the signal. The other part, conversely, is shown to contain information on refractive indices. Also, non sphericity, as well as inhomogeneity, information could possibly be recovered (Onofri et al., 1994-a).

4 Connected topics

We end this paper by briefly mentioning a few connected topics relevant to particle characterization.

4.1 Laser sheets

Although phase-Doppler instruments rely on the use of Gaussian beams, there may be advantages in using laser sheets. Laser sheets may be obtained by transforming Gaussian beams with the aid of a cylindrical lens. Laser sheets are for instance used in particle image velocimetry, for polarization ratio measurements , for measuring wall shear stress in a turbulent boundary layer , for particle sizing, or for studying droplet distribution effects on planar laser imaging of sprays (Adrian, 1984, 1992, Ryan et al. 1990, Naqwi et al., 1986, 1990, 1992, Hess, 1990). Therefore, GLMT has been specified to the case of laser sheets and used to investigate some optical characterization problems (Ren et al., 1992-b, 1993, 1994-a, 1994-b, Mroczka et al. 1993, Gréhan et al., 1993-b, 1993-c, 1994-a). The use of laser sheets in phase-Doppler instruments might be an interesting prospect.

4.2 Stratified spheres

In the previous subsection, GLMT has been specified to the case of laser sheets. It does not change the GLMT framework which is insensitive to the nature of the incident beam (GLMT is an arbitrary beam theory). Considering a new kind of beam however requires some extra-work associated with the knowledge of the electromagnetic description of the beam, to be used as an input in GLMT. Instead of changing the nature of the beam, we may consider other cases than the classical homogeneous sphere assumed in LMT and GLMT. Then, for each of these other cases, a new theory must be built.

The simplest case is the stratified sphere which preserves the spherical symmetry enjoyed by GLMT. The theory of interaction between an arbitrary incident beam and a stratified sphere has indeed been recently completed (Onofri et al., 1994-b). Relying on the plane wave case, we may forecast many applications of this theory to optical characterization, in particular in combusting systems which induce temperature gradients inside droplets (see for instance Kai et al, 1993). The development of phase-Doppler instruments to the study of stratified spheres might then be another interesting prospect.

4.3 Cylinders

Many particles are not spheres nor even stratified spheres. It is therefore somewhat of an unpleasant state of affair that most of our experimental techniques have to assume that we deal with spherical particles, even if they are not. This attitude may lead to the so-called spherical chicken syndrome as defined by B.H.Kaye , 1989,i.e. to misleading measurements and physical models. A possible way to escape of this problem is to develop the phase-Doppler instruments in such a way that they could provide non sphericity information as previously discussed. Another way is to develop other instruments less sensitive to non sphericity. There, a good candidate is the so-called top-hat beam technique which allowed rather successful measurements on irregular particles, such as sand particles (Allano et al., 1984, Gréhan and Gouesbet, 1986, Gréhan et al., 1986, Kleine et al., 1982, Corbin et al. 1991, Gouesbet et al., 1994, and references therein). Other techniques are the Visible Infra-red Double Extinction (VIDE) technique to investigate the case of large optical thicknesses, or quasi-elastic light scattering spectroscopy devoted to ensemble measurements on submicronic particles (Guidt et al. 1990, Lhuissier et al. 1989, and references therein).

Another approach is to start a systematic study of well defined shapes. After spheres, and stratified spheres, the best candidate to investigate seems to be the infinite cylinder. Many potential applications can be expected. Examples are the control of the production of glass wool for thermal isolation, of optical fibers for data transmission, or the study of pulverization processes when droplets are produced by instabilities acting on an initial liquid jet. Another promising remark is that many particles, even if they are not infinite cylinders, may locally look like an infinite cylinder when they are illuminated by focused beams. Therefore, this special case of a regular particle might also help to study a class of irregular particles, including sinuous fibers.

Concerning phase-Doppler instruments, it appears pretty sure that the extension of this technique from spheres to cylinders should lead to significant results. On the theoretical side, let us mention that the theory of interaction between an arbitrary incident beam and an infinite cylinder is under construction. However, the problem appears much more complicated than expected as testified by (Gouesbet and Gréhan, 1993, 1994-a, 1994-b). In particular, the basic so-called separability theorem used in designing both LMT and GLMT looks like failing for the cylinder problem. A pure mathematical work concerning this theorem is therefore likely to be compulsory, such as discussed in (Gouesbet, 1994-b). The construction of the GLMT basically required one decade. It is hoped that the completion of the cylinder problem will be less time demanding. In any case, we may forecast future interesting developments in this topic, both from an experimental and a theoretical point of view.

5 Conclusion

One of the major recent news concerning phase-Doppler anemometry has been the uncovering of the existence of trajectory ambiguity effects (Gaussian beam defects) which may lead to dramatic measurement errors if the optical design and the data processing are not set up in an appropriate way. These effects, first revealed by theoretical investigations, have been experimentally confirmed. At first sight, this is bad news. In particular, it is quite possible that some data displayed in the literature are erroneous due to the lack of knowledge on these effects.

A strong and systematic effort has therefore been devoted to the analysis of these effects, and to their elimination. Several new optical designs, possibly associated with refined data processing, actually allow one to cure them. May be more interesting, TAE, first considered as a nuisance, may also be considered as a chance because they carry extra information that we may recover, therefore opening new possibilities.

Finally, related topics have been considered, namely the specification of GLMT to the case of laser sheets,its generalization to the case of stratified spheres, and the challenging problem of interaction between arbitrary shaped beams and infinite cylinders.

6 References

Adrian R. J., 1984, Scattering particle characteristics and their effect on pulsed laser measurements of fluid flow : speckle velocimetry vs. particle image velocimetry. Applied Optics, 23, 1690-1691.

Adrian R. J., 1992, The role of particle image velocimetry in fluid mechanics, in : optical methods and data processing in heat and fluid flow, London, April 1992, pp 1-6..

Allano D., Gouesbet G., Gréhan G., Lisiecki D. 1984, Droplet sizing using a top-hat laser beam technique, Journal of Physics : Applied Physics, 17,1,43-58.

Al-Chalabi S. A. M. , Hardalupas Y., Jones A. R. , Taylor A. M. K. P., 1988, Calculation of the calibration curves for the phase-Doppler technique, comparison between Mie theory and geometrical optics, in Gouesbet G.,Gréhan G. (edts), optical particle sizing : theory and practice, Plenum Press, New-York, pp 107-120.

Aizu Y., F. Durst, Gréhan G., Onofri F. , Xu T. H., 1993, PDA-systems without Gaussian beam defects, Proceedings of the third international congress on optical particle sizing, August 23-26, Yokohama.

Aizu Y., Domnick J., Durst F., Gréhan G., Onofri F., Qiu H.H., Sommerfeld M., Xu T.H., Ziema M., 1994-a, New generation of phase-Doppler instruments for particle velocity, size, and concentration measurements, Particle and Particle Systems Characterization, 11, 43-54.

Aizu Y., Durst F., Gréhan G., Onofri F., T. H. Xu, 1994-b, PDA-systems without Gaussian beam defects, To be published in ASME journal.

Bachalo W. D., Houser M. J., 1984, Phase-Doppler spray analyzer for simultaneous measurements of drop size and velocity distributions, Optical Engineering. 23, 583-590.

Bachalo W. D., Sankar S. V., 1988, Analysis of the light scattering interferometry for spheres larger than the light wavelength, Proceedings of the fourth international symposium on applications of laser anemometry to fluid mechanics, Lisbon.

Corbin F., Gréhan G., Gouesbet G., 1991, Top-hat beam technique : improvements and application to bubble measurements, Particle and Particle Systems Characterization, 8, 222-228.

Domnick J., Ertel H., Tropea C., 1989, Processing of phase-Doppler signals using the cross-spectral density function, in applications of laser anemometry to fluid mechanics, R. J. Adrian, T. Asanuma, D. F. G. Durao, F. Durst, J. H. Whitelaw (eds), Springer-Verlag.

Durst F., Zaré M., 1975, Laser Doppler measurements in two-phase flows. Proceedings of the LDA-symposium, Copenhagen, pp 403-429.

Gouesbet G., 1994-a, Generalized Lorenz-Mie theory and applications, Plenary lecture, Third international congress on optical particle sizing, August 23-26,1993, Yokohama, Japan, Proceedings, republished in Particle and Particle Systems Characterization, 11, 22-34.

Gouesbet G., 1994-b, The separability dogma revisited with applications to light scattering theory, submitted to Journal of Optical Society of America.

Gouesbet G., Gréhan G., 1982, Sur la généralisation de la théorie de Lorenz-Mie, Journal of Optics (Paris), 13,2,97-103.

Gouesbet G., Gréhan G., 1993, Recent advances in generalized Lorenz-Mie theory and applications. Third European aerosol conference 93, Universitt Duisburg, Oct 4-8, seminar : trends in aerosol research. Proceedings of the seminar of the Sonderforschungbereich 209, Universitt Duisburg Gesamthochschule, A. Schmidt-Ott, ed, pp 17-23.

Gouesbet G., Gréhan G., 1994-a, Interaction between shaped beams and an infinite cylinder, including a discussion of Gaussian beams, Particle and Particle Systems Characterization, 11, pp. 299-308.

Gouesbet G., Gréhan G., 1994-b, Interaction between a Gaussian beam and an infinite cylinder, using non-sigma separable potentials, submitted to Journal of Optical Society of America A.

Gouesbet G., Lock J. A., 1994, A rigorous justification of the localized approximation to the beam shape coefficients in generalized Lorenz-Mie theory, II. Off-axis beams, To be published in Journal of Optical Society of America A.

Gouesbet G., Maheu B., Gréhan G., 1988, Light scattering from a sphere arbitrarily located in a Gaussian beam, using a Bromwich formulation, Journal of Optical Society of America A, 5,9,1427-1443.

Gouesbet G., Gréhan G., Maheu B., 1990, A localized interpretation to compute all the coefficients gnm in the generalized Lorenz-Mie theory, Journal of Optical Society of America A, 7,6,998-1007.

Gouesbet G., Berlemont A., Desjonquères P., Gréhan G., 1993, On modelling and measurement problems in the study of turbulent two-phase flows with particles, Sixth workshop on two-phase flow predictions, Erlangen, March 23-26, 1992. Proceedings : Forschungszentrum Jülich GmbH, bilateral seminar of the international bureau, 1993, pp 365-387, M. Sommerfeld, editor.

Gouesbet G., Lock J. A., Gréhan G., 1994, Do you know what a laser beam is?, Proceedings of the seventh workshop on two-phase flow predictions, Erlangen, Germany, April 11-14.

Gréhan G., Gouesbet G., 1986, Simultaneous measurements of velocities and sizes of particles in flows using a combined system incorporating a top-hat beam technique. Applied Optics, 25,19,3527-3538.

Gréhan G., Gouesbet G., Rabasse C., 1980, The computer program Supermidi for Lorenz-Mie theory and the research of one-to-one relationships for particle sizing. Proceedings of the symposium on long range and short range optical velocity measurements, Institut franco-allemand de Saint-Louis, 15-18th sept.

Gréhan G., Maheu B., Gouesbet G., 1986-a, Scattering of laser beams by Mie scatter centers : numerical results using a localized approximation, Applied Optics, 25,19,3539-3548.

Gréhan G., Gouesbet G., Kleine R., Renz V., Wilhelmi I.,1986-b, Corrected laser beam techniques for simultaneous velocimetry and sizing of particles in flows and applications. Third international symposium on applications of laser anemometry to fluid mechanics, Lisbon, July 7-9. Proceedings.

Gréhan G., Gouesbet G., Vannobel F., Dementhon J. B., 1990, Generalized Lorenz-Mie theory : an application to phase-Doppler technique. Proceedings of the fifth international symposium on applications of laser anemometry to fluid mechanics, Lisbon, Portugal, July 9-12.

Gréhan G., Gouesbet G., Naqwi A., Durst F., 1991, Evaluation of phase-Doppler systems using generalized Lorenz-Mie theory, international conference on multiphase flows, university of Tsukuba, Japan, Sept 24-27. Proceedings.

Gréhan G., Gouesbet G., Naqwi A., Durst F., 1992-a, On elimination of the trajectory effects in phase Doppler systems, 5th European symposium on particle characterization, Nürnberg, March 1992, Proceedings, pp 309-318.

Gréhan G., Gouesbet G., Naqwi A., Durst F., 1992-b, Trajectory ambiguities in phase-Doppler systems : use of polarizers and additional detectors to suppress the effect. Sixth international symposium on applications of laser techniques to fluid mechanics, Lisbon, July 20-23, proceedings.

Gréhan G., Gouesbet G., Naqwi A., Durst F., 1993-a, Particle trajectory effects in phase-Doppler systems : computations and experiments, Particle and Particle Systems Characterization, 10,332-338.

Gréhan G., Ren K. F., Gouesbet G., Naqwi A., Durst F., 1993-b, Application of the generalized Lorenz-Mie theory to multiphase flows experimental techniques, Ninth symposium on turbulent shear flows, Kyoto, Japan, August 16-18, Proceedings, paper 27-1-1.

Gréhan G., Ren K. F., Gouesbet G., Naqwi A., Durst F., 1993-c, Evaluation of particle sizing technique based on laser sheets, Third international congress on optical particle sizing, August 23-26, Yokohama, Japan, Proceedings.

Gréhan G., Ren K. F., Gouesbet G., Naqwi A., Durst F., 1994-a, Evaluation of a particle sizing technique based on laser sheets, Particle and Particle Systems Characterization, 11, 101-106.

Gréhan G., Gouesbet G., Naqwi A., Durst F., 1994-b, Trajectory ambiguities in phase-Doppler systems : study of a near-forward and
a near-backward geometry, Particle and Particle Systems Characterization, 11, pp. 133-144.

Gréhan G., Onofri F., Girasole T., Gouesbet G., Durst F., C. Tropea,
1994-c, Measurement of bubbles by phase-Doppler technique and
trajectory ambiguity, Seventh International Symposium on applications of laser techniques to fluid mechanics, paper 18-2.

Guidt J.B., Gouesbet G., Le Toulouzan J. N., 1990, An accurate validation of visible infra-red double extinction simultaneous measurements of particle sizes and number-densities by using densely
laden standard media, Applied Optics, 29, 7, 1011-1022.

Hess C.F., 1990, Poster at the 2nd international congress on optical particle sizing, Tempe, Arizona, USA, March 1990. Company
Metro-laser.

Kai L., Massoli P., D'Alessio A., 1993, Studying inhomogeneities of
spherical particles by light scattering, 3rd international symposium on optical particle sizing, Yokohama, Japan, pp 135-143.

Kaye B.H., 1989, A random walk through fractal dimensions, VCH.

Kleine R., Gouesbet G., Rück, 1982, Lokale, optische Messungen
von Teilchengeschwindigkeiten, Teilchengrssenverteilungen und
Teilchenkonzentrationen. Sensor 82. Essen, January 12-14. Proceedings.

Lhuissier N., Gouesbet G., Weill M.E., 1989, Extensive measurements
on soot particles in laminar premixed flames by quasi-elastic light
scattering spectroscopy, Combustion science and technology, 67,
17-36.

Lock J.A., 1993, The contribution of high-order rainbows to the scattering of a Gaussian laser beam by a spherical particle, Journal of
Optical Society of America A, 10, 4, 693-706.

Lock J.A., 1994, An improved Gaussian beam scattering algorithm,
under preparation.

Lock J.A., Gouesbet G., 1994, A rigorous justification of the localized approximation to the beam shape coefficients in generalized
Lorenz-Mie theory, I. On-axis beams, To be published in Journal
of Optical Society of America A.

Naqwi A., Reynolds W.C., Carr L. W., 1986, Dual-cylindrical wave
laser Doppler method for measurement of wall shear stress, in
Adrian, Durao, Mishina, Whitelaw (eds) : laser anemometry in
fluid mechanics-II, Ladoan-Instituto superior tecnico, Lisbon, 105-
122.

Naqwi A., Liu X. Z., Durst F., 1990, Dual-cylindrical wave method for particle sizing, Particle and Particle Systems Characterization, 7, 45-53.

Naqwi A., Liu X. Z., Durst F., 1992, Evaluation of the dual-cylindrical wave laser technique for sizing of liquid droplets, Particle and Particle Systems Characterization, 9, 44-51.

Naqwi A., Ziema M., Gréhan G., Gouesbet G., 1993, Accuracy considerations in the optical design of phase Doppler systems, in the study of turbulent two-phase flows with particles, Sixth workshop on two-phase flow predictions, Erlangen, March 23-26, 1992. Proceedings : Forschungszentrum Jülich GmbH, bilateral seminar of the international bureau, pp 412-420, M. Sommerfeld, editor.

Maheu B., Gouesbet G., Gréhan G., 1988, A concise presentation of the generalized Lorenz-Mie theory for arbitrary location of the scatterer in an arbitrary incident profile, Journal of Optics (Paris), 19,2, 59-67.

Mroczka J., Ren K. F., Gréhan G., Gouesbet G., 1993, Particle sizing by using polarisation ratios, on the use of laser sheets, first international conference-workshop on modelling in measurement processes, June 7-9, Technical University of Wroclaw, Szklarska-Poreba, Poland, proceedings.

Onofri F., Rozé C., Gréhan G., 1992, Traitement des signaux phase-Doppler et ADL sujets aux effets de trajectoire, analyse par ondelettes, Proceedings des 9èmes journées sur les aérosols, Paris, Dec 8-9.

Onofri F., Gréhan G., Gouesbet G., Xu T. H., Brenn G., Tropea C., 1994-a, Phase-Doppler anemometry with dual-burst technique for particle refractive index measurements, Seventh International Symposium on applications of laser techniques to fluid mechanics, paper 24-1.

Onofri F., Gréhan G., Gouesbet G., 1994-b, Electromagnetic scattering from a multilayered sphere located in an arbitrary beam, submitted to Applied Optics.

Ren K.F., Gréhan G., Gouesbet G., 1992-a, Localized approximation of generalized Lorenz-Mie theory. Faster algorithm for computations of the beam shape coefficients, Particle and Particle Systems Characterization, 9, 2, 144-150.

Ren K.F., Gréhan G., Gouesbet G., 1992-b, Laser sheet scattering by spherical particles, Proceedings of the sixth international symposium on applications of laser techniques to fluid mechanics, Lisbon, July 20-23.

Ren K.F., Gréhan G., Gouesbet G., 1993, Laser sheet scattering by spherical particles, Particle and Particle Systems characterization, 10, 332-338.

Ren K.F., Gréhan G., Gouesbet G., 1994-a, Symmetry relations in generalized Lorenz-Mie theory, Journal of Optical Society of America A., 11, 6, 1812-1817

Ren K.F., Gréhan G., Gouesbet G., 1994-b, Evaluation of laser sheet beam shape coefficients in generalized Lorenz-Mie theory by using a localized approximation, Journal of Optical Society of America A., 11, 7, 2072-2079

Ryan C.H., Pal A., Lee W., Santoro R. J., 1990, Droplet distribution effects on planar laser imaging of sprays, Atomisation and sprays, 2, 155-177.

Saffman M., 1986, The use of polarized light for optical particle sizing, in R. J. Adrian, T. Asanuma, D. F. G. Durao, Durst F., J. H. Whitelaw (eds) : Laser anemometry in fluid mechanics-III. Selected papers from the third international symposium on applications of laser anemometry to fluid mechanics.

Saffman M., Buchhave P., Tanger H., 1984, Simultaneous measurement of size, concentration and velocity of spherical particles by a laser Doppler method, in R. J. Adrian, D. F. G. Durao, F. Durst, Mishina, J. H. Whitelaw (edts), laser anemometry in fluid mechanics-II, Ladoan- Lisbon, pp85-104.

Sankar S.V. and W.D. Bachalo, 1991, Response characteristics of phase Doppler particle analyzer for sizing spherical particles larger than the wavelength, Applied Optics, vol. 30, n 12, 1487-1496.

Sankar S.V., Inenaga A. and Bachalo W. D. 1992, Trajectory dependent scattering in phase doppler interferometry: minimizing and eliminating sizing error, Proceedings of the sixth international symposium on applications of laser anemometry to fluid mechanics, Lisbon, Portugal, July 20-23.

Tropea C., Xu T., Gréhan G., Onofri F., Haugen P., 1994, Dual-mode – phase-Doppler anemometry, Seventh International Symposium on applications of laser techniques to fluid mechanics, paper 18-3

Xu T.H.,Tropea C., 1994, Improving performance of two-component Phase Doppler anemometers, accepted for publication in Measurement Science Technology.

Quantization of Doppler Signals: how many bits are needed?

Knud Andersen and Anders Høst-Madsen

Dantec Measurement Technology A/S
Tonsbakken 16-18, 2740 Skovlunde, Denmark.

Abstract. In this paper we evaluate the effect of quantization of Doppler signals. We do this by calculating Cramér-Rao lower bounds for one bit quantization and by doing computer simulations. Both approaches show that there are serious problems with one bit quantization around certain critical frequencies. Contrasting, it is shown that 4 bit quantization performs nearly as good as no quantization. The conclusion is that 4 bits are adequate while fewer number of bits give unreliable results.

1. Introduction

Many aspects are crucial to obtain accurate and reliable LDA measurements. Good optics and lasers are necessary to obtain high quality Doppler signals; advanced accurate processing of the Doppler signal is needed to estimate the Doppler frequency with high precision; and, finally, the data analysis techniques applied to the measurements should give optimal and reliable results. Processing of the Doppler signal plays a pivotal role, since this determines the accuracy in any subsequent processing. This processing should use the information in the Doppler signals to the fullest extent, otherwise optimization in other links of the chain is wasted. In this paper we will consequently concentrate on this part of the LDA measurement.

The accurate determination of the Doppler frequency is dependent on a number of factors: burst detection and determination, filters, the processing method, windowing, interpolation etc. However, again we can point at one essential link in this chain: the sampling of the Doppler signal. The Doppler signal is a continuous signal, $s(t)$. In order to process this digitally, it must be sampled at certain instants to obtain $s(t_i)$. Further this value should be represented by a digital number. This is called quantization. Thus, the infinitely many possible states of the signal is reduced to a finite number of bits, and of course information of the signal is lost in this process. This information loss should be kept at a negligible minimum; once the information is lost, it is lost forever and can never be restored. Therefore, a proper sampling method is essential to Doppler processing.

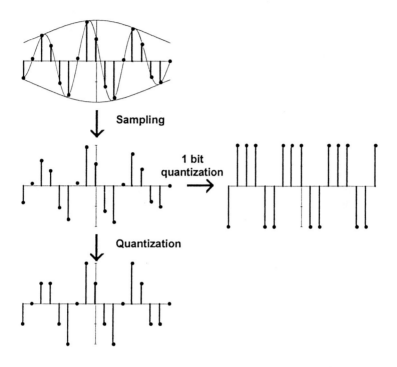

Fig. 1.1 Sampling and quantization.

The contents of this paper is the determination of the requirements to the sampling in order to keep the information loss at a minimum.

The first parameter of the sampling process: when the signal should be sampled is readily determined by the well-known Shannon sampling theorem (alias the Nyquist criterion): the signal should be sampled at a frequency twice the maximal frequency in the Doppler signal. If this condition is fulfilled Shannon says that no information is lost in the sampling. Thus the information loss occurs in the quantization of the signal. In this paper we will investigate this information loss in the quantization and determine how this can be kept at a negligible minimum, answering the question: *Quantization of Doppler Signals: how many bits are needed?*

2. Quantization Of Doppler signals

In order to process the Doppler signal digitally it is converted to a digital value in an A/D-converter. This quantization can be done by a differing number of bits. The two extreme cases are 1 bit quantization and quantization by an infinite number of bits, corresponding to no quantization. For the number of bits converging towards infinity the effect of quantization will converge towards the optimal minimum of no quantization. In principle the more bits the better, but in a practical implementation the number of bits has to be limited, and the question at issue is how low this number can be in order for the effects of quantization to be minimal.

It should be noted that we are discussing the quantization of the primary Doppler signal, not the resolution of the frequency output of an LDA processor. However, the number of bits in the quantization of the primary Doppler signal will limit the accuracy of the estimated Doppler frequency, thereby also limiting the number of *significant* bits in the estimated Doppler frequency. To output more bits than is significant considering the accuracy of the Doppler frequency makes little sense, and as we will see a low quantization will considerably limit the accuracy of the Doppler frequency.

One bit quantization (see Ibrahim et al. (1990)) is equivalent to determining the sign of the signal

$$s_{\text{one bit}}(t_i) = \begin{cases} +1 & s(t_i) \geq 0 \\ -1 & s(t_i) < 0 \end{cases}$$

From an immediate point of view, it might seem that one bit is too little to be useful. However, one bit quantization means determining the zero crossing of the signal, and to determine the frequency of the signal this is sufficient information, as known from the old counter principle, see also Ibrahim et al. (1990). However, it *will* limit the accuracy.

We will make a quantitative evaluation of one bit quantization, and compare this with no quantization and quantization with a limited number of bits. The accuracy of the Doppler signal is characterized by bias and variance. If the Doppler frequency is f then

$$\text{bias} = \overline{f_{\text{estimated}}} - f_{\text{true}}$$

$$\text{variance} = \overline{\left(f_{\text{estimated}} - \overline{f_{\text{estimated}}} \right)^2}$$

If the Doppler signal has bias, then measured mean values will be wrong, and normally one will try to keep bias at a minimum. The variance on the measurements will directly add to the turbulence intensity and also this should be limited.

We will calculate lower bounds for the variance of the estimated Doppler frequency. For no quantization this has been done in Rife and Boorstyn (1974) and in Lading and Edwards (1993) and it should be noticed that this bound does not apply to one bit quantization, in the sense that it gives too low bounds. We will calculate lower bounds that do directly apply to one bit quantization. Some work in this direction has previously been done in Ibrahim et al. (1990), but the bounds given in this reference are much too optimistic and do not give an impression of the problems with one bit quantization.

3. The Cramér-Rao bound

In the appendix we calculate the so called Cramér-Rao lower bound for one bit quantization. We will briefly discuss the significance of the Cramér-Rao lower bound. For a more complete discussion see, e.g. Kendal and Stuart (1993) and Papoulis (1988). If estimating a statistical quantity such as the Doppler frequency, this cannot be done with infinite accuracy, not even theoretically. There is a lower bound on the variance of the measurement, expressed in the Cramér-Rao inequality.

No *unbiased* estimator can do better than the Cramér-Rao (CR) bound. The CR bound thus plays the same role as Heisenberg's uncertainty relation in quantum mechanics or the speed of light in relativity theory. This does not mean that the CR bound can be reached. There might not exists any unbiased estimator that has as low variance as predicted by the CR bound. On the other hand, a biased estimator might have lower variance than the CR bound. An estimator that gives a constant value independently of the input will for example have zero variance but unlimited bias

The CR bound is an information theoretic criterion. This is important in the context of quantization. As mentioned in the introduction, quantization means information reduction and the CR bound is an evaluation of how this information reduction will manifest as variance on estimated value. If the reduced information is used optimally, the CR bound may be reached, but to repeat: it can never be surpassed.

4. Cramér-Rao bounds for Doppler signals

4.1 Doppler signal characteristics

To derive CR bounds for Doppler frequency estimation we will assume that the Doppler signal is a sinusoidal embedded in additive Gaussian white noise. This is only a partial truth. The noise in Doppler signals both consists of signal independent noise and Poisson noise due to the random arrival of photons (shot-noise), which is dependent on the amplitude of the signal. An investigation of the influence of this latter noise has been done in the papers Lading (1993) and Lading and Edwards (1993). To get the full picture both kinds of investigations have to be carried through. However, in the most difficult and critical measurement situations, the signal independent background noise will be dominating. Further, when taking into consideration the many different noise sources of the Doppler signal and using the central limit theorem, the assumption of additive Gaussian white noise seems very reasonable. Thus the Doppler signal can be expressed as

$$s(t) = b(t)\sin(\omega t + \theta) + n(t)$$

The factor $b(t)$ is the envelope of the signal. The exact form of the envelope will depend on the optical configuration, particle sizes etc. Two cases are of special interest. The envelope can be Gaussian due to the profile of the laser beams, or the envelope can be approximately rectangular (top-hat) due to apertures in the system or if only a part (i.e. the central part) of the burst is processed. We will consider both cases, but especially the latter. It can be proved that a rectangular envelope will give the best frequency estimation of all envelopes with a given power, so that a lower bound for a rectangular envelope will give a lower bound for any envelope.

4.2 Cramér-Rao bound for non-quantized sampling

The CR-bound for non-quantized sampling was found in Rife and Boorstyn (1974) for a rectangular envelope to be

$$\mathrm{var}\left[\hat{\omega}\right] \geq V_{CR,\text{non-quantized}}$$

$$= \frac{12}{(SNR)^2 T^2 N(N^2 - 1)}$$

Where T is the sampling interval and N is the number of samples.

For arbitrary envelopes this expression can be easily generalized to (see also appendix)

$$\mathbf{I}_n = \left(\frac{b(t_n)}{\sigma}\right)^2 \begin{bmatrix} t_n^2 & t_n \\ t_n & 1 \end{bmatrix}$$

$$V_{CR,\text{non-quantized}} = \left(\sum_{n=0}^{N} \mathbf{I}_n\right)^{-1}_{11}$$

(1)

where t_n are the sampling times and σ the noise amplitude. On the second line, the matrices should be summed, then the sum should be inverted as matrix; the lower bound for the variance on the signal frequency is then the element in row 1, column 1.

4.3 Cramér-Rao bound for one-bit quantization

In the appendix we have derived the CR-bound for one bit quantization, given by the expression

$$I_n(\omega,t_n,\theta) = \frac{2}{\pi}\left(\frac{b}{\sigma}\right)^2 \begin{bmatrix} t_n^2 & t_n \\ t_n & 1 \end{bmatrix} \Pi(\omega,t_n,\theta)$$

$$V_{CR,\text{one-bit}}(\omega,\theta) = \left(\sum_{n=0}^{N} I_n(\omega,t_n,\theta)\right)^{-1}_{11} \tag{2}$$

$$V_{CR,\text{one-bit}}(\omega) = \frac{1}{2\pi}\int_0^{2\pi} V_{CR,\text{one-bit}}(\theta)d\theta$$

The second line of (2) expresses the CR bound for one bit quantization depending on both SNR ($= b / \sigma$) frequency ω and phase θ. Since the Doppler bursts have random phase with any value of the phase equally probable, we have averaged over the phase to derive at the SNR and frequency dependent CR bound on line 3, which is the expression we will use.

The function $\Pi(\omega,t_n,\theta)$ is given by the rather complicated expression

$$\Pi(\omega,t_n,\theta) = \frac{e^{-\left(\frac{b}{\sigma}\right)^2 \cos^2(\omega t_n+\theta)}}{\left(1-\operatorname{erf}\left(\frac{1}{\sqrt{2}}\frac{b}{\sigma}\cos(\omega t_n+\theta)\right)\right)^2}\sin^2(\omega t_n+\theta)$$

$$+\frac{e^{-\left(\frac{b}{\sigma}\right)^2 \sin^2(\omega t_n+\theta)}}{\left(1-\operatorname{erf}\left(\frac{1}{\sqrt{2}}\frac{b}{\sigma}\sin(\omega t_n+\theta)\right)\right)^2}\cos^2(\omega t_n+\theta)$$

The first thing to notice about (2) is that it is similar to the CR bound for no quantization (1) except for the factors $2/\pi$ and $\Pi(\omega,t_n,\theta)$. Since it can be proven that $\Pi(\omega,t_n,\theta)\le 1$ (se note in appendix) we arrive after some considerations to the conclusion

$$V_{CR,\text{one-bit}}(\omega) \geq \frac{\pi}{2} V_{CR,\text{non-quantized}}$$

or put in words

The variance for one-bit quantization always is at least 50 % higher than the variance for no quantization.

This statement, however, does not take into consideration the frequency dependency of the factor $\Pi(\omega, t_n, \theta)$, which gives a very remarkable behavior. Unfortunately (2) is so complicated that an analytical evaluation is impossible, and consequently one has to turn to a numerical calculation of it. Also this is tricky, but with care it is possible, and we have done so for a selection of frequencies and SNR in the results section below.

The CR bound will not be calculated for quantization of an arbitrary number of bits. Although this could be done in theory, the expressions would be so complicated that they would be without any interest. Instead we have done the evaluation of quantization for a higher number of bits by computer simulations.

It should be remembered that the CR bound calculated for one bit quantization is *independent of the signal processing method, it be FFT or correlation processing.*

5. Computer simulations

To relate the theoretical calculation of the CR bound in the previous section, we have also made computer simulations showing the effect of quantization. The computer simulations have been done for a Doppler signal with rectangular envelope embedded in additive Gaussian white noise, quadrature mixed to give a complex signal. The processing has been done by a FFT analysis with 4 times zero padding followed by a peak search after Brent's method (see Press et al. (1992)). This is a very slow method which is not at all feasible for real-time implementation, but as the subject of this paper is quantization effects we have chosen this method to avoid the influence of other factors like interpolation.

For no quantization the FFT method (or rather Fourier method) is maximum likelihood, i.e. the (asymptotically) optimal method for frequency estimation (as proved in Rife and Boorstyn (1974). *For one-bit quantization the Fourier method is not maximum likelihood,* contrary to some belief (see, e.g. Ibrahim et al. (1990)). The maximum likelihood method for one-bit quantization can be derived by examining the likelihood function in the appendix. Thus, it is indeed possible that a better method than FFT can be found for one-bit quantization, but in this paper we will not proceed further in this direction.

The simulations will be done for one bit quantization and no quantization corresponding to the CR bounds calculated above, and in addition for 4 bit (complex) quantization, such as used by the DANTEC BSA.

6. Results.

In this section we will present the results of the calculation of CR bounds for one-bit quantization, and the simulation results for no quantization, 4 bit quantization and 1 bit quantization, mainly in graphical form. All calculations and simulations show here are for 32 (complex) samples, which applies to a typical burst. The behavior for a larger number of samples is qualitatively identical but of course with different numbers. In all graphs frequency and standard deviations of frequencies are relative to sampling frequency (i.e., bandwidth). The numbers therefore easily may be scaled.

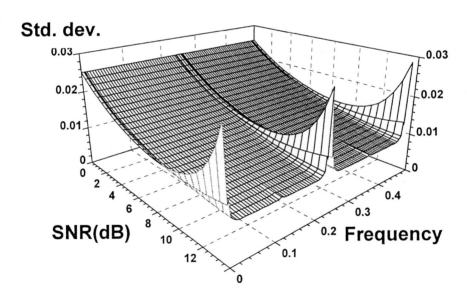

Fig. 6.1. CR lower bound for one bit quantization; 32 complex samples. Standard deviation versus frequency and SNR. Standard deviation is relative to the sampling frequency

Standard deviation of frequency

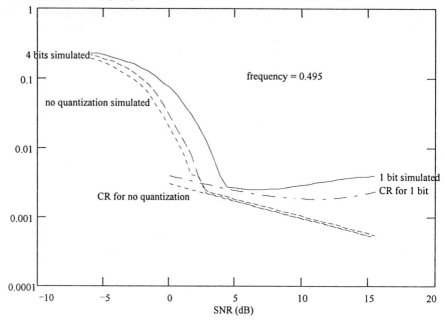

Fig. 6.2. Simulation results and CR bounds as function of SNR. The frequency is 0.495. Standard deviation is relative to sampling frequency.

The CR bound for one-bit quantization depends on both frequency and SNR. Figure 6.1 shows this dependency in a 3D-plot. The remarkable feature is that at the frequencies that are a multiple of a quarter of the sampling frequency, $nf_s/4$, the variance does, as expected, decrease with increasing SNR, *until* it reaches a minimum, and then it increases dramatically. Thus, the paradoxical thing happens that adding noise to the measurements increases the accuracy. This is a strange, but to signal processing well-known, effect of quantization (see e.g. Bilinskis and Mikelsons (1992)). In the terms of the introduction one could say that there is a terrible waste of information around these frequencies.

In figure 6.2 and 6.3 we show the results of simulations. It is seen that there is a close agreement between the simulations and the CR bounds, and that 4 bit quantization is very close to non-quantized sampling. There is a disagreement between the CR bound and simulation for one-bit quantization just at the critical frequencies $nf_s/4$. Here the variance of the simulations are less than the CR bound. This seems to contradict that the CR bound is a lower bound, but one should bear in mind that this is only true for unbiased estimators, and FFT for one-bit quantization is *not unbiased*.

Standard deviation on frequency

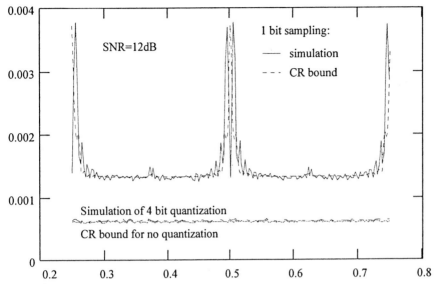

Fig. 6.3. Simulation results and CR bounds as function of frequency. The CR bound for no quantization, the simulations results for no quantization and the simulation results for 4 bit quantization are indistinguishable.

We have investigated the question of bias further. In figure 6.4 we have input a range of frequencies distributed Gaussian as typically in turbulence. This was done for both one bit and 4 bit quantization, producing quite different results. The result for 4 bit quantization is almost indistinguishable from the input data. However, one bit quantization produces an histogram with a quite different form from the input. It is seen that he frequencies $nf_s/4$ acts as "black holes" that swallow every frequencies in their neighborhoods. Therefore the variance is low exactly at $nf_s/4$, lower than predicted by CR, but at the expense of bias.

Obviously, the behavior of one bit quantization will pose problems particularly when measuring turbulence. Here, an unwary user who is not aware of the problems around these critical frequencies may suddenly encounter some strange experimental results. Since one of the critical frequencies is 0, there will inevitably be problems when measuring reversing flows. The only way to avoid this is to stay within a frequency band of a quarter of the sampling bandwidth.

Finally figure 6.2 indicates that the threshold is higher for one-bit quantization than for no quantization with 4 bit quantization being identical to no

270

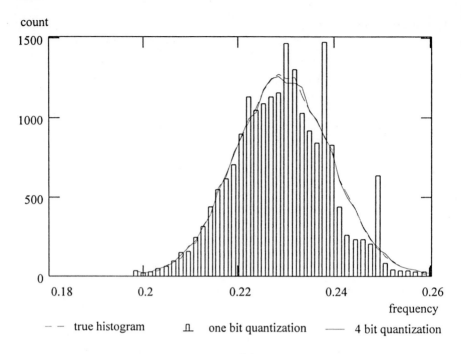

count

Fig. 6.4. Bias effect around the frequencies $nf_s / 4$.

quantization. This effect however has to be investigated further; we do not have any theoretical explanation for this.

From the computer simulations it appears that 4 bits quantization is adequate for an accurate determination of the Doppler frequency. However, to get the full benefit from the 4 bits, the signal need to be adjusted to the dynamic range. In the Dantec BSA an amplitude normalizing circuit in front of the analog to digital converter ensures that the dynamic range of the 4 bits is always used fully also when the Doppler signal amplitude varies over a large dynamic range. We have not made investigations for a fewer number of bits to see if 2 or 3 bits would also be adequate since we estimate that this normalization of the signal would be virtually impossible with the very limited dynamic range of 2 or 3 bits. Therefore, 2 or 3 bit quantization will in reality perform as badly as 1 bit quantization.

The effects of quantization was also investigated in Rogers et al. (1991). In this paper, however, the behavior around the critical frequencies was not investigated, which makes a direct comparison of the results with our results difficult.

7. Conclusion.

We have investigated the influence of quantization on the estimation of the Doppler frequency. This has been done by calculating the Cramér-Rao lower bound for one-bit quantization, which gives a lower limit to the variance of the estimated Doppler frequency, and by computer simulations. Based on the investigations the following can be concluded:

- *The variance for one-bit quantization always is at least 50 % higher than the variance for no quantization.*
- *around the frequencies which are mulipla of one quarter of the sampling frequency the variance and bias for one bit quantization explodes.*
- *The variance and bias at these frequencies can seriously influence turbulence measurements*
- *The threshold (lower limit) SNR for one bit quantization is around 3 dB higher than for no quantization.*
- *Quantization with 4 bits gives an accuracy very close to no quantization without the problems around certain frequencies.*

It should be noticed that these are fundamental problems of one-bit quantization, not problems of a concrete implementation. Based on this, it can be recommended that the following precautions be taken if one bit quantization is to give reliable results

- *The range of Doppler frequencies should be kept within one quarter of the sampling bandwidth for one bit quantization.*

The answer to the question posed is the title can now be answered: in general one bit is to little, while 2 or 3 bits are problematic due to adjustment problems. A quantization of 4 bits is adequate, and there is no gain in using a higher number of bits.

Finally it should kept in mind that once then quantization problem has been solved satisfactory many other factors of the signal processing are important in obtaining accurate LDA measurements.

Appendix: derivation of CR bound

Let the signal be given by

$$s(t) = b_q(t)\cos(\omega t + \theta)$$

here b_q is the envelope of the signal, which depends on some parameters in the parameter vector q (such as signal amplitude and duration), which might be know or unknown. We will primarily concentrate on the case where b_q is constant, so that $q = b_0$ and $b_q(t) = b_0$, and in the derivations below we will not explicitly note the dependencies of b and accordingly write b for $b_q(t)$.

We will assume that sampling is done on the complex signal, i.e., on the signal and its Hilbert transform, which may be obtained from the real signal by quadrature mixing. Thus, the samples (before quantization) can be written as

$$X_n = s(t_n) + W(t_n)$$
$$Y_n = \tilde{s}(t_n) + \tilde{W}(t_n)$$

where $\tilde{\ }$ denotes Hilbert transform and $\tilde{s}(t) = b\sin(\omega t + \theta)$. W is an additive noise contribution, which is assumed to be white Gaussian of zero mean and variance σ^2. Thus, the samples are uncorrelated, and the probability density function (pdf) of a single sample is given by

$$p(X_n; \mathbf{p}) = \frac{1}{\sqrt{2\pi}\sigma} \exp\left(-\frac{1}{2\sigma^2}(X_n - b\cos(\omega t_n + \theta))^2\right) \tag{3}$$

where we have explicitly noted the dependency of p on the parameter vector $\mathbf{p} = [\omega, \theta, q]$. The pdf for Y_n is given equivalently. Since the noise is uncorrelated from sample till sample, the total probability distribution is give as a product:

$$p(\mathbf{X}, \mathbf{Y}; \mathbf{p}) = \prod_{i=1}^{N} p(X_i; \mathbf{p}) \prod_{i=1}^{N} p(Y_i; \mathbf{p})$$

From this expression we can calculate the CR bound for non-quantized sampling, as done in Rife and Boorstyn (1974).

One bit quantization corresponds to obtaining the sign of the samples. Thus we obtain a new series of samples, which can assume only the values +1 and -1,

$$x_n = \text{sign}(X_n)$$
$$y_n = \text{sign}(Y_n)$$

The pdf for x_n is found from (3)

$$p(x_n = -1; \mathbf{p}) = P(X_n < 0; \mathbf{p}) = \int_{-\infty}^{0} \frac{1}{\sqrt{2\pi}\sigma} \exp\left(-\frac{1}{2\sigma^2}(X_n - b\cos(\omega t_n + \theta))^2\right) dX_n$$

$$p(x_n = +1; \mathbf{p}) = P(X_n > 0; \mathbf{p}) = \int_{0}^{\infty} \frac{1}{\sqrt{2\pi}\sigma} \exp\left(-\frac{1}{2\sigma^2}(X_n - b\cos(\omega t_n + \theta))^2\right) dX_n$$

We can write this as the common expression

$$p_n^x(s) = p(x_n = s; \mathbf{p}) = \int_{0}^{\infty} \frac{1}{\sqrt{2\pi}\sigma} \exp\left(-\frac{1}{2\sigma^2}(X_n - sb\cos(\omega t_n + \theta))^2\right) dX_n$$

The elements of the Fisher matrix, due to the real component, are accordingly

$$I_{ij}^x = \sum_{n=1}^{N} \sum_{s=\pm 1} \frac{1}{p_n(s)} \int_{0}^{\infty} \frac{1}{\sqrt{2\pi}\sigma} \exp(-k_n(x)) \frac{\partial k_n}{\partial \varphi_i}(x) dx \int_{0}^{\infty} \frac{1}{\sqrt{2\pi}\sigma} \exp(-k_n(x)) \frac{\partial k_n}{\partial \varphi_j}(x) dx$$

where $\varphi_1, \ldots$ are the unknown parameters to be estimated, and

$$k_n(x) = \frac{1}{2\sigma^2}(x - sb\cos(\omega t_n + \theta))^2$$

We find an equivalent expression for the Fisher matrix I_{ij}^y of the imaginary component of the signal, and the total Fisher matrix is

$$\mathbf{I}_{ij} = \mathbf{I}_{ij}^x + \mathbf{I}_{ij}^y$$

In general this is a rather complicated expression. To get a principal insight, we will therefore at first consider the simplified case where we assume that phase and amplitude of the signal is known. In this case there is only one unknown parameter to be estimated, the frequency ω, and the Fisher matrix is a scalar. After some calculation we find

$$I_n = \frac{2}{\pi}\left(\frac{b}{\sigma}\right)^2 t_n^{\,2} \cdot$$

$$\left(\frac{e^{-\left(\frac{b}{\sigma}\right)^2 \cos^2(\omega t_n + \theta)}}{\left(1 - \mathrm{erf}\left(\frac{1}{\sqrt{2}}\frac{b}{\sigma}\cos(\omega t_n + \theta)\right)\right)^2}\sin^2(\omega t_n + \theta)\right.$$

$$\left.+\frac{e^{-\left(\frac{b}{\sigma}\right)^2 \sin^2(\omega t_n + \theta)}}{\left(1 - \mathrm{erf}\left(\frac{1}{\sqrt{2}}\frac{b}{\sigma}\sin(\omega t_n + \theta)\right)\right)^2}\cos^2(\omega t_n + \theta)\right)$$

$$V_{CR,\text{one bit}} = \frac{1}{\displaystyle\sum_{n=0}^{N} I_n}$$

The factor in the parenthesis is a periodic, highly oscillating function depending on both phase and frequency

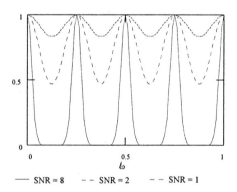

Fig. A.1. One bit factor

For non-quantized sampling the CR bound is in comparison (see Rife and Boorstyn (1974)):

$$I_n = \left(\frac{b}{\sigma}\right)^2 t_n^2$$

$$V_{CR,\text{non-quantized}} = \frac{1}{\displaystyle\sum_{n=0}^{N} I_n}$$

We can make the following evaluation of the CR bound for 1 bit quantization:

$$I_n \leq \frac{2}{\pi}\left(\frac{b}{\sigma}\right)^2 t_n^2 \frac{1}{1-\text{erf}\left(\dfrac{1}{\sqrt{2}}\dfrac{b}{\sigma}\right)}$$

$$V_{CR,\text{one bit}} \geq \frac{\pi}{2}\left(1-\text{erf}\left(\frac{1}{\sqrt{2}}\frac{b}{\sigma}\right)\right) V_{CR,\text{non-quantized}}$$

This is not the best inequality, however. We are certain without doubt that the factor in the parenthesis is always ≤ 1 (see figure 6), although we have not made a rigorous mathematical proof of this. Therefore we can give the inequality

$$V_{CR,\text{one-bit}} \geq \frac{\pi}{2} V_{CR,\text{non-quantized}}$$

For both frequency and phase unknown, the CR bound can be found to

$$\mathbf{I}_n =$$

$$\frac{2}{\pi}\left(\frac{b}{\sigma}\right)^2 \begin{bmatrix} t_n^2 & t_n \\ t_n & 1 \end{bmatrix} \cdot$$

$$\left(\frac{e^{-\left(\frac{b}{\sigma}\right)^2 \cos^2(\omega t_n + \theta)}}{\left(1-\text{erf}\left(\frac{1}{\sqrt{2}}\frac{b}{\sigma}\cos(\omega t_n + \theta)\right)\right)^2} \sin^2(\omega t_n + \theta) \right.$$

$$\left. + \frac{e^{-\left(\frac{b}{\sigma}\right)^2 \sin^2(\omega t_n + \theta)}}{\left(1-\text{erf}\left(\frac{1}{\sqrt{2}}\frac{b}{\sigma}\sin(\omega t_n + \theta)\right)\right)^2} \cos^2(\omega t_n + \theta) \right)$$

$$\mathbf{V}_{CR,\text{one bit}} = \left(\sum_{n=0}^{N} \mathbf{I}_n\right)^{-1}$$

and for non-quantized sampling the CR bound is (see Rife and Boorstyn (1974)):

$$\mathbf{I}_n = \left(\frac{b}{\sigma}\right)^2 \begin{bmatrix} t_n^{\,2} & t_n \\ t_n & 1 \end{bmatrix}$$

$$V_{CR,\text{non-quantized}} = \left(\sum_{n=0}^{N} \mathbf{I}_n\right)^{-1}$$

For both frequency, phase and amplitude unknown we get the somewhat more complex:

$$\mathbf{I}_n = \frac{2}{\pi}\frac{1}{\sigma^2}$$

$$\left(\begin{bmatrix} b^2 t_n^{\,2}\sin^2(\omega t_n) & -\frac{1}{2}b t_n\sin(2\omega t_n) & b^2 t_n\sin^2(\omega t_n) \\ -\frac{1}{2}b t_n\sin(2\omega t_n) & \cos^2(\omega t_n) & -\frac{1}{2}b\sin(2\omega t_n) \\ b^2 t_n\sin^2(\omega t_n) & -\frac{1}{2}b\sin(2\omega t_n) & b^2\sin^2(\omega t_n) \end{bmatrix} \frac{e^{-\left(\frac{b}{\sigma}\right)^2\cos^2(\omega t_n)}}{\left(1 - \text{erf}\left(\frac{1}{\sqrt{2}}\frac{b}{\sigma}\cos(\omega t_n)\right)\right)^2} + \right.$$

$$\left.\begin{bmatrix} b^2 t_n^{\,2}\cos^2(\omega t_n) & \frac{1}{2}b t_n\sin(2\omega t_n) & b^2 t_n\cos^2(\omega t_n) \\ \frac{1}{2}b t_n\sin(2\omega t_n) & \sin^2(\omega t_n) & \frac{1}{2}b\sin(2\omega t_n) \\ b^2 t_n\cos^2(\omega t_n) & \frac{1}{2}b\sin(2\omega t_n) & b^2\cos^2(\omega t_n) \end{bmatrix} \frac{e^{-\left(\frac{b}{\sigma}\right)^2\sin^2(\omega t_n)}}{\left(1 - \text{erf}\left(\frac{1}{\sqrt{2}}\frac{b}{\sigma}\sin(\omega t_n)\right)\right)^2}\right)$$

$$V_{CR,\text{one bit}} = \left(\sum_{n=0}^{N} \mathbf{I}_n\right)^{-1}$$

Notice that in all these expressions the variance depends on both phase and frequency for the signal. For a given frequency, the phase of the Doppler burst will be random and any phase will be equally possible. Thus, to get the CR lower bound for a given phase, the variance should be averaged over all phases (a minor consideration will show that still gives a lower bound). Thus

$$V_{CR,\text{non-quantized}}(\omega) = \int_0^{2\pi}\left(\sum_{n=0}^{N} \mathbf{I}_n\right)^{-1} d\theta \tag{4}$$

since the integration is to be done after the matrix inversion, it is virtually impossible to arrive at an analytical expression. The evaluation of (4) must therefore be done numerical, which is also non-trivial due to the strange behavior of the one bit factor (figure A.1).

References

Bilinskis, I. and Mikelsons, A.: Randomized signal processing, Prentice Hall, London 1992.

Ibrahim, K. M., Werthimer, G. D. and Bachalo, W. D.: "Signal Processing Considerations for Laser Doppler and Phase Doppler Applications", 5th International Symposium on Applications of Laser Techniques to Fluid Mechanics, Lisbon 1990, Springer-Verlag 1991, pp. 291-316.

Kendal, M. and Stuart, A.: "The advanced theory of statistics, vol. 2" London 1963.

Lading, L. and Edwards, R., "Laser velocimeters; lower limits to uncertainty", Applied Optics, vol. 32, No. 21, July 1993, pp. 3855-3866.

Lading, L.: "Signal processing for time resolved measurements", Summer School on optical diagnostics for flow processes, Risø, 1993.

Papoulis, A.: Probability, Random Variables and Stochastic Processes, McGraw-Hill, New York, 1988.

Press, H. et al.: Numerical receipies in C, Cambridge University Press, 1992.

Rife, D. and Boorstyn, R.: "Single-Tone Parameter Estimation from Discrete Time Observations", IEEE Transactions on Information Theory, vol. IT-20, No. 5, Sep. 1974, pp. 591-598.

Rogers, P., Blancha, B. and Murphy, R. J.: Effects of A/D Resolution on the Performance of Frequency Domain Processors, 4th International Conference on Laser Anemometry, Advances and Applications, Cleveland, Ohio 1991, ASME 1991, pp. 603-608.

Direction Sensitive LDA
Using a Single High-Frequency Pulsed Laser Diode
And a Single Photodetector Without Frequency Shifting

H. Wang, H. Müller, V. Strunck and D. Dopheide

Physikalisch-Technische Bundesanstalt (PTB)
Laboratory for Fluid Flow Measuring Techniques
Bundesallee 100, D-38116 Braunschweig
Federal Republic of Germany

Abstract. Conventional directional LDAs are based on the frequency shifting technique. In this paper a technique has been proposed for the realization of a direction sensitive HF-pulsed LDA on the basis of switch demultiplexing and quadrature signal processing. This technique allows one to take advantage of HF-pulsed LDAs and to use only one high-frequency pulsed laser diode and one detector for directional flow velocity measurements without frequency shifting. The first experimental verification of this technique has been made. Improvements to the experimental set-up and practical miniaturized realizations of a directional HF-pulsed LDA using light fibres and integrated optical devices are described.

Keywords. Laser Doppler anemometer, fluid measurement technique, high frequency technique, demultiplexing technique, quadrature signal processing

1 Introduction

It has been reported that high-frequency pulsed laser diode LDAs (HF-pulsed LDA) for simultaneous multi-component flow velocity measurements can be realized by using only one detector, one data processing chain and laser diodes at an identical wavelength (Dopheide et al., 1990). In addition to this, the single pulse emitted from the HF-pulsed laser diodes has a high peak power, giving a better signal-to-noise ratio (SNR). Experimental investigations on these systems have been described in detail by Dopheide et al. (1993) and by Wang et al. (1994a). At present, the HF-pulsed multi-component LDAs need more than one laser diode and are not capable of discriminating the direction of velocity components.

Frequency shifting is generally the most convenient method used in realizations of direction sensitive LDAs. In principle, this technique can also be applied in HF-pulsed LDAs for direction discrimination. A technique for the realization of a directional LDA without frequency shifting has been described by Köhler et al. (1993), but, two detectors and a variable attenuator are required.

Since the peak power of laser diodes can be enhanced by the high-frequency pulsing technique, it becomes possible to divide the output beam of the HF-pulsed laser diode into several beams to realize:

1)   a multi-component LDA as described by Wang et al. (1994b);
2)   a directional HF-pulsed LDA.

In this case, only a single laser diode is required to measure more than one component of a directional flow velocity simultaneously.

This paper describes the technique used to realize such a direction sensitive system with a single laser diode and a single photodetector and without frequency shifting. The first experimental results are reported. Improvements made to the experimental set-up and practical realizations of this technique, for example, miniaturized directional LDAs, using fibre optics and integrated optical devices are also outlined.

2   Method

2.1   Experimental Set-up

As shown in Fig. 1, the PTB's two-dimensional HF-pulsed LDA has been modified to consist of a single laser diode (Dopheide et al., 1990 & 1993; Wang et al., 1994b). The two equidistant fringe patterns are parallel in the measuring volume instead of being placed orthogonally like the fringe patterns used for multi-component measurements.

Fig. 2 shows the light output pulses of the laser diode, delayed pulse trains through the by-pass and their superposition measured in the measuring volume. The clearly separated pulse trains allow the two fringe patterns to be generated in the measuring volume.

As shown in Fig. 3, the fringe patterns do not coincide exactly but are locally shifted relatively to each other in the direction parallel to the flow direction. A particle flying through the measuring volume now results in two individual Doppler signals whose phase difference indicates the direction of the flow velocity. In order to use the quadrature signal processing technique described by Czarske et al. (1993), the two equidistant fringe systems are locally shifted relatively to each other by a quarter fringe spacing. The experimental realization of such an arrangement will be discussed later in more detail.

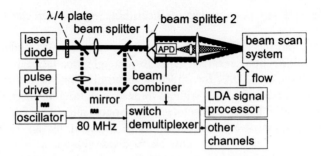

Fig. 1. Set-up of the direction sensitive HF-pulsed LDA with only one laser diode and one photodetector. One 80 MHz electrical pulse train is applied to the laser diode driver to trigger the laser diode, giving the first laser pulse train at 80 MHz (black line, main beam). By means of the beam splitter and mirrors, a second laser pulse train which is phase-delayed to the first one is obtained (dashed black line, by-pass beam). The phase difference between the first and the second laser pulse train can be adjusted the mirrors, so that two 80 MHz pulse trains are observed in sequence in the measuring volume (see Fig. 2). The switch demultiplexer is explained in Fig. 4

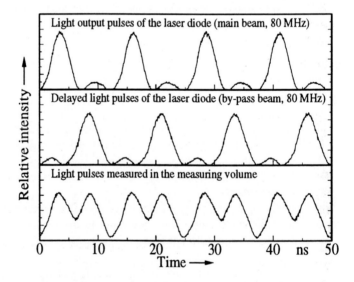

Fig. 2. Light pulses measured in the measuring volume

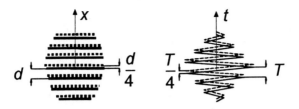

Fig. 3. Cross sections of successive fringe patterns in the measuring volume and the generation of a quadrature signal pair by two equidistant fringe systems locally shifted relatively to each other by a quarter fringe spacing d. Such an arrangement allows the quadrature signal processing technique to be used (see Figs. 5 and 6)

2.2 Switch Demultiplexing

The scattered light is collected on an APD. The pulse signals from the amplifier are connected to a switch demultiplexer which converts the pulse signals modulated by the Doppler signals into two standard Doppler signals for the laser pulse trains 1 and 2. By optimizing the phase shift of these two reference signals through the delay, either channel 1 or channel 2 can be switched on and off, thus separating the Doppler signal of every component. The separated Doppler signals can be processed using a conventional LDA signal processor (see Fig. 4, Wang et al., 1994a).

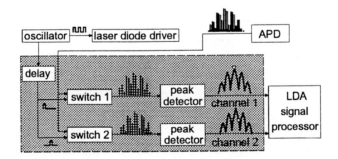

Fig. 4. Working principle of a switch demultiplexer. The received electrical pulse trains are switched into channels 1 and 2 synchronously by two reference signals at 80 MHz. The peak detector is optimized at the peak of the signal pulses at 80 MHz (coherent sampling) which converts the burst signal behind the switch into the standard Doppler signals. This allows the conventional LDA signal processor to be used for signal processing in HF-pulsed LDAs

282

2.3 Quadrature Signal Processing

If the conditions in Fig. 3 are satisfied, the Doppler signals after switch demultiplexing are actually a quadrature signal pair and can be processed using the quadrature signal processing technique (Czarske et al., 1993, Müller et al. 1994). In such a case, the two Doppler signals have a 90° phase difference and the same amplitude $A(t)$:

$$a_1 = A(t)\sin \Phi(t) \qquad (1)$$
$$a_2 = A(t)\cos \Phi(t) \qquad (2)$$

The phase angle $\Phi(t)$ is given by the quotient of signals a_1 and a_2 and becomes independent of the amplitude $A(t)$:

$$\Phi(t) = \arctan(a_1 / a_2) \qquad (3)$$

Based on the measured time series of phase angle values during a burst, an estimation of the Doppler frequency can be made by using the least-squares method. A linear regression by

$$\Phi(t) = \Phi_0 + 2\pi f_D t \qquad (4)$$

allows a direction analysis of the averaged tracer particle velocity in the measuring volume. This velocity is proportional to the center frequency f_D in Eq. (4).

The pair of quadrature signals can be considered as the real part and the imaginary part of a complex rotating vector where the rotation describes the time dependence of the phase angle $\Phi(t)$. As shown in Figs. 5 and 6, the rotation direction of the Lissajous figure gives the sign of the flow velocity and the rotation number of the spiral is equal to the fringe number in the measuring volume.

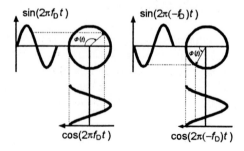

Fig. 5. Phase angle measurement by a quadrature signal pair. The signal pair $\sin(2\pi f_D t)$ and $\cos(2\pi f_D t)$ curve in the left hand shows one direction, while the signal pair $\sin(2\pi(-f_D)t)$ and $\cos(2\pi(-f_D)t)$ in the right hand represents the opposition direction

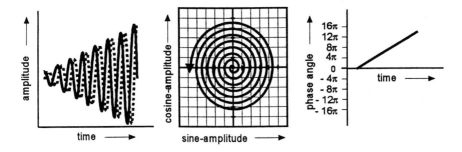

Fig. 6. First cycles of a measured LDA quadrature signal pair (left), Lissajous figures of the signal pair (middle) and time sequence of the phase angle values which gives both direction and magnitude of the flow velocity (right) (Müller et al., 1994)

3　Results

Referring to Fig. 1, the direction sensitive HF-pulsed LDA consists of practically two dual-beam LDAs which have two distinguishable fringe systems with identical fringe spacings. These two fringe systems are relevant to the main and by-pass beams, and are illuminated alternatively because of the optical path length difference between the main and by-pass beams. During experiments, an optical path length of about 1.5 m is used due to the pulse frequency of 80 MHz. In such a case, the main and by-pass beams become two conjugate pulse trains and allow the switch demultiplexing technique to be used (Wang et al., 1994a). A shorter optical path length is required if a higher pulse frequency is used. For example, an optical path length difference of about 0.3 m is required if a pulse frequency of 1 GHz is used.

In principle, the main and by-pass beams have the same beam diameter in the measuring volume if the diode laser output is collimated to have a very small divergence. During experiments, however, divergent laser beams have been obtained due to the limitation of the experimental conditions. The main and by-pass beams have therefore different beam widths in the measuring volume due to their different optical path lengths, resulting in two Doppler signals with different numbers of the period (Fig. 7). This makes the use of the quadrature signal processing technique a little bit difficult, but, the direction and the magnitude of the flow velocity can be easily determined.

In order to show the influence of optical path length differences between main and by-pass beams on the shape of fringe patterns, experiments have been done after the optical path length difference is reduced from 1.5 m to 0.3 m. As shown in Fig. 8, such main and by-pass beams allow to generate two similar fringe patterns and to use the quadrature signal processing technique.

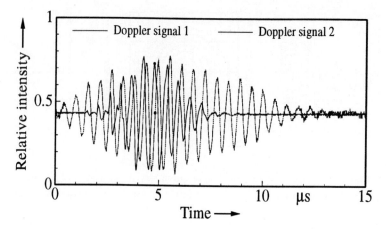

Fig. 7. The Doppler signals after switch demultiplexing. The sign of the flow velocity is determined by the sign of the phase difference between the Doppler signals 1 and 2 and its magnitude by the power spectrum of either of the two Doppler signals. An inverse flow direction will result in an opposite phase difference. In order to obtain better SNR, the forward scattering has been detected

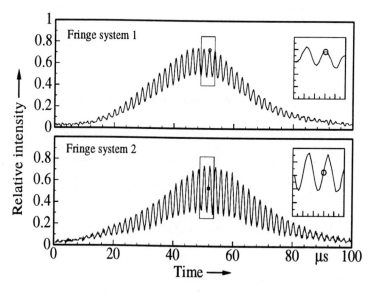

Fig. 8. Fringe patterns measured in the measuring volume when the optical path length between the main and by-pass beams for the fringe systems 1 and 2 is reduced to about 0.3 m. The laser beam of fringe system 1 is blocked when fringe system 2 is measured, and vice versa. The phase shift is obvious if the parts of enlarged fringe patters shown in the insets are compared

4 Improved Experimental Set-up Using Light Fibres

Referring to Figs. 9 and 10 the mirrors and beam splitters are replaced by light fibres in order to produce the optical path length difference between the diode laser beams. The HF-pulsed laser beam is coupled into the light fibre, which in turn is split into two light fibres of different lengths. The short light fibre is used to obtain one dual-beam LDA, while the other long light fibre produces the second fringe system. The time difference between these two parallel fringe patterns can be controlled by the length of the light fibres. In such a case, the two similar equidistant fringe systems can be generated by adjusting collimators to ensure that the fringe patterns have the same properties and allow the quadrature signal processing technique to be used.

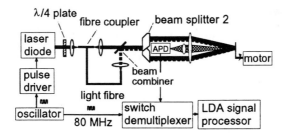

Fig. 9. Set-up of the direction sensitive HF-pulsed LDA with only one laser diode and light fibres

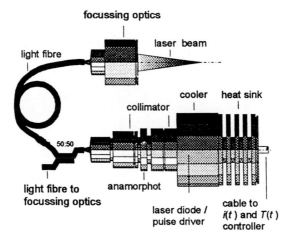

Fig. 10. Pigtailed laser diode with integrated focusing optics used in the direction sensitive HF-pulsed LDA

Fig. 11 shows the Doppler signals obtained using the set-up in Fig. 9. It is obvious that these Doppler signals form a quadrature signal pair and can be processed using the quadrature signal processing technique described in section 2.3.

It should be pointed out that:

1) Referring to Fig. 9, there is a coupling loss of more than 50% if a conventional beam combiner is used. The use of polarizing beam splitters and combiners, and of polarization maintaining light fibres, supplies also two orthogonal linearly polarized beams in the measuring volume, reduces a coupling loss and gives a better SNR

2) As an alternative to the polarization optics described above, the experimental set-up in Fig. 9 can be further improved if, instead of the beam splitter, a mirror is used. As seen in Fig. 12, the use of the mirror can reduce the coupling loss. Such an arrangement can also be applied to the experimental set-up in Fig. 1 to reduce the coupling loss and to enhance the SNR.

3) The use of pigtailed laser diodes would be an advantage to reduce the requirements for adjusting the optical system in Figs. 10 and 12.

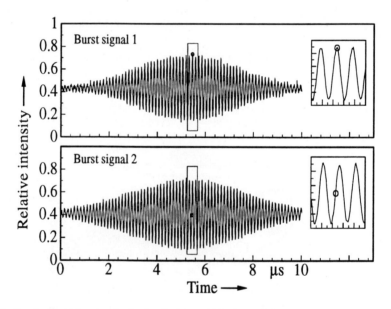

Fig. 11. Quadrature signal pair obtained after switch demultiplexing in Fig. 9. The insets show parts of the enlarged burst signal

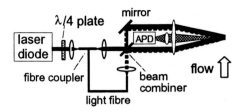

Fig. 12. Set-up of the direction sensitive HF-pulsed LDA with only one laser diode and light fibres. Instead of the beam splitter, a mirror is used to reduce the coupling loss and enhance the SNR. The integration of the mirrors and light fibres would ensure the stability of the shifted fringe systems

4)   Difficulties have been met in keeping the stability of two equidistant fringe systems in the set-up shown in Figs. 1 and 9. Better results have been obtained if a short optical path length difference is used. Such instabilities can surely be reduced using integrated optics.

5)   The "coherent sampling" technique proposed by Dopheide et al. (1990) can be used for data processing in the experimental set-up. This could be advantageous if a high speed waveform digitizer for PCs is used instead of a transient recorder (up to 1 GHz is now commercial available, for example, the *STR ★ 81G   High Speed Waveform Digitizer* from the Keithley Instrument GmbH).

As described by Wang et al. (1994b) and Stieglmeier et al. (1993), the cw laser can be used as the light source in HF-pulsed LDAs if an optical amplitude modulator is applied. When a control voltage is applied onto the central plate of a hybrid Mach-Zehnder modulator (Fig. 13), an incoming cw light beam can be modulated up to 20 GHz and used as an HF-pulsed laser source for the realization of LDAs.

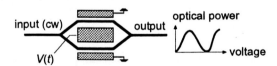

Fig. 13. Mach-Zehnder modulator described by Power. When a voltage $V(t)$ is applied to the central electrode, the cw light input is modulated to the high-frequency pulse output in accordance with the electro-optical effect (Powell, 1993)

288

5 Discussion

It has been experimentally verified that only one HF-pulsed laser diode and one detector are required to measure directional flow velocities without applying frequency shift techniques. Consequently, only one temperature stabilization, one current stabilization and one pulse driver are required for the direction sensitive HF-pulsed LDA.

Such a LDA can also be realized by using light fibres and integrated optical devices for the miniaturized design of directional HF-pulsed LDAs without frequency shifting. The use of the quadrature signal processing technique for example, the miniaturized signal processor developed by von Wnuck et al. (1994), would simplify data processing of the direction sensitive HF-pulsed LDA.

References

Czarske, J.; Hock, F.; Müller, H (1993) Quadrature demodulation - a new LDV-burst signal frequency estimation. Proc. of the 5th International Conference, Laser Anemometry - Advances and Applications. SPIE Vol. 2052, pp 79-86 (Konigshof, Veldhoven, The Netherlands)

Dopheide, D.; Strunck, V.; Pfeifer, H. J (1990) Miniaturized multi-component laser Doppler anemometers using high-frequency pulsed laser diodes and new electronic signal acquisition systems. Experiments in Fluids. 9: 309-316

Dopheide, D.; Rinker, M.; Strunck, V (1993) High-frequency pulsed laser diode application in multi-component laser Doppler anemometry. Optics and Lasers in Engineering. 18: 135-145

Köhler, R.; Stetter, M.; Schroff, G (1993) Laserdoppler-Richtungserkennung - ohne Braggzellen, ganz anders. In: Tagungsband 2. Workshop Lasermethoden in der Strömungsmesstechnik: Aktueller Stand und neue Anwendungen (Ed Dopheide, D.; Leder, A.; Ruck, B). Paper 1, Achen: Verlag Shaker

Müller H.; Czarske, J.; Kramer, R.; Többen, H.; Arndt, V.; Wang, H.; Dopheide, D (1994) Heterodyning and quadrature signal generation: Advantageous techniques for applying new frequency shift mechanisms in the laser Doppler velocimeter. In: The Proceedings of The Seventh International Symposium on Application of Laser Techniques to Fluid Mechanics. Vol. I, pp 23.3.1 - 23.3.8 (Lisbon, Portugal)

Powell, M. A (1993) An ally for high speed and the long haul: Integrated optical modulator applications in communication systems. Photonics Spectra. Special CLEO issue: 102-108

Stieglmeier, M.; Theis, F.; Tropea, C.; Weber, H.; Rasch, A (1993) Zwei Komponenten LDA mit integriert-optischem X-Schalter und kohärenter Abtastung. In: Tagungsband 2. Workshop Lasermethoden in der Strömungsmesstechnik: Aktueller Stand und neue Anwendungen (Ed Dopheide, D.; Leder, A.; Ruck, B). Paper 6, Achen: Verlag Shaker

von Wnuck, J.; Strunck, V.; Dopheide, D (1994) Universelle Miniatur-Signalauswertung für die Laser Doppler Anemometrie, PTB Bericht No. PTB-MA-32, Braunschweig, January 1994

Wang, H.; Dopheide, D.; Müller, H.; Strunck, V. (1994a) Optimized signal separation in a high-frequency pulsed laser diode Doppler anemometer for multi-component measurements, PTB Mitteilungen. 104: 83-88

Wang, H.; Strunck, V.; Müller, H.; Dopheide, D (1994b) New technique for multi-component flow velocity measurements using only one laser diode and integrated optical devices, Paper presented at The Conference on Lasers and Electro-Optics (CLEO® '94), May 8-13, 1994, Anaheim, USA

Measurement of bubbles by Phase Doppler technique and trajectory ambiguity

G. Gréhan[1], F. Onofri[1], T. Girasole[1], G. Gouesbet[1]
F. Durst[2] & C. Tropea[2]

[1] Laboratoire d'Energétique des Systèmes et Procédés,
INSA de ROUEN, URA.CNRS.230, CORIA.
BP08, 76131, Mont-Saint-Aignan Cédex, France.
Tel: (33) 35 52 83 87, Fax: (33) 35 52 83 90

[2] Lehrstuhl für Strömungsmechanik
University of Erlangen-Nürnberg
Cauerstr. 4, 91058 Erlangen, Germany

Abstract. That paper is devoted to the analysis of the importance of the finite size of the laser beam on the quality of bubble measurements by a Phase Doppler Anemometer (PDA) (trajectory ambiguity or Gaussian beam effect). Detection at 30 ° and 70° are studied for a standard PDA, a non-standard PDA and dual burst configurations. The possibility to sort bubbles and perfectly reflecting particles is discussed.

Keywords. Particle sizing, Generalized Lorenz-Mie theory, light scattering.

1 Introduction

To measure particle velocity/particle size correlations, the phase Doppler Anemometer (PDA) represents one of the most versatile and accurate techniques available today.

Nevertheless, like any technique, the PDA is not without limitations:

- When the size of the particle under investigation is of the same order as the incident wavelength, the phase shift/diameter relationship is no longer linear because of Mie resonances. This leads to ambiguity in the size determination, see Sankar et al. (1990) and Naqwi et al. (1992).

- When the particle size is not small compared to the focused beam diameter, the phase shift/diameter relationship is also nonlinear because of the nonuniform incident intensity across the surface of the particle. The nonuniform illumination leads to a change in the ratio of refracted and reflected light received by the detectors and thus modifies the intended phase shift/diameter relationship. This effect is known as trajectory ambiguity or Gaussian beam defect, since the arising error depends on the particle positions or particle trajectory inside the measurement control volume.

The latter effect has been studied in detail by using geometrical optics (Bachalo & Sankar (1988), Sankar & Bachalo (1991)) and Generalized Lorenz-Mie Theory (GLMT) (Gréhan et al. (1991,1994)). Various solutions to remove or at least minimize the trajectory ambiguity have been proposed for classical geometries (Gréhan et al. (1992)) or using more original designs such as the Planar geometry (Aizu et al. (1993), Modified Standard geometry (Xu & Tropea (1994)) or the dual burst technique (Onofri et al. (1994)). Nevertheless, these studies have been focused on refractive particles, i.e. water droplets and PDA configurations in which refraction is the dominating scattering mechanism to be considered. In the present paper the trajectory ambiguity will be examined for reflective dominated systems, typically systems for the measurement of particles with a relative refractive index less than one, i.e. air bubbles in water.

This paper describes a study of trajectory effects in the phase Doppler technique taking into account all the properties of the optical signal (phase shift, visibility, and pedestal) and is organized as follows. Section 2 reviews the GLMT and simulation model as well as the geometry under study. Then section 3 displays computational results for the classical geometry and discusses similarities and differences compared with the droplet case. Section 4 is devoted to modified standard geometries while section 5 is devoted to the dual burst technique. Section 6 is a conclusion.

2 GLMT, simulation model and geometry under study

2.1 GLMT

The classical Lorenz-Mie Theory (LMT), which is about one century old, describes the interaction between a plane wave and a homogeneous sphere. It has been widely used to design instruments and to analyze data in optical sizing. Unfortunately, most techniques use laser sources which produce finite cross-section light beams so that when the particle diameters are not small enough with respect to the beam width, LMT is inevitably misleading. Therefore, the research team at Rouen worked out a generalization of LMT. A description of the analytical work is given in Gouesbet et al. (1991), where earlier references may also be found.

The physics behind GLMT is very simple. It solves Maxwell's equations with the appropriate boundary conditions at the surface of the sphere and at an infinite distance from the sphere. GLMT introduces two new series of beam shape coefficients $g_{n,TM}^m$ and $g_{n,TE}^m$, which describe the incident beam. For practical computations special algorithms have been developed for the case of circular and elliptical Gaussian beams. The more powerful technique is based on the so-called localization principle (Gouesbet et al. (1990)), which has been recently rigorously demonstrated (Lock and Gouesbet (1994), Gouesbet and Lock (1994)). Only codes based on the localization principle are used to carry out computations in the present paper.

2.2 Simulation Model

The scattered light signal from a dual-beam system may be expressed in the following form (Naqwi & Durst (1991)):

$$P(t) = P_s [1 + \nu \, cos(\Delta\omega \, t + \Phi)] \tag{1}$$

where P_s, ν, Φ and $\Delta\omega$ are the signal pedestal, visibility, phase and heterodyne Doppler frequency, respectively. The parameters P_s, ν, and Φ are functions of the complex scattering amplitudes S_1 and S_2, which are provided by GLMT (Gréhan et al. (1994)).

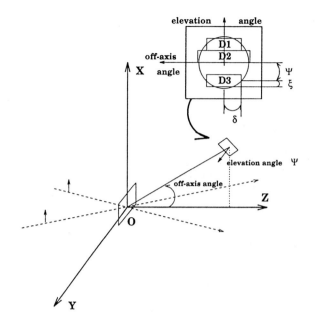

Figure 1: *Geometry of phase Doppler*

2.3 PDA setup

In order to be able to compare with results previously published, the geometry selected is as close as possible to the layout studied in Bachalo & Sankar (1988), Sankar & Bachalo (1991), using geometrical optics or Gréhan et al. (1991,1994), using GLMT. The optical layout is illustrated in Fig. 1, where various angles are defined, the size and location of the rectangular receiving apertures used for detectors D_1, D_2 and D_3, are also shown.

Two laser beams, polarized parallel to the OX axis, propagate in the (YOZ) plane. The two beam waists are centered at point O. The results presented here pertain to an angle of 1.35° between the two-beams. The beam waist diameter is 80 μm and the wavelength is 0.4747 μm. At the centre of the measuring volume, each beam has an intensity of 10^7 W/m^2.

Each detector is defined by: its center location, an off-axis angle θ in the plane (ZOX) and an elevation angle Ψ, and its aperture size. Parallel to the fringes, the receiving aperture subtends a half-angle δ,

Detector	Ψ	δ	ξ
D_1	4.1°	4.39°	1.57 °
D_2	1.27°	5.71°	1.27 °
D_3	-4.1°	4.39°	1.57 °

Table 1: *Elementary detector parameters*

and normal to the fringes the half-angle is denoted by ξ. Table 1 gives the values used in this paper. Air bubbles in water with a complex refractive index of $0.75 - 0.0i$ are considered as the scattering particles.

2.4   Computation domain

The computation domain in this paper is limited to the plane (XOY), where the particle center was moved in steps of 16 μm parallel to the OX and OY directions, describing a square with each side of 160 μm using 121 grid points. For each of these 121 locations, the pedestal, visibility and phase recorded by each of the three detectors were computed.

Each component of this 11x11 matrix was interpolated onto a 25x25 grid and used to draw maps as in Gréhan et al. (1994).

3   Standard geometry

3.1   30° detection

The collection unit is in the plane (XOZ), with an off-axis angle of $\theta = 30°$. The main particle trajectory is parallel to the Y axis.

Figures 2-a and 2-b display iso-pedestal lines (more exactly iso-level of $\log_{10}(P_s)$, where P_s is expressed in Watt), corresponding to detector D_1, for bubbles of 1 and 40 μm diameter, respectively.

Figures 3-a and 3-b display iso-phase shift lines between detector pairs 1-2, Φ_{12} and 1-3, Φ_{13} respectively, for a bubble of 1 μm diameter. Figures 3-c and 3-d show Φ_{12} and Φ_{13} but for a 40 μm diameter bubble. Maps of visibility are not presented because the visibility is always very good.

Figures 2-a and 2-b show that the pedestal is a strong function of the bubble diameter and location in the control volume. Furthermore, the

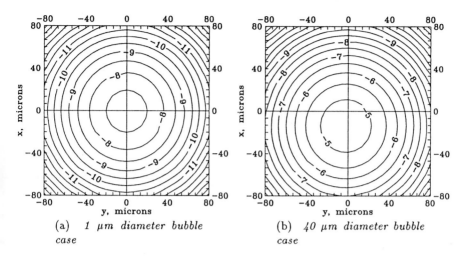

(a) *1 μm diameter bubble case*

(b) *40 μm diameter bubble case*

Figure 2: *Iso-pedestal lines for detector D_1.*

location of the maximum value of the pedestal depends on the particle size (and detector location). For example, for the 40 μm diameter bubble (Fig. 2-b), the maximum is located at $(Y = 0, x \approx -16 \ \mu m)$. *The location of the maximum of the pedestal moves opposite than for the droplet case (Gréhan et al. (1994)).*

Figures 3-a and 3-b, for 1 μm diameter bubble, show that the phase shift is nearly independent on the particle location in the control volume. The gradient of illuminating light intensity onto the particle is very small. Hence the assumption of a uniform illumination of the particle is verified and the classical Lorenz-Mie theory could be used successfully (the phase shifts predicted by LMT are $\Phi_{12} = 0.468°$ and $\Phi_{13} = 1.338°$).

Figures 3-c and 3-d, for the 40 μm diameter bubble, show that the phase shift is strongly dependent on the particle location.

If we want to measure at the same time 1 μm and 40 μm diameter bubbles, the level of minimum pedestal must be -7.5 (see Fig. 2-a). Then from figure 2-b and figures 3-c and 3-d, the measured phase shift for a 40 μm diameter bubble evolves (i) from 20° to 28° for Φ_{12} (ii) from

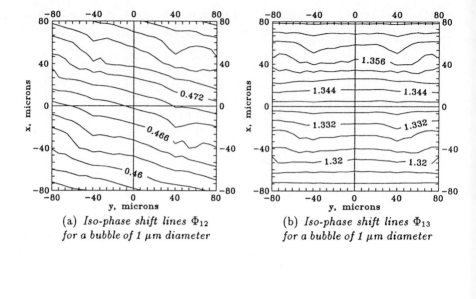

(a) *Iso-phase shift lines* Φ_{12}
for a bubble of 1 μm diameter

(b) *Iso-phase shift lines* Φ_{13}
for a bubble of 1 μm diameter

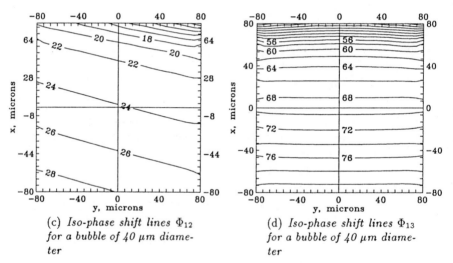

(c) *Iso-phase shift lines* Φ_{12}
for a bubble of 40 μm diame-
ter

(d) *Iso-phase shift lines* Φ_{13}
for a bubble of 40 μm diame-
ter

Figure 3: *Iso-phase shift lines*

$58°$ to $82°$ for Φ_{13}. Note that the plane wave prediction for a 40 μm bubble are $\Phi_{12,LMT} = 24.10°$ and $\Phi_{13,LMT} = 69.17°$.

Then for this standard geometry, as the mean particle trajectory is parallel to the Y axis, the measured diameter for a 40 μm bubble could vary from about 37 to 53 μm, if the measured phase shift is interpreted with the plane wave relationship Phase shift/diameter ($d_p = 0.658\,\Phi_{13}$). *For bubbles, the trajectory ambiguity <u>overestimates</u> the bubble diameter in contrast with the droplet case where it corresponds to an <u>underestimate</u> of the droplet diameter (Gréhan et al. (1994)).*

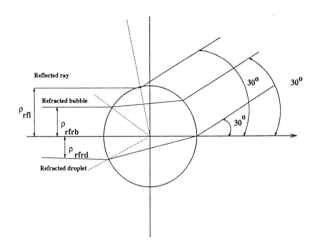

Figure 4: *Rays contributing to the scattering process at an off-axis of $30°$, for a water droplet in air or an air bubble in water*

Nevertheless, the trajectory effect is less important than for the droplet case. The difference can be explained as follows.
At $30°$, for a water droplet as well as for an air bubble the two main contributions are due to refraction and reflection.

For example for a 40 μm diameter particle, by reflection, the ray which is collected corresponds to an impact parameter ρ_{rfl} of $\approx$ 19.3 μm for both cases (droplet or bubble). But by refraction, the ray which is collected corresponds to an impact parameter ρ_{rfrb} of $\approx$ 10.4 μm for the bubble, and ρ_{rfrd} of $\approx$ -9.7 μm for the droplet (see figure 4). Then the modification of the relative importance of rays reflected and refracted involves rays separated from $\approx$ 9 μm for the bubble, but $\approx$ 29 μm for the droplet.

3.2   70° detection

Nevertheless, a 30° detection is really only an attractive solution when the optical access is limited, it is not the best choice for measuring bubbles. When possible a location of the detector unit at ≈ 70° is recommended.

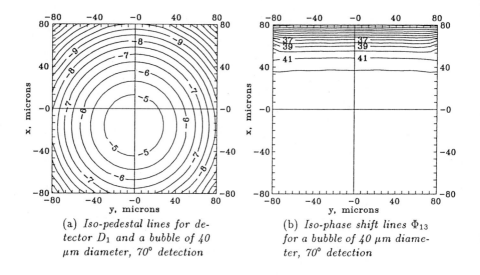

(a) *Iso-pedestal lines for detector D_1 and a bubble of 40 µm diameter, 70° detection*

(b) *Iso-phase shift lines Φ_{13} for a bubble of 40 µm diameter, 70° detection*

Figure 5: *70° Detection*

Figures 5-a and 5-b correspond to such a detection angle of 70°. From Figure 5-a the shift of the location of the maximum of the pedestal toward negative X is close to the one predicted for a 30° detection. Comparing Figs 3-d and 5-b, it is clear that the trajectory effect is much less severe for the 70° detection: contour lines are not present in the bottom of the figure due to the invariance of the phase shift. The phase shift predicted by LMT for a 40 µm bubble is of 42.79°. As the phase shift/particle diameter relationship is $d_p = 1.058\ \Phi_{13}$, the measured diameter of the 40 μm bubble varies from 42 μm to 45 μm across the measurement volume.

4 Modified standard Geometry

The modified standard geometry retains the same optical design that was previously computed, but the main particle trajectory is now parallel to the X-axis, and the signal processing is only carried out on the part of the signal corresponding to the maximum of the pedestal (Xu & Tropea (1994)).

Re-interpreting the series of Figures 2-a to 3-d, we conclude that when the mean particle trajectory is parallel to X, and the signal processing is limited to the maximum of the pedestal, the bubble diameter is measured with a good accuracy, including detection at an angle of 30°.

5 Dual burst technique

The aim of the dual burst technique is to measure the size, velocity, and refractive index of a particle by using only one detection unit and a continuous analysis of the phase shift evolution during each burst. The principle of the technique is fully described, in the case of refracting droplets, by Onofri et al.(1994). The aim of this section is to discuss the similarities and differences with the droplet case, in order to describe the applicability of the method to liquid flow with solid (refractive or strictly reflecting) particles and bubbles.

Fig. 6-a displays the evolution of the phase shift and pedestal for three bubbles of different size travelling parallel to OX. Taking into account a variation of the intensity limited to 10^4, the presence of a second burst is not clear, in contrast to the refracting droplet case.

Nevertheless, from Fig. 6, GLMT phase shift predictions for $X = -80\mu m$ are 46.6, 95.4 and 192.7 ° , and for $X = 0$, 34.3, 64.6 and 119.2 ° for the bubbles of 20, 40 and 80 μm, respectively. To compare, the phase shifts predicted by geometrical optics for such bubbles wit h this geometry are (i) by reflection 48, 98, and 195 °, (ii) by refraction 30, 60, and 117 ° for the 20, 40, and 80 μm bubbles, respectively.

Then, for example with the 80 μm bubble, the phase evolves between approximately the value predicted by geometrical optics for a reflective

300

scattering process to a refractive scattering process. The gradient of light on the particle is strong enough to shift the "refracted burst" and the "reflected burst" but not enough to separate them. This evolution is limited to the negative abscissa, for the positive abscissa another scattering process is prevailing, but not discussed here because the level of the corresponding pedestal is very low.

Fig. 6-b displays the same evolution as Fig. 6-a but for a perfectly reflecting particle (m=1.5-5i). The particle is so absorbing that no light can cross through it. Again, we limit our attention to the negative abscissa, as for Fig. 6-a. The main behaviour is that now the phase is exactly constant, corresponding to geometric optics predictions for a reflecting particle. Then, especially for big particles, the difference between bubbles and strictly reflecting particles can be extracted from the continuous evolution of the phase in the burst of bubbles.

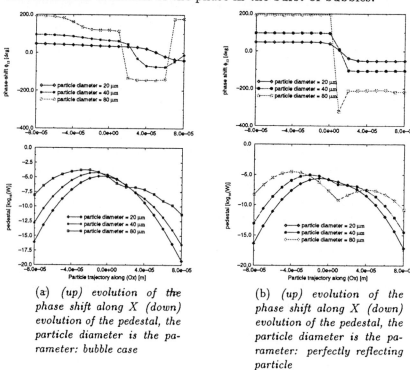

(a) *(up) evolution of the phase shift along X (down) evolution of the pedestal, the particle diameter is the parameter: bubble case*

(b) *(up) evolution of the phase shift along X (down) evolution of the pedestal, the particle diameter is the parameter: perfectly reflecting particle*

Figure 6: *Dual burst geometry, 2 $w_0 = 40\mu m$*

6 Conclusion

In the present paper, the influence of the trajectory effect on size measurements of bubbles by phase Doppler technique has been discussed for a classical geometry, a modified standard geometry, as well as for the Dual Burst technique. In the latter case, possibility of distinction between bubbles and perfectly reflecting particles has been dicussed. These results will help to understand the physical effects which are at the origin of the measurements of ghost bubbles in some experimental set-ups (Moechti, 1993).

7 Acknowledgments

The authors are grateful to the Commission of the European Communities, the French government (PROCOPE program), the German Ministry of Research and Technology, and the Deutsche Forschungsgesellschaft for providing financial support for this work.

8 References

Aizu Y., Durst F., Gréhan G., Onofri F. & Xu T.H. 1993, A PDA system without Gaussian beam defects, *3rd Int. Cong. on Optical Particle Sizing, Yokohama, Japan*, pp. 461-470.

Bachalo W.D. & Sankar S.V. 1988, Analysis of the light scattering interferometry for spheres larger than the light wavelength, *Proc. 4th Intl. Symposium on Applications of Laser Anemometry to Fluid Mechanics, Lisbon* paper 1.8.

Gouesbet G., Maheu B. & Gréhan G. 1988, Light scattering from a sphere arbitrarily located in a Gaussian beam, using a Bromwich formulation, *J.O.S.A. A*, vol. 5, pp. 1428-1443.

Gouesbet G., Gréhan G. & Maheu B. 1990, Localized interpretation to compute all the coefficients g_n^m in the generalized Lorenz-Mie theory, *J.O.S.A. A*, vol. 7, pp. 4003-4018.

Gouesbet G., Gréhan G. & Maheu B. 1991, Generalized Lorenz-Mie theory and applications to optical sizing, in *Combustion Measurements*, ed. Norman Chigier, pp.339-384, Hemisphere Publishing Corporation, New York.

Gouesbet G. & Lock J.A. A rigorous justification of the localized approximation to the beam shape coefficients in generalized Lorenz-Mie theory. II: Off-axis beams, *to be published by J.O.S.A. A,*

Gréhan G., Gouesbet G., Naqwi A. & Durst F. 1991, Evaluation of a phase Doppler system using generalized Lorenz-Mie theory, *Int. Conf. on Multiphase flows'91, Tsukuba, Japan,* pp. 291-296.

Gréhan G., Gouesbet G., Naqwi A. & Durst F. 1992, Trajectory ambiguities in phase Doppler systems: Use of polarizers and additional detectors to suppress the effect, *6th intl. symp. on applications of laser techniques to fluid mechanics, Lisbon,* paper 12-1

Gréhan G., Gouesbet G., Naqwi A. & Durst F. 1994, Trajectory ambiguities in phase Doppler systems: study of a near-forward and a near-backward geometry, *To be published by Part. Part. Syst. Charact.*

Lock J.A. & Gouesbet G. 1994, A rigorous justification of the localized approximation to the beam shape coefficients in generalized Lorenz-Mie theory. I: On-axis beams, *to be published by J.O.S.A. A;*

Moechti T.Y. 1993, Etude expérimentale des écoulements de Poiseuille à bulles plans, Thése, Grenoble le 20 Décembre 1993, Chapitre V.

Naqwi A. & Durst F. 1993, Analysis of laser light-scattering interferometric devices for in-line diagnostics of moving particles, Applied Optics, vol. 32, pp. 4003-4018.

Naqwi A., Ziema M. & Durst F. 1992, Fine particle sizing using an extended phase Doppler anemometer, *5th European Symp. Particle Characterization (PARTEC92), Nürnberg, Germany,* pp. 267-279.

Onofri F., Gréhan G., Gouesbet G., Xu T.H., Brenn G. & Tropea C. 1994, Phase-Doppler anemometer with dual-burst technique for particle refractive index measurements, *7th Intl. Symposium on Applications of Laser Anemometry to Fluid Mechanics, Lisbon*

Sankar S.V. & Bachalo W.D. 1991, Response characteristics of the phase Doppler particle analyzer for sizing spherical particles larger than the light wavelength. *Appl. Optics,* vol. 30, pp. 1487-1496.

Sankar S.V., Weber B.J. & Bachalo W.D. 1990, Sizing fine particles with the phase Doppler interferometric technique, *Proc. of the 2nd Int. congress on Optical particle sizing, Tempe, USA,* pp. 277-287.

Xu T.H. & Tropea C. 1994, Improving performance of two-component Phase Doppler anemometer, *To be published by Meas. Sci. Techn.*

Examination of the Rainbow Position of Optically Levitated Droplets for the Determination of Evaporation Rates of Droplets

N. Roth, K.Anders, A. Frohn

ITLR Institut für Thermodynamik der Luft- und Raumfahrt
Universität Stuttgart Pfaffenwaldring 31, D-70550 Stuttgart,
Germany

Abstract. In the present paper theoretical and experimental studies of the radiation pressure and of the intensity distribution in the rainbow region are presented. It has been shown that both radiation pressure and rainbow position show an oscillatory behaviour in their dependence on droplet radius. This property has been used to detect changes in radius of optically levitated droplets. From the theoretical calculations it follows that the period of these oscillations is practically independent of size and refractive index for a wide range of both quantities. Furthermore, it could be shown theoretically and experimentally that the oscillations of radiation pressure and rainbow position are strongly correlated and have the same period.

1 Introduction

In the past six years rainbow refractometry has become an important tool to determine refractive indices or temperatures of liquid droplets. The method at the first time described by Roth et al. (1988) has been treated in different additional papers by the same authors (Roth et al. 1989, 1990, 1991, 1992, 1993) and by other groups for instance by Massoli et al. (1993), Sankar et al. (1993) or Schneider et al. (1993). The knowledge of refractive indices is important for the study of basic phenomena such as droplet combustion or droplet evaporation as well as for the correction of droplet sizing methods. According to a first order approximation based on geometrical optics the rainbow angle is only a function of the wavelength λ and of the real part m of the refractive index of the liquid. At a closer look the problem becomes more complicated. For instance a size dependence of the rainbow position has to be taken into account for small droplets. Walker (1976) has given a correction factor to account for this effect. With this correction one obtains for instance for a water droplet with a radius of $10\,\mu$m an increase of approximately 4° in the rainbow angle compared to the approximation of geometrical optics. Internal refractive index gradients may influence the accuracy of the method as described by Schneider et al. (1993). On the other hand the influence caused by internal gradients may offer the opportunity to get information on these gradients according to Kai et al. (1993).

The determination of the rainbow position from a measured intensity distribution is not straight forward. The angular position of the highest intensity in the rainbow region θ_P as a function of droplet radius r shows an oscillatory behaviour leading to an uncertainty for the determination of the refractive index. Details on this problem have been described by Anders et al. (1993). The influence of these uncertainties can be decreased applying a correlation algorithm to the intensity distribution in the vicinity of the rainbow itself, thus taking into account the whole intensity distribution in this range rather than the peak value of the intensity. In this algorithm the correlation of the intensity distribution and one period of a $\sin^2$-function has been evaluated. The free parameter of the $\sin^2$-function has been choosen in such a way that the area between the function and the abscissa is the same as the area between the intensity distribution and the abscissa. The maximum of the correlation function has been regarded as the rainbow position θ_{RC} . The oscillations of the rainbow position are smoothed by this procedure but do not vanish completely. Depending on the value of the refractive index the oscillations in a θ_{RC}-r-plot vary between almost sinusoidal functions and functions with sharp peaks. The purpose of the present paper is to examine the relation between the lowest characteristic frequency of the θ_{RC}-r-relation and its dependence on size and refractive index. In addition the correlation between this frequency and the lowest frequencies of the oscillations of radiation pressure forces will be studied.

2 Theoretical Results

As a basis for the evaluation of the experiments theoretical calculations were performed. In the experiments described in the next chapter droplets are levitated and stabilized by radiation pressure forces of a vertical laser beam. The light of the laser beam is scattered by the droplet and can be used for measuring properties of the levitated droplet. In the present paper the interest is focussed on the angular position of the first rainbow and on the vertical position of the droplet relative to the focus of the levitating laser beam. The vertical position of the droplet along the z-axis is determined by the vertical component of the light pressure forces acting on the droplet. Therefore calculations to determine the rainbow position θ_{RC} and calculations of the radiation pressure cross-sections $C_{pr,z}$, which are proportional to the radiation pressure forces, have been performed.

2.1 Rainbow Position

To determine the position of the first rainbow the correlation algorithm described in the introduction has been applied to theoretical intensity distributions calculated for the rainbow region. For the calculations of the intensity distributions in the rainbow region a computer program based on Mie-theory

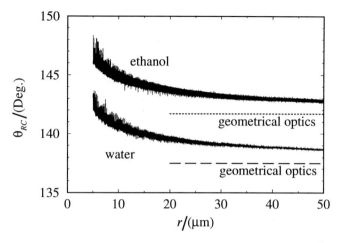

Fig. 2.1. Rainbow angle θ_{RC} as a function of droplet radius r. The increment between subsequent radii at which θ_{RC} has been calculated, was $0.01\,\mu$m. Results are shown for water and ethanol, which have the refractive indices $m = 1.33$ and $m = 1.36$. This and all following calculations were performed for the wavelength $\lambda = 514.5\,$nm.

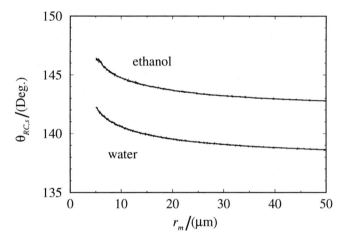

Fig. 2.2. Mean rainbow angle $\theta_{RC,s}$ as a function of mean droplet radius r_m. Results are shown for water and ethanol, which have the refractive indices $m = 1.33$ and $m = 1.36$. The results have been obtained averaging over a sample of 20 droplets in a narrow radius interval.

and proposed by Bohren and Huffman (1983) was used in a slightly modified version. These and all following calculations were performed for the wavelength $\lambda = 514.5\,$nm used in the experiments. In all calculations absorption in the droplet was neglected. Results for the rainbow angle θ_{RC} are shown in

Fig. 2.1 for two different liquids in a range of droplet radius from $r = 5\,\mu m$ to $r = 50\,\mu m$. With increasing droplet radius the rainbow angle θ_{RC} approaches the rainbow angle obtained from geometrical optics. The oscillations of the rainbow angle, which increase with decreasing droplet radius, result in uncertainties for the determination of the refractive index. It can be seen that the amplitudes of the oscillations for water are smaller than those obtained for ethanol. A method for smoothing the oscillations of the rainbow by averaging over a sample of droplets covering a small range of droplet radii has been described by Roth et al. (1992). A typical result of this averaging technique is shown in Fig. 2.2. In this case the intensity distributions of 20 droplets with different radii in the very narrow interval from $r = r_m - 0.1\,\mu m$ to $r = r_m + 0.1\,\mu m$ have been summed up. Here r_m is the mean radius of the interval. The correlation algorithm has been applied to the calculated mean intensity distribution. As a result the rainbow angle of the sample $\theta_{RC,s}$ has been obtained as a function of the mean radius in the interval. From such results the refractive index of the droplets can be determined with high accuracy.

2.2 Radiation Pressure Cross-Sections

For the calculations of radiation pressure cross-sections a computer code based on the generalized Lorenz-Mie theory (GLMT) was used (Gouesbet et al. 1988, Ren et al. 1994). In Fig. 2.3 the evolution of the radiation pressure

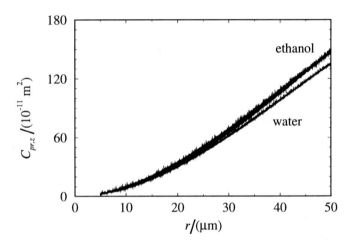

Fig. 2.3. Radiation pressure cross-sections $C_{pr,z}$ as a function of droplet radius r. The increment between subsequent radii at which $C_{pr,z}$ has been calculated, was $0.01\,\mu m$. Results are shown for water and ethanol, which have the refractive indices $m = 1.33$ and $m = 1.36$. Here and for all following calculations of radiation pressure cross-sections the waist radius of the laser beam was $w_0 = 100\,\mu m$ and the wavelength was $\lambda = 514.5\,nm$.

cross-sections as a function of the droplet radius is shown. Results are plotted for water and ethanol, assuming refractive indices $m = 1.33$ and $m = 1.36$. These and all following results have been obtained for the beam waist radius $w_0 = 100\,\mu$m. As can be seen from Fig. 2.3 the radiation pressure forces do not increase monotonically with increasing droplet radius, one observes an increasing behaviour with superimposed oscillations.

2.3 Comparison Between Rainbow Position and Radiation Pressure Cross-Section

For a comparison of the oscillations of the radiation pressure forces with the oscillations of the rainbow position θ_{RC} calculations with a high resolution in droplet radius were performed for both $C_{pr,z}$ and θ_{RC}. Thirteen different values of refractive indices have been chosen, which correspond to hydrocarbons from pentane to hexadecane and water. With this selection a wide range from $m = 1.333$ to $m = 1.4345$ is covered. In the following two figures the rainbow angle θ_{RC} and the radiation pressure cross-section $C_{pr,z}$ are shown as a function of the droplet radius r. Results for two different radius intervals are presented for water, hexane, and pentadecane. The rainbow angle θ_{RC} is plotted in the upper plot of each figure, whereas the radiation pressure cross-section is shown in the three lower plots. Figure 2.4 shows results for the radius range from $r = 5\,\mu$m to $r = 6.5\,\mu$m. In Fig. 2.5 results for the radius range from $r = 15\,\mu$m to $r = 16.5\,\mu$m are shown. The results of $\theta_{RC}(r)$ and $C_{pr,z}(r)$ should be compared for the same refractive index. For the low frequency oscillations of both the rainbow angle and the radiation pressure cross-sections a similar behaviour is found. The period of the lowest frequency of these oscillations -neglecting the sharp peaks- seems to be in the same order of magnitude for θ_{RC} and $C_{pr,z}$ and for different size ranges and refractive indices. A detailed examination of the period of these oscillations was made using a fast-Fourier transform algorithm (FFT). The FFT algorithm has been applied subsequently to packages of 1024 neighbouring points of the droplet radius for which θ_{RC} or $C_{pr,z}$ was calculated. After each FFT the window of 1024 neighbouring points has been shifted by one point. As a results of this procedure one obtains the period p_R for the oscillations of θ_{RC} and the period p_P for the oscillations of $C_{pr,z}$ as a function of the mean radius r_m associated to the 1024 points. Results of this procedure are shown in Fig. 2.6 and in Fig. 2.7. There the radius range from $r_m = 10\,\mu$m to $r_m = 11.5\,\mu$m has been evaluated. Again results for water, hexane, and pentadecane are shown. In Fig. 2.6 the period p_R of the oscillations of the rainbow angle and in Fig. 2.7 the period p_P of the oscillations of the radiation pressure cross-section is presented. As can be seen from these diagrams the values of period p_R and period p_P are practically the same for equal values of refractive index. Variations with the droplet radius are negligible. A small dependence of the periods p_R and p_P on the refractive index can be found. For the three size ranges from $r_m = 5\,\mu$m to $r_m = 6.5\,\mu$m, from $r_m = 10\,\mu$m

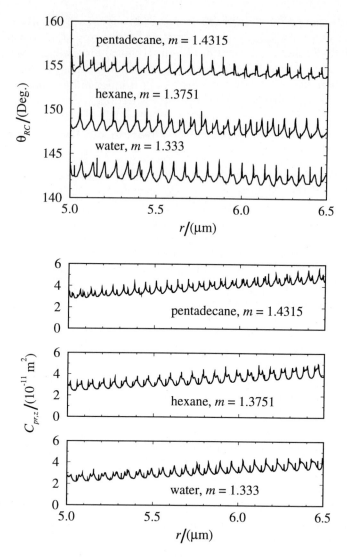

Fig. 2.4. In the upper diagram the rainbow angle θ_{RC} as a function of droplet radius r is shown. In the three lower diagrams the radiation pressure cross-sections $C_{pr,z}$ are presented in the same size range from $r = 5.0\,\mu$m to $r = 6.5\,\mu$m. Results are given for water, hexane, and pentadecane. The increment between subsequent radii at which the values have been calculated, was $0.001\,\mu$m.

to $r_m = 11.5\,\mu$m, and from $r_m = 15\,\mu$m to $r_m = 16.5\,\mu$m the FFT algorithm with the shift of the packages was applied for the data obtained for the refractive indices of the hydrocarbons from pentane to hexadecane and for the refractive index of water. For each size range and for each refractive

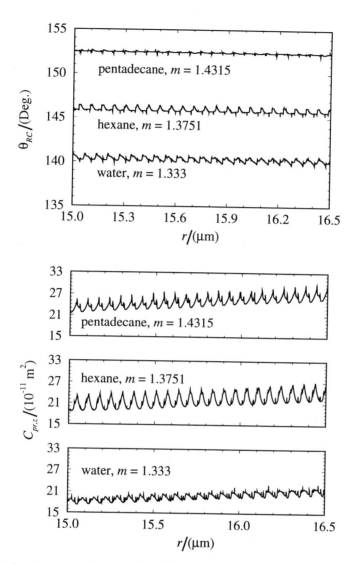

Fig. 2.5. In the upper diagram the rainbow angles θ_{RC} as a function of droplet radius r is shown. In the three lower diagrams the radiation pressure cross-sections $C_{pr,z}$ are presented for the same size range from $r = 15.0\,\mu m$ to $r = 16.5\,\mu m$. Results are given for water, hexane, and pentadecane. The increment between subsequent radii at which the values have been calculated, was $0.001\,\mu m$.

index the mean values $p_{R,av}$ and $p_{P,av}$ of the periods were calculated. The results are shown in Fig. 2.8 and Fig. 2.9 for the average periods $p_{R,av}$ of the oscillations of $\theta_{RC}(r_m)$ and for the average periods $p_{P,av}$ of the oscillations $C_{pr,z}(r_m)$ as a function of the refractive index. It can clearly be seen that the

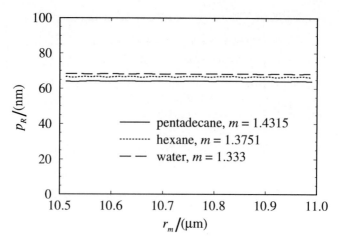

Fig. 2.6. Period p_R of the oscillations of the rainbow position θ_{RC} as a function of mean radius r_m. Result for water, hexane, and pentadecane are shown in the size range from $r = 10.0\,\mu$m to $r = 11.5\,\mu$m.

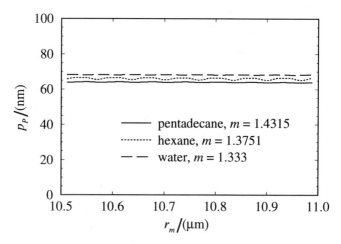

Fig. 2.7. Period p_P of the oscillations of the radiation pressure cross-sections $C_{pr,z}$ as a function of mean radius r_m. Result for water, hexane, and pentadecane are shown in the size range from $10.0\,\mu$m to $r = 11.5\,\mu$m.

average periods $p_{R,av}$ and $p_{P,av}$ are practically independent of the droplet radius within the studied size ranges. The values of $p_{R,av}$ and $p_{P,av}$ change very weakly with the refractive index from approximately 68 nm for water with $m = 1.333$ to approximately 64 nm for hexadecane with $m = 1.4345$. The transition from the higher to the lower value becomes more evident in the enlarged view shown in Fig. 2.10. Here the values of the periods $p_{R,av}$

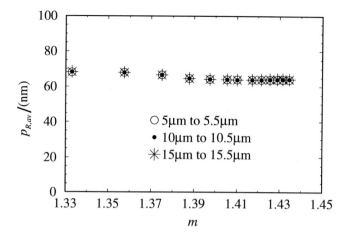

Fig. 2.8. Averaged period $p_{R,av}$ of the oscillations of the rainbow angle θ_{RC} for the thirteen different refractive indices representing water and the hydrocarbons from pentane to hexadecane. Results are shown for three different size ranges. The averages have been taken over each individual radius interval.

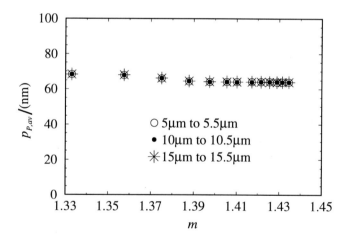

Fig. 2.9. Averaged period $p_{P,av}$ of the oscillations of the radiation pressure cross-sections $C_{pr,z}$ for the thirteen different refractive indices representing water and the hydrocarbons from pentane to hexadecane. Results are shown for three different size ranges. The averages have been taken over each individual radius interval.

and $p_{P,av}$ are compared for one radius interval. As can be seen in Fig. 2.10 for refractive indices below $m = 1.36$ a constant value close to 68 nm has been found. For further increasing refractive indices following a transition a constant value close to 64 nm for refractive indices above 1.39 is obtained.

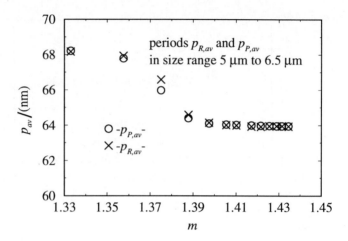

Fig. 2.10. Averaged period $p_{P,av}$ of the oscillations of the radiation pressure cross-sections $C_{pr,z}$ in comparison with averaged period $p_{R,av}$ of the oscillations of the rainbow angle θ_{RC} for the thirteen different refractive indices representing water and the hydrocarbons from pentane to hexadecane. Results are shown for one size range in an enlarged view.

Furthermore, it should be emphasized that the values of p are identical for both rainbow positions and radiation pressures. In an experiment with evaporation or condensating droplets both the oscillations of the rainbow or the oscillations of the radiation pressure can be associated directly with the rate of change of droplet radius $dr/d\tau$ for a range in which $p(m) = const.$ In this case one has

$$\left| \frac{dr}{d\tau} \right| = p_i f_i \quad , \quad i = R, P \tag{1}$$

where f_P and f_R are the frequency of the oscillations of radiation pressure or rainbow position respectively. To obtain these values the temporal evolution of one of these values has to be recorded in an appropriate manner. All hydrocarbons from octane to hexadecane have refractive indices above 1.39. This gives the basis to examine the evaporation of binary mixtures of these hydrocarbons as p_R and p_P in this range is independent of size and composition. This theoretical study and the studies of Anders et al. (1993) and Roth et al. (1992) show that it is possible to measure the refractive index, the droplet size and the evaporation rate in observing and evaluating the scattered light in the region of the first rainbow. This study is the theoretical basis for experimental investigations of evaporation rates of optically levitated droplet. The experimental setup and some experimental results are presented in the following paragraph.

3 Experiments

Optical levitation is an appropriate tool to study individual particles. In the work of this paper liquid droplets were levitated optically and stabilized in a vertical laser beam. The technique of optical levitation has been as described in previous papers for instance by Ashkin and Dziedzic (1971). The droplet is observed in a chamber, whose temperature can be adjusted in an appropriate range. Depending on temperature and saturation in the chamber the droplet will evaporate or grow by condensation. The droplet is illuminated by the levitating laser beam of an Ar^+-laser with a wavelength of 514.5 nm.

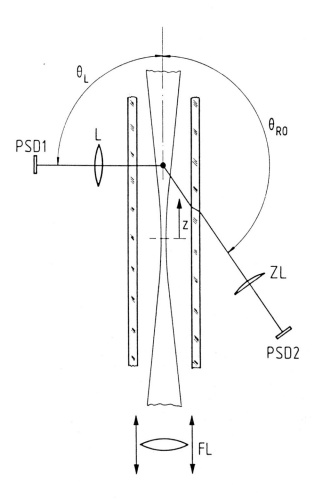

Fig. 3.1. Schematical view of the experimental setup. The symbols are explained in the text.

This allows to apply optical non-intrusive measuring techniques to study the droplet behaviour during the evaporation or condensation process.

The optical setup used in the experiments of the present paper is shown in Fig. 3.1. The rate of radius change of a droplet corresponding to the evaporation rate or condensation rate can be determined in evaluating the temporal evolution of the droplet position relative to the focus of the levitating laser beam. This is due to the fact that a droplet with decreasing or increasing radius performs an ocsillatory motion along the axis of the laser beam due to the oscillatory dependence of the light pressure forces on the droplet radius as described in the previous chapter. Light scattered at an angle of 90° is focussed by lens L on the surface of the position sensing detector PSD1. The output signal of this PSD is proportional to the first moment of the light distribution on the sensor surface and is therefore proportional to the position of the light spot on the sensor, which of course corresponds to the position of the droplet. The optical setup is designed to allow the determination of even small changes of the droplet position. An evaporating droplet levitated in a laser beam performs oscillatory motions. These oscillations are detected with PSD1 and recorded by a transient recorder. An example of a record from an evaporating pentadecane droplet is shown in Fig. 3.2. There the output signal of PSD1 is plotted versus time. The calculations described in the previous chapter showed that to some extend the period length of the oscillations of the radiation pressure as a function of radius are independent of the droplet radius and the refractive index. Therefore the change of droplet radius with time $dr/d\tau$ of a levitated droplet can be determined by evaluating a temporal evolution of the droplet position. In order to determine the real part of the refractive index the angular position of the first rainbow has to be recorded as described for example by Roth et al. (1991,1993). In the experiments of the present paper the optical arrangement was designed that the rainbow position recorded is not influenced by changes of the positon of the droplet in the laser beam. The light scattered in the rainbow region was imaged by the cylindrical lens ZL on the surface of the position sensing detector PSD2 positioned in the focal plane of the lens. In this case the cylindrical lens performs an optical Fourier transform of the scattered light. The application of a PSD to measure rainbow positions was described by Roth et al. (1993) for monodisperse droplet streams. In the present paper the temporal evolution of the rainbow position for an individual droplet has been examined in detail. The output signal of PSD2, which detects the rainbow position is shown in Fig. 3.3. This signal has been obtained for the same pentadecane droplet whose position in the laser beam is shown in Fig. 3.2.

Comparing Fig. 3.2 and Fig. 3.3 one can see that the rainbow position shows peak values occuring with the same frequency as the oscillations of the droplet position, the characteristic of the signal however is quiet different at this particular refractive index. Theoretical calculations with high resolution in radius of the rainbow position applying the correlation algorithm show the

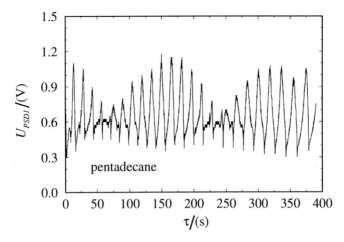

Fig. 3.2. Output signal U_{PSD1} of the PSD1 as a function of time τ, which represents the vertical droplet position.

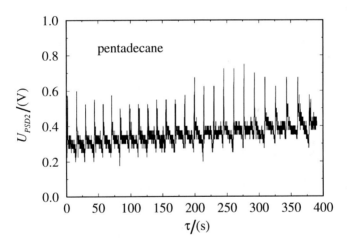

Fig. 3.3. Output signal U_{PSD2} of the PSD2 as a function of time τ, which represents the position of the first rainbow.

same behaviour. This can be clearly seen by comparing Fig. 3.3 with Fig. 2.4. By carefully examining the slope between the peaks one can immediately recognize that the droplet is evaporating rather than growing in this case. In order to get information on the temporal change of droplet radius $dr/d\tau$, which corresponds to the evaporation or condensation rate of the droplet, frequencies of the oscillatory motion of the droplet or of the oscillations of the rainbow position have to be determined. This can be done by applying the FFT-procedure described in chapter 2.3 to the experimental results shown in

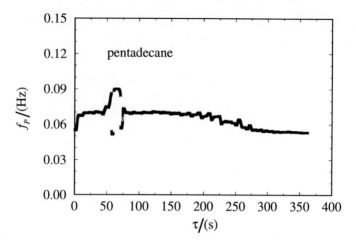

Fig. 3.4. Temporal evolution of the frequncy f_P of the oscillations of the droplet position. This result has been obtained from the data shown in Fig. 3.2.

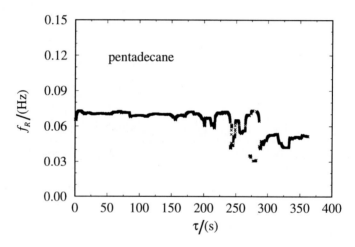

Fig. 3.5. Temporal evolution of the frequncy f_R of the oscillations of the angular position of the first rainbow. This result has been obtained from the data shown in Fig. 3.3.

Fig. 3.2 and in Fig. 3.3. Results of the FFT procedure are shown in Fig. 3.4 and in Fig. 3.5. As expected the frequencies f_P and their temporal evolution obtained for the oscillatory motion shown in Fig. 3.4 are the same as those obtained for the rainbow positions denoted by f_R and shown in Fig. 3.4. Only minor deviations can be seen. These can be explained by poor signal quality, for example low amplitude, of the input signal. Applying Eq. (1) the rate $dr/d\tau$ as a function of time can be determined immediately. A slight decrease of the frequencies and therefore for the rate of radius change with increasing time can be observed in both figures. This can be explained by an increasing vapour pressure of pentadecane in the observation chamber in this particular experiment.

4 Conclusions

In the present paper it has been shown that the rainbow signal can be used to examine the rate of change of droplet radius, corresponding to the evaporation or condensation rate. It has been shown that combining this technique with techniques described in previous papers allows to determine size, refractive index, and evaporation or condensation rate from the intensity distribution in the rainbow region. This gives the possibility to study droplets consisting of binary mixtures. The transient behaviour of the evaporation process of mixtures can be examined by recording the angular rainbow position, which is a measure for the refractive index and therefore for the temperature and composition, and by recording the periodic oscillations, which are a measure for the change of radius of the droplet. Furthermore, it has been shown theoretically and experimentally that there is a direct relation between fluctuations of the radiation pressure and fluctuations in the rainbow position and that the period of these fluctuations is in a wide range practically independent of droplet size and refractive index.

Acknowledgements

We gratefully acknowledge the assistance of Dr. G. Gréhan of INSA Rouen/ France, who has provided the computer program to calculate radiation pressure cross-sections by GLMT.

References

Anders K., Roth N., Frohn A. 1993. Light scattering at the rainbow angle: Information on size and refractive index. Proc. 3rd Int. Congr. on Optical Particle Sizing, Yokokama (Japan), 237-242.

Ashkin A., Dziedzic J.M. 1971. Optical levitation by radiation pressure. Appl. Phys. Lett., **19**, 283-285

Bohren C.F., Huffman D.R. 1983. Absorption and scattering of light by small particles. John Wiley & Sons, New York.

Gouesbet G., Maheu B., Gréhan G. 1988. Light scattering from a sphere arbitrarily located in a Gaussian beam, using Bromwich formulation. J.Opt. Soc. Am. A **5**, 1427-1443.

Kai L., Massoli P., D'Alessio A. 1993. Studying inhomogeneities of spherical particles by light scattering. Proc. 3rd Int. Congr. on Optical Particle Sizing, Yokokama (Japan), 135-143.

Massoli P., Beretta F., D'Alessio A., Lazzaro M. 1993. Temperature and size of single transparent droplets by light scattering in the forward and rainbow regions. Appl. Opt., **30**, 3295-3301.

Ren K.F., Gréhan G., Gouesbet G. 1994. Radiation pressure forces exerted on a particle arbitraily located in a Gaussian beam by using the generalized Lorenz-Mie theory, and associated resonance effects. (accepted Opt. Commun.)

Roth N., Anders K., Frohn A. 1988. Simultaneous measurement of temperature and size of droplets in the micrometer range. Proc. 7. Int. Con. on Optical Methods in Flow and Particle Diagnostics ICALEO 88, L.I.A. **67**, 294-304.

Roth N., Anders K., Frohn A. 1989. Temporal evolution of size and temperature measurements of burning ethanol droplets for different initial temperatures. Proc. Joint Meeting of the German and Italian Sections of the Combustion Institute, Ravello (Italy), 2.3.

Roth N., Anders K., Frohn A. 1990. Simultaneous measurements of temperature and size of droplets in the micrometer range. J. Laser Applications, **2**, No.1, 37-42.

Roth N., Anders K., Frohn A. 1991. Refractive-index measurements for the correction of particle sizing methods. Appl. Opt., **30**, No. 33, 4960-4965.

Roth N., Anders K., Frohn A., 1992. Simultaneous determination of refractive index and droplet size using Mie-theory. Proc. 6th Int. Symp. on Application of Laser Techniques to Fluid Mechanics, Lisbon, 15.5.1

Roth N., Anders K., Frohn A., 1993. Measurement of the rainbow positon using a PSD-Sensor. Proc. 3rd Int. Congr. on Optical Particle Sizing, Yokohama (Japan), 183-187.

Roth N. Anders K., Frohn A.,1994. Determination of size, evaporation rate, and freezing of water droplets using light scattering and radiation pressure. Part. Part. Syst. Charact., **11**, 207-211.

Sankar S.V., Ibrahim K.M., Buermann D.H., Fidrich M.J., Bachalo W.D. 1993. An integrated phase Doppler/rainbow refractometer system for simultaneous measurements of droplet size, velocity and refractive index. Proc. 3rd Int. Congr. on Optical Particle Sizing, Yokokama (Japan), 275-284.

Schneider M., Hirleman E.D., Saleheen H., Chowdhary D.Q., Hill S.C. 1993. Rainbows and radially-inhomogeneous droplets. Proc. 3rd Int. Congr. on Optical Particle Sizing, Yokohama (Japan), 323-326.

Walker J.D. 1976. Multiple rainbows from single drops of water and other liquids. Am. J. Phys., **44**, 421-433.

A New Technique to Measure Refractive Index with Phase Doppler Anemometry

H.H. Benzon[1] , T. Nonn[2] & P.Buchhave[1]

[1] The Technical University of Denmark, Physics Dept.
Lyngby, Denmark

[2] Dantec Measurement Technology
16-18 Tonsbakken, Skovlunde 2740, Denmark.

Abstract. A standard phase-Doppler system was modified to determine the feasibility of refractive index measurements utilizing new optical techniques. With the use of two laser beam systems of different wavelength and a single receiving unit different optical configurations can be explored by varying the scattering, elevation and crossing angles independently for each system. The index of refraction of a droplet can be extracted from the phase-shift ratio of two distinct pairs of apertures. Selecting the proper optical parameters yields a sensitive relationship between phase ratio and refractive index.

For measurements an LDA configuration with two transmitting optics was chosen because of its widespread use and standard components. The two probes were situated 40° relative to each other. This was done to maximize the relationship between the ratio of phase shifts and the refractive index. Tests were conducted on water-glucose mixtures with varying refractive index.

Keywords. Particle sizing, phase-Doppler Anemometry, refractive index.

1 Introduction

The fundamental concept behind Phase-Doppler Anemometry (PDA) is that for carefully selected scattering angles a linear relationship can be derived between the phase shift of a particle measured as it passes through an LDA measurement volume and its size. This relationship holds for so long as certain restrictions are enforced (i.e., size range of particles relative to LDA probe volume, dominant scattering mode, polarization) and that the physical properties of the system should remain constant (i.e., refractive index, wavelength of light). In multiphase flows and in sprays where varying temperature environments are experienced, as in combustion, the refractive index of droplets can clearly vary. This variation within the normal PDA operation can add bias to measurements. Several methods have been developed to extend the operation of PDA to include the measurement of refractive index with some success. A two-receiver configuration explored by Pitcher *et al.* [1] and Naqwi *et al.* [2] allude to this. The strength of their method lies in that the dependence of the phase shift on refractive index varies more in the near forward off-axis regions than in the far off-axis regions. Alternatively, Sankar *et al.* [3] utilized the clear dependence of the rainbow angle with refractive index in another instrument configuration. This technique, however, is complex and as indicated by the authors the system is sensitive to high frequency oscillations.

An appealing side to refractive index measurement is the ability to confirm different particle types within the same particle distribution. A persistent problem is the existence of multiple modes of light scattering in sprays. Droplets can reflect as well as refract and therefore present a problem when calculating spray statistics. In the standard PDA implementation there is a region of overlap as indicated in Nonn *et al.* [4] between reflecting and refracting particles that lead to ambiguity of some results. Having additional optical information can perhaps help resolve this issue.

2 Theory

The implementation of [1] utilizes the fact that the Doppler-phase is a linear function of diameter and that the refractive index is contained within the proportionality constant for first and higher orders of refraction. This means that the ratio of two phases will be independent of diameter and only dependent on the refractive index of the particle and the parameters that describe the optical configuration. The measurement was accomplished by using two receivers positioned at two different scattering angles. The angles were chosen so that refraction dominated at both angles.

This system is equivelant to one where there are two laser beam systems, for example green and blue, and a single receiver. The light received from this system can then be subsequently separated and analyzed. The advantage of this system is that the three apertures of the receiver can be used for both laser systems. In Figure 1, the reference coordinate system for the green beam system is shown. If we superimpose the blue system and rotate it about the x-axis we get a system that is equivalent to the system suggested by [1]. The advantage of this method is that one can change the elevation angle and scattering of one beam system independently. The purpose in all this is to maximize the change in the phase ratio.

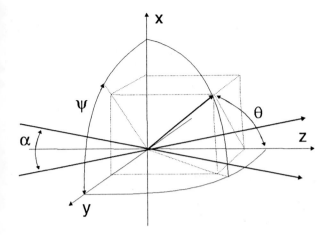

Figure 1. Definition of beam crossing angle α, scattering angle θ and elevation angle ψ.

Rotating the blue system around the z-axis in the reference system has the effect of placing the receiving unit at a different elevation angle. The standard PDA system contains three receiving apertures. In the new system there are three pairs of angularly placed apertures. The index of refraction can be determined from the ratio between the same two aperture pairs for the different beam systems. Having additional apertures also means having more calibration curves for the determination of refractive index, hence data can be verified with different combinations.

The following section will describe the selection of appropriate values for parameters to maximize measurement of the refractive index. To produce a sensitive calibration curve for the refractive index the phase shifts of one of the systems must change more rapidly. The correct choice for optical parameters can be seen in the Table 1. Of the parameters listed the elevation angle is the most

	Rapidly varying Doppler phase	Slowly varying Doppler phase
Wavelength λ	small (blue system)	large (green system)
Scattering angle θ	small	large
Crossing angle α	large	small
Elevation angle ψ	large	small

Table 1. Optimum selection of optical parameters

sensitive with respect to the phase of the Doppler signal and the wavelength the least.The overall dynamic range and resolution can be improved by selecting the appropriate calibration curve from the two beam system.

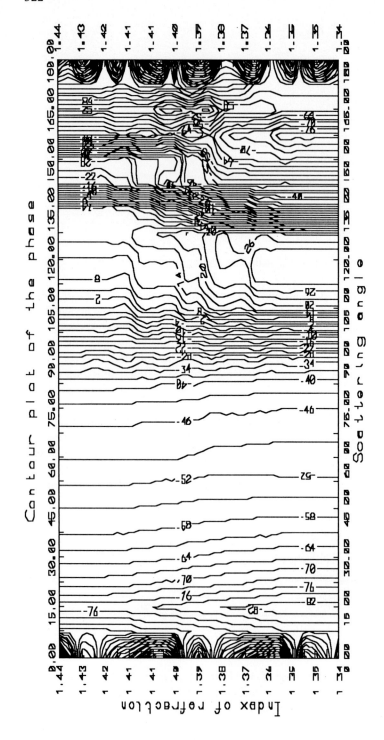

Figure 2. Lorenz-Mie based iso-phase plot as a function of scattering angle and index of refraction.

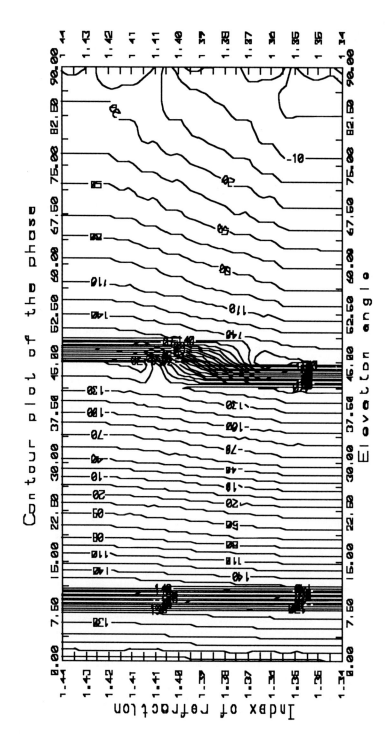

Figure 3. Lorenz-Mie iso-phase plot as a function of elevation angle and refractive index. Scattering angle 30°.

One gets an idea how the Doppler-phase varies with elevation angle, scattering angle and the refractive index from the iso-plots in Figures 2 and 3. The plots are calculated for a standard PDA system with a crossing angle (α, Fig. 1) of 12.52°, parallel polarization and particle size of 12 μm using Lorenz-Mie theory. The Doppler signal is integrated over the PDA aperture. The scattering angle in the elevation plot is 30°.

From Figure 2 it can be seen that the Doppler phase exhibits more complicated behavior for the back scatter region, as is expected. One could imagine using the second-order refraction to measure in regions where the index of refraction varies little. In the near forward region the Doppler-phase varies rapidly, reaching a maximum change at 20°. As we increase scattering angle the sensitivities decreases. Around a scattering angle of 70° refraction stills dominates but the sensitivity is small. It can be also be inferred from the plots that the Doppler -phase is more sensitive to changes of elevation angle then scattering angle.

3 Experimental Layout

As indicated earlier a standard Dantec LDA system was used as the test bed for refractive index measurement. To allow for these measurements the standard 57X10 PDA Receiving Optics was modified with three 55X35 color separators. The five photomultipliers were connected to the color separators and routed to a modified Dantec 58N10 PDA signal processor.

The optical layout of the present experiment consisted of the two FiberFlow probes (green and blue) situated at two scattering angles to the receiver, as in Figure 4. The green and blue system were placed 40° with respect to each other, thus making the effective scattering angle with the receiver for the blue and green systems 30° and 70°, respectively. The focal lengths for the transmitting probes were 400 mm, with a beam separation 40 mm, creating a probe volume of diameter 260 μm. The receiver used had a focal length of 310 mm.

In the standard PDA arrangement three phase detectors are used to measure the phase differences between three apertures. An additional phase detector was employed to measure the phase shift from one pair of the apertures of the blue beam system. More detectors could be employed, up to six, to make a more robust validation. An advantage of using two colors is that the intensity of the beams can be varied independently relative to the receiver. This is useful since the intensity can vary considerably as one goes from the near forward to side scatter.

To ensure proper measurements coincidence between the two pairs of beams is paramount. However, since this configuration also allows for 3-D velocity measurements, this requirement would anyway have to be fulfilled. Alignment was accomplished by first using a 100 μm pinhole, optimizing, and then repeating the procedure with a 50 μm pinhole. The pinholes by themselves were not sufficient to guarantee coincidence, therefore, to maximize overlap of the probe volumes a light sensing diode was placed behind the 50 μm pinhole and the beams manipulated until the maximum beam power was attained at the crossing. The slightest deviation in the alignment had the effect of making the refractive index measurements unusable. In order to get a valid measurement the phases measured from the separate beam systems must correspond to the same particle. Thus, this method makes an excellent test of coincidence for 3D LDA systems.

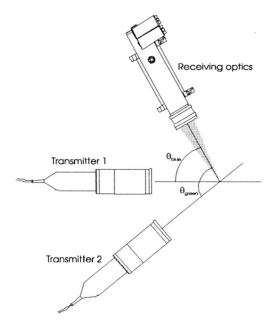

Figure 4. Optical layout with two transmitters situated at 30° and 70° with respect to the receiver.

The test particles were spray droplets made from solutions of water with varying amounts of dissolved glucose. By adding incremental amounts by weight of solid glucose into distilled water one can change the refractive index. The uncertainty in the measurement of glucose dissolved was limited to within 0.01 %.

Figure 5 illustrates the system calibration curve for a 11.9 μm and 12.5 μm particle. The curve is based on the U1-2 phase shift ratio for the green and blue systems. It must be emphasized that the Lorenz-Mie calculations for refractive index are size dependent, that is, each particle size has its own characteristic oscillation. For small particles the oscillations are more pronounced than for larger particles. It was stressed in [1,2] that such oscillatory behavior in small particles would make refractive index measurement difficult if not impossible. A very small spread of particle size will effectively randomize the phase oscillation

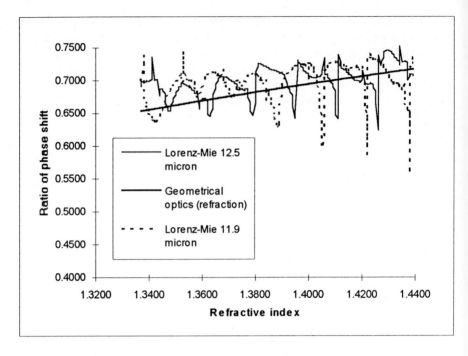

Figure 5. Comparison of Lorenz-Mie and refraction calibration curve for different refractive indexes. Particle diameters were 11.9 µm and 12.5 µm.

and average the index measurement. The result will then coincide with the geometric optics calibration curve. To test the robustness of this setup a spray was measured having an effective size distribution of 2-30 µm. It is thought, that a distribution of particles over a range of sizes would tend to average out the effect of the oscillations.

4 Measurements

Measurements were made using water droplets issued from an airbrush mixed with glucose in controlled proportions to vary the refractive index. At 20 °C water exhibits a refractive index of 1.334. Adding by percent weight glucose from 0.5% up to 60 % increases the refractive index from 1.337 to 1.439 [5]. Figure 6 displays a typical run with the characteristic peak indicating the dominant phase ratio, corresponding to the refractive index measured.

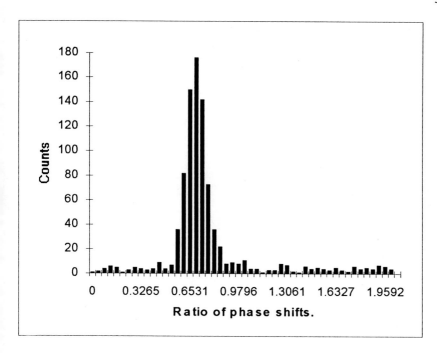

Figure 6. Histogram of phase shift ratio using glucose 40 % by weight.

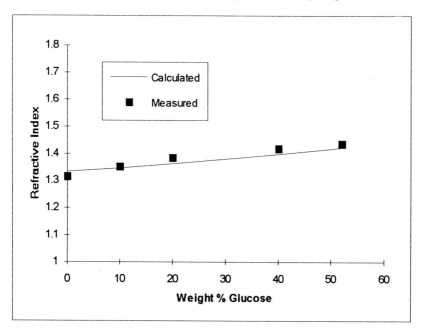

Figure 7. Measured and estimated refractive index as a function of % weight glucose.

The accuracy of the measurements over the range of refractive indices tested are listed in table 2 and displayed in Figure 7. The maximum error in the measurement of refractive index was under 2 %. Considering the difficulties implied by the scattering relations and alignment of the system this is a positive result. Though it was not attempted in these tests, an improvement to the sensitivity could be accomplished by choosing an alternate phase ratio; i.e., the ratio between the U1-3 and U1-2 for the two systems. Alternatively, the elevation angle and/or the crossing angle in one of the two systems can be changed.

% weight glucose	Actual index	Measured index	Deviation (%)
0	1.334	1.3167	1.2964
10	1.3477	1.3530	0.395
20	1.3635	1.3876	1.7683
40	1.3986	1.4222	1.687
52	1.4222	1.4394	1.216

Table 2. Comparison between measured and actual values of refractive index.

5 Conclusion

In this paper we have described a new method for measuring the relative refractive index. This technique has several advantages in that there are many possible permutations of the optical configuration. That is, the elevation angle, scattering angle and crossing angles can be selected independently for each beam system. This means that the phase ratio can be made more sensitive to the index of refraction.

The implementation in this paper shows that a standard PDA system can be easily extended to include the measurement of refractive index. Measurements were made with sprays consisting of droplets of substances with varying index of refraction. The accuracy on average was within 2 % of the estimated value determined from geometrical optics. Furthermore, the refractive ratios that are measured are excellent indicators of coincidence between the beam systems.

References

1. Saffman M., G.Pitcher and G.Wigley. Sensitivity of Dropsize Measurements by Phase Doppler Anemometry to Refractive Index Changes in Combusting Fuel Sprays. In Proceedings of The International Symposium on applications of Laser Anemometry to Fluid Mechanics, Lisbon, Portugal, 1990.
2. Naqwi A., F. Durst and X.-Z.Lui. An Extended Phase-Doppler System for Characterization of Multiphase Flows. In Proceedings of The International Symposium on Applications of Laser Anemometry to Fluid Mechanics, Lisbon, Portugal, 1990.

3. Sankar S., K.M.Ibrahim, D.H:Buermann, M.J.Fidrich, and W.D. Bachalo. An Integrated Phase Doppler/Rainbow Refractometer System for Simultaneous Measurement of Droplet Size, Velocity, and Refractive Index. In Proceedings of 3rd International Congress on Optical Particle Sizing, Yokohama, Japan, 1993.
4. Thomas Nonn and Zeev Dagan. Analysis of Particle Mixtures Exhibiting Different Scattering Modes by Phase Discrimination. In Proceedings of 3rd International Congress on Optical Particle Sizing, Yokohama, Japan, 1993.
5. Weast. Handbook of Chemistry and Physics. 53rd edition, pg. D-211, CRC Press.

Simultaneous Determination of Temperature and Size of Droplets from the Rainbow using Airy Theory

J.P.A.J. van Beeck and M.L. Riethmuller

von Karman Institute for Fluid Dynamics, Chaussée de Waterloo 72, B-1640 Rhode-Saint-Genèse, Belgium, Fax: +32 2 3582885 E-mail: vanbeeck@vki.ac.be

Abstract. The anisotropy of laser light scattered by a droplet exhibits a strong dependence on droplet size and temperature. Around the geometrical rainbow angle this dependence is such that it can be used to determine these two parameters. This determination is done with the help of the Airy theory for the rainbow to avoid an extensive calibration of the angular scattered light intensity. A comparison between the Airy and the Mie theories shows that one has to be careful in applying the Airy theory for this purpose. The Airy theory has also been compared with the experimental first order "monochromatic rainbow" created by single falling droplets crossing a laser beam. There is a qualitative good agreement between experiment and theory after the measured rainbow pattern has been properly smoothed.

Keywords. Droplet size, droplet temperature, rainbow, Airy theory for the rainbow, Mie theory

1   Introduction

Measurements of droplet velocity and droplet size are nowadays widely performed by optical techniques such as the Laser Doppler Velocimetry and the Phase Doppler Particle Analysis. Similar well-established non-intrusive techniques do not yet exist for the droplet temperature. Roth et a. (1990) proposed a laser technique which, as for the aforementioned successful techniques, is based on the detection of the scattered laser light. The scattering angle, at which the first order "monochromatic rainbow" is seen, depends on the droplet temperature and the droplet size. The droplet size has been determined from the spatial intensity distribution of the light scattered in the forward hemisphere (see König et a. (1986)). Once this quantity is known, the temperature can be deduced from the position of the rainbow. In this way two CCD-cameras have to be used, one in the forward and one in the backward hemisphere. The present paper will show that it is possible to deduce the droplet size from the

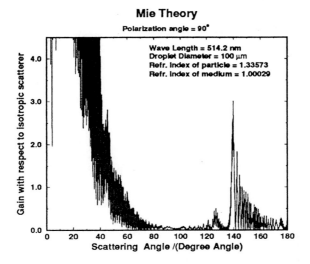

Fig. 1: Gain factor with respect to an isotropic scatterer as a function of the scattering angle according to the Mie theory. The droplet diameter is $100\,\mu m$, the refractive index of the particle is 1.33573 and of the medium is 1.00029.

interference pattern that can be seen near the geometrical rainbow angle, thus requiring only one CCD-camera. Therefore, the Airy theory for the rainbow is applied. The applicability of this theory for this new method will be shown by comparing it with the "correct" Mie theory. Finally so-called rainbow patterns, coming from single falling droplets, have been recorded on the CCD-camera and will be presented.

2   The rainbow as non-intrusive measurement technique

One could state that by using the rainbow for technical purposes this beautiful phenomenon is deromanticized. However, making research on the rainbow shows that also scientifically seen it is a beauty. Just look to figure 1. Herein is depicted the gain factor with respect to an isotropic scatterer as a function of the scattering angle for a droplet being hit by a planar monochromatic electromagnetic wave front. The solution for this problem is given by the Mie theory (see for instance van de Hulst (1957) or Bohren and Huffman (1983) for elaborate descriptions) and is presented here for the particular case of a droplet diameter of $100\,\mu m$, a wave length of the incident light of $514.2\,nm$ and a real refractive index of the particle of 1.33573. The refractive index of the medium is set to 1.00029 and the polarization is perpendicular to the scattering plane.

Near a scattering angle of 140° an interference pattern can be seen that resembles the first order rainbow. Figure 2 shows a close-up of this rainbow

332

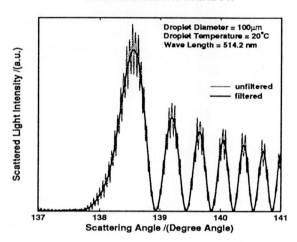

MONOCHROMATIC RAINBOW

Droplet Diameter = 100μm
Droplet Temperature = 20°C
Wave Length = 514.2 nm

—— unfiltered
—— filtered

Scattered Light Intensity /(a.u.)

137   138   139   140   141
Scattering Angle /(Degree Angle)

Fig. 2: Filtered and unfiltered first order "monochromatic rainbow".

pattern for a droplet diameter of $1\,mm$. The pattern in figure 2 for which the ripple structure (discussed by van Beeck and Riethmuller (Poitiers 1994)) is filtered out, is applied for the optical technique described below.

Before explaining this technique it is useful to look to the results of the Airy theory, which is able to predict the filtered interference pattern of figure 2 as well. The advantage of the Airy theory is its simplicity compared to the very complex and computationally expensive Mie theory.

The Airy theory is an extension of geometrical optics based on the Huygens' principle (see Hecht (1987)). Concerning the first order rainbow, the theory deals with rays that have suffered one internal reflection in the drop as

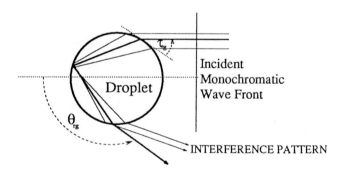

Fig. 3: The geometrical optical rays that form the first order rainbow.

Airy Theory

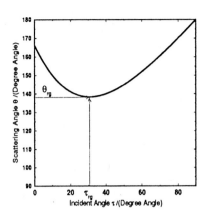

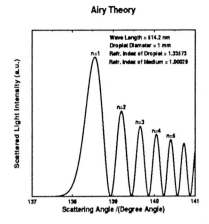

Fig. 4: Scattering angle θ versus incident angle τ for a geometrical ray suffered one internal reflection in a droplet. (Particle and medium refractive index are 1.33573 and 1.00029, respectively)

Fig. 5: The rainbow pattern according to the Airy theory.

is sketched in figure 3. The relationship between the scattering angle θ and the incidence angle τ is given by geometrical optics and is presented in figure 4. The geometrical rainbow angle θ_{rg} is situated at the minimum of the curve. This θ_{rg} depends only on the refractive index of the liquid. The high intensity of the rainbow is due to the fact that a large part of the incident light emerges at a small range of scattering angles θ.

Until so far geometrical optics. The subsidiary maxima are explained as follows. Near τ_{rg}, which is the incident angle of the ray that leaves the droplet at θ_{rg}, there exist rays with different τ's that leave the droplet at the same θ (see sketch in figure 3). These rays are parallel and therefore interfere at infinity, yielding a minimum or maximum in intensity according to their phase shift. The Airy theory computes this phase shift by assuming a cubic wave front near the rainbow (see for instance van de Hulst (1957) about this); thus a pattern as in figure 5 is obtained. It has to be noted that also for the Airy theory the polarization is chosen perpendicular to the scattering plane. For the polarization parallel to the scattering plane the rainbow is hardly visible because the internal angle corresponding (by means of Snell's law) to τ_{rg} is close to the Brewster angle. This is why the polarization perpendicular to the scattering plane is chosen for the optical technique discussed here.

The shape of the Airy's rainbow pattern does not only depend on the refractive index of the liquid but also on the droplet size and the wave length

334

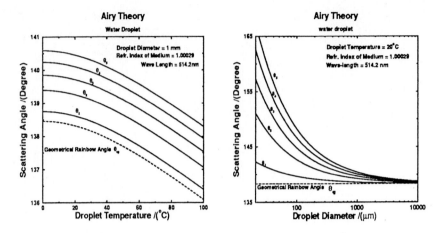

Fig. 6a: Angular positions of the first five extrema in the rainbow pattern as a function of the droplet temperature.

Fig. 6b: Angular positions of the first five extrema in the rainbow pattern as a function of the droplet diameter.

of the laser light. If one uses a laser beam as light source and if the relationship between the refractive index of the liquid and the temperature is well known (see Thormählen et a. (1985) for the refractive index of water), then the rainbow pattern depends on the droplet temperature and size. This dual dependence can be seen in figures 6a and 6b. In figure 6a the scattering angle correponding to the first five maxima in the rainbow pattern, denoted by $n = 1$ to $n = 5$, have been plotted as a function of the temperature for a fixed droplet diameter of $1mm$. Figure 6b depicts the same as a function of the droplet size for a temperature of $20°C$. Note, as mentioned above, that the geometrical rainbow angle does not exhibit any dependence on the droplet size. From these graphs the experimental method can be understood. As the angular difference between the maxima does not depend on the temperature (figure 6a), this spacing can serve to determine the diameter without knowing the temperature before. With this knowledge about the droplet size, the temperature can then be deduced from the scattering angle at which one of the peaks in the rainbow pattern is positioned. In this way, both parameters can be determined from one signal, i.e. the rainbow.

3   Comparison between Airy and Mie computations

In this section the applicability of the Airy theory for the proposed non-intrusive measurement technique will be discussed. Therefore it will be compared with the "correct" Mie theory. A comparison between the rainbow given by Mie computations and by the Airy approximation has been carried out by

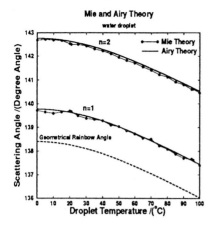

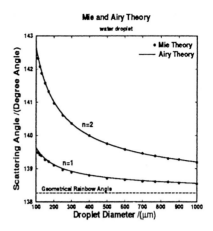

Fig. 6c: Angular positions of the first two extrema in the rainbow pattern as a function of the droplet temperature according to Airy and Mie calculations. ($\lambda = 514.2\,nm$, $D = 100\,\mu m$ and the refractive index of the medium is 1.00029.)

Fig. 6d: Angular positions of the first two extrema in the rainbow pattern as a function of the droplet diameter according to the Mie and Airy theory. ($\lambda = 514.2\,nm$, $T = 20°C$ and the refractive index of the medium is 1.00029.)

Wang and van de Hulst (1991). Their conclusion was that for the first order rainbow the theories agree with each other for diameters of water droplets down to $0.02\,mm$. Nevertheless, another comparison will be given here which is more suitable to the envisaged application. Namely, similar curves as in figures 6a and 6b have been reproduced by means of the Mie theory. As these calculations are very time consuming, only the first two maxima have been considered. For the comparison the high frequencies in the Mie theory have been filtered out as was done for the thick curve in figure 2. This leads to the desired comparison between the theories in figures 6c and 6d.

Figure 6c depicts the dependence of the peak positions on the droplet temperature for a droplet diameter of $100\,\mu m$. Remarkable are the wiggles in the curves of the Mie theory. For the first peak this feature even causes an ambiguity for the temperature measurement around $20\,°C$. One could suggest to use the second extremum of the rainbow for the determination of the temperature because for this peak the wiggles have a smaller amplitude. But for the second extremum an angular shift of about $0.1°$ between both theories can be seen; such a shift corresponds to a temperature change of $2\,°C$. The results of Wang and van de Hulst (1991) indicate that for higher order subsidiary maxima this shift becomes even larger; thus one can also exclude the use of these maxima for the discussed experimental technique. Therefore figure 6c tells us that for

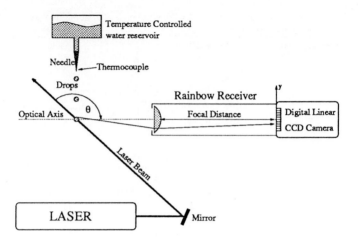

Fig. 7: Experimental setup for the detection of the "monochromatic rainbow".

droplet sizes around $100\,\mu m$ the use of the Airy theory can lead to errors of several degree Celsius concerning the droplet temperature. Roth et a. (1992) have demonstrated that the amplitude of the wiggles increases for smaller droplets, thus implying that for droplet diameters in the micrometer range certainly the Mie and not the Airy theory has to be used. This, in addition to the fact that for these small droplets more ambiguities occur concerning the droplet temperature determination, makes this technique unfavourable for droplets in the micrometer range.

Figure 6d shows more positive results. Herein is depicted the dependence on the droplet size. The dots produced by the Mie theory coincide well with the curves of the Airy theory except for the droplets near $D = 100\,\mu m$ as has been noticed in the previous paragraph. This means that the wiggles damp out and the angular shift disappears for droplets exceeding $D = 100\,\mu m$! Therefore the Airy theory can applied without any problem for droplets in the millimeter range; the measurements presented in the next section have been performed in this range. But it has to be noted that close study of the presented calculations shows that even for these big droplets the (still present) wiggles lead to an error of several percent concerning the droplet sizing and an error of a fraction of a degree Celsius concerning the temperature determination. If more accurate results are required, the Mie theory itself has to be applied.

4   Experimental Results

The experimental results shown here consist of experimental rainbow patterns arising from single falling droplets in the millimeter range. Such droplets are produced with the setup, outlined in figure 7. The droplets are initiated

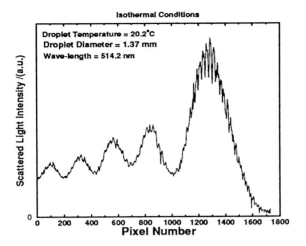

Fig. 8: (Unsmoothed) calibration rainbow pattern recorded in order to relate each pixel number to a certain scattering angle.

from the needle of a medical syringe. The hydrodynamic pressure (of several $1000\ Pa$) in combination with a valve ensures a range of droplet diameters from $1\ mm$ to $4\ mm$. The initial temperature of the droplet is measured with a thermocouple mounted inside the thermally isolated needle. The falling droplet crosses the laser beam which has a diameter of about $5\ mm$ at this crossing point. The "monochromatic rainbow" is projected on the digital linear CCD-array. The integration time of the camera is chosen such that only one single rainbow pattern is recorded at a time.

A calibration is carried out to relate each pixel number to a certain scattering angle. It avoids the use of a geometrical measurement of the scattering angle. The procedure consists of the detection of a rainbow pattern issued from a droplet at ambient temperature. This calibration rainbow pattern is depicted in figure 8. The droplet size is deduced from the fringe spacing to be $1.37\ mm$. The ambient temperature is given by the thermocouple and equals $20.2°C$. The Airy theory delivers for these parameters the corresponding rainbow pattern so that the desired relation between pixel number and scattering angle can be established. This calibration procedure suffers from the evaporation of the droplet along its path, resulting in a slight change in temperature. This unknown change plus the uncertainty in the determination of the peak positions leads to a total error in the temperature of less than $\pm 1°C$.

After the calibration the temperature and diameter can be determined for any heated droplet. Figure 9 shows a typical experimental low pass filtered signal yielding a "rainbow temperature" of $49°C$. The thermocouple indicated

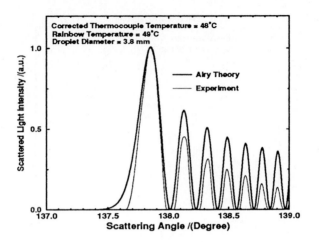

Fig. 9: Typical low pass filtered rainbow pattern and the comparison with the Airy theory.

$51°C$ but after a numerical correction for the cooling of the droplet before it crosses the laser beam (given graphically by van Beeck and Riethmuller (Applied Optics 1994)), it equals the "rainbow temperature" within $1°C$. The reconstruction of the theoretical rainbow pattern demonstrates a remarkable agreement of the positions of all the peaks which was expected for this droplet size in section 3. As those positions determine the desired parameters the disagreement concerning the peak intensities is just a minor defect.

A final remark concerns the shape of the droplets. For rainbows coming from droplets bigger than $2\,mm$, the "rainbow temperature" suffers largely from the non-sphericity of the oscillating droplets; it is well known that the amplitude of the oscillation increases with the droplet size (see for instance Becker et a. (1991)). The presented Airy theory is only valid for spheres thus yielding incorrect quantities for non-spherical droplets. The reason for the perfect agreement in figure 9 was probably because of the accidental event that an oscillating droplet happened to momentarily have a spherical shape. This has been pointed out by van Beeck and Riethmuller (Applied Optics 1994) and further studied by the same authors in (Poitiers 1994).

5   Conclusions

The art of using the first order rainbow for the measurement of the droplet temperature has been improved. The droplet size, necessary for this rainbow method, can be deduced from the fringes, visible in the rainbow itself. A comparison between Airy and Mie computations shows that for a droplet diameter bigger than $100\,\mu m$ the Airy theory for the rainbow can be used for the determination of the temperature and diameter provided the signal is properly low pass filtered; the accuracy is several percent concerning the droplet size and a fraction of one degree Celsius concerning the temperature. If higher precision is required, the Mie theory has to be applied.

References

Roth N., Anders K., A. 1990. Simultaneous Measurement of Temperature and Size of Droplets in the Micrometer Range, Journal of Laser Applications, vol. 2, number 1.

König G., Anders K., Frohn A. 1986. A New Light Scattering Technique to Measure the Size of Periodically Generated Moving Particles, J. Aerosol Sci., **17**, 157-167.

Bohren C.F., Huffman D.R. 1983. Absorption and Scattering of Light by Small Particles, Wiley, New York.

van de Hulst H.C. 1957.
Light Scattering by Small Particles, Wiley, New York.

van Beeck J.P.A.J, M.L. Riethmuller 1994. Determination Non Intrusive de la Dimension et de la Temperature des Gouttes dans une Pulverisation, 4e Congrès Francophone de Vélocimétrie Laser", September 1994, Poitiers, France.

Hecht Eugene 1987. Optics, Addison Wesley Publishing Company.

Thormählen I., Straub J., Grigull U. 1985. Refractive Index of Water and its Dependence on Wavelength, Temperature and Density, J. Phys. Chem. Ref. Data, vol. 14, number 4.

Wang Ru T., van de Hulst H.C. 1991. Rainbows: Mie Computations and the Airy Approximation, Appl. Opt., vol. 30, number 1.

Roth N., Anders K., Frohn A. 1992. Simultaneous Determination of Refractive Index and Droplet Size Using Mie Theory, Proc. Sixth International Symposium on Applications of Laser Techniques to Fluid Mechanics, Lisbon.

van Beeck J.P.A.J., Riethmuller, M.L. 1994. Non-intrusive Measurements of Temperature and Size of Raindrops, to be published in Applied Optics.

Becker E., Hiller, W.J. & Kowalewski, T.A. 1991. Experimental and Theoretical Investigation of Large-amplitude Oscillations of Liquid Droplets, J. Fluid Mech., **231**, 189-210.

A rigorous procedure for design and response determination of phase Doppler systems

Amir Naqwi and Rajan Menon

Fluid Mechanics Instrument Division, TSI Inc., St. Paul, MN 55164, USA

Abstract. A three-step procedure is described for selecting the critical parameters of a phase Doppler system and for determining the transfer function of the instrument. This procedure covers all possible values of the material properties for homogeneous spherical particles and has the potential to be transformed into an *expert system* for setting up phase Doppler instruments and preparing look-up tables to convert measured phases into particle diameters.

Keywords. Phase Doppler, Particle Sizing, Optical Instrumentation

1. Introduction

A rigorous procedure for designing optical layouts of phase Doppler systems is provided. The first step of the design is based on geometrical optics. For this purpose, the phase Doppler designers' chart introduced earlier [1] is divided into 15 domains representing certain combinations of particle refractive index and scattering angle. Within each domain, scattering characteristics are fairly consistent for a given state of polarization and a given level of light attenuation through the particle. Charts are provided, so that the phase Doppler user can select a state of polarization and a range of scattering angles for his/her application. At this stage, an estimate of the required beam angle and elevation angle is available for a standard layout. Also, the beam angle for a planar layout can be estimated at this level.

In the second step, due attention is given to the particle trajectory effects. As shown by Naqwi [2], the trajectory effect becomes more and more severe with the increasing dynamic range and concentration of the particles and leads to narrower choices of scattering angles. For transparent particles in a rarer medium, charts are provided to select the scattering angles that would minimize the trajectory effect.

In the third step of design, a choice between planar and standard (or orthogonal) optical layout [2,3] is made. The newly introduced Adaptive

Phase/Doppler Velocimeter (APV) can be configured as a standard or a planar system.

Subsequently, size and shape of the receiving apertures are determined. After selection of the optical parameters, Mie scattering theory is used to compute the phase-diameter relationship, esp. if the level of confidence regarding the proposed geometrical scattering mechanism is low. Usually, this relationship is not strictly monotonic and may exhibit oscillations with high (submicron) or low periodicity. Two "monotonizing" methods are outlined that transform the Mie response curve to the nearest nonlinear monotonic function. These procedures result in smaller errors and uncertainties than the commonly used linear regression technique. A look-up table is subsequently generated that allows conversion of phase into particle diameter. The authors prefer to use look-up tables as opposed to conversion factors, even if the phase-diameter relationship is strictly linear. The look-up tables enable faster processing of the data, by eliminating a multiplication operation.

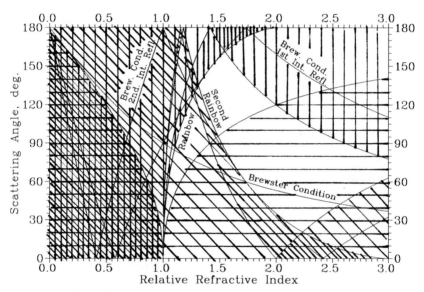

Figure 1: Critical scattering angles as a function of refractive index

2. Identification of the Operating Domain

Phase Doppler systems were originally analyzed using the geometrical theory of light scattering, which indicates that the relationship between the phase shift and the particle diameter is linear, provided that a single geometrical scattering mode is dominant. One of the first three modes of geometrical scattering is usually employed for particle sizing. These modes may be described as (i) reflection of light from particle surface, (ii) refraction of light through the particle, and (iii) emergence of light rays from the particle after one internal reflection.

The conditions under which one of these scattering modes are applicable, have been discussed in literature, but only with reference to special cases. A considerable number of cases are discussed by Naqwi & Durst [1 & 4], who also introduced phase Doppler designers' chart, shown in Fig. 1. In this chart, areas covered by horizontal, vertical and oblique lines represent the presence of refracted, singly internally-reflected and doubly internally-reflected light respectively. The spacing between the vertical lines is reduced to half, one-third and one-fourth in some regions in order to indicate the presence of double, triple and quadruple contributions of the internally reflected rays. Furthermore, several critical scattering angles, such as those representing the Brewster conditions for surface reflections and internal reflections as well as rainbow angles are included.

The governing equations for some higher order angles are not given in Ref. 1. These mathematical relationships are provided below:

Brewster condition for the first internal reflection:

$$\cos \theta = (m^2 - 1)(m^4 - 14m^2 + 1)/(m^2 + 1)^3 , \tag{1}$$

where θ and m represent the scattering angle and the refractive index of the particle relative to the surrounding medium.

The scattering angle of the doubly internally-reflected light corresponding to the edge ray is given as,

$$\cos \theta = 2\left(\frac{4 - 3m^2}{m^3}\right)^2 - 1, \quad 1 \leq m < \infty . \tag{2}$$

The second rainbow is represented by

$$\cos \theta = (m^8 + 4m^6 + 270m^4 - 972m^2 + 729)/32m^6 , \quad 1 \leq m \leq 3, \tag{3}$$

and the Brewster condition for second internal reflection corresponds to the scattering angles expressed by

$$\cos \theta = (-m^8 + 28m^6 - 70m^4 + 28m^2 - 1)/(m^2 + 1)^4 . \tag{4}$$

The above critical scattering angles follow directly from the geometrical formalism of light scattering given by Van de Hulst [5].

The introduction of the designers' chart has made it possible to devise *generalized criteria* for determining the applicability of individual geometrical scattering modes, instead of restricting the discussion to a few special cases. In this context, four independent variables are considered in the present work, i.e. refractive index of the particle relative to the surrounding medium, attenuation coefficient of the particle material, scattering angle of the receiving optics and polarization of the incident light.

The designers' chart is divided into 15 domains based on the values of scattering angle and the refractive index, as shown in Fig. 2. The rationale for the domains in Fig. 2 is provided by the critical angles depicted in Fig. 1, which strictly represents non-absorbing particles. Subsequently a measure of light attenuation is incorporated in the above scheme of classification. An attenuation parameter is introduced as below:

$$a = \chi d_{p\,max} = 4\pi n\kappa d_{p\,max}/\lambda_0 ,$$ (5)

where $d_{p\,max}$ is the largest particle diameter and χ is 1/e-attenuation coefficient, which is related to the imaginary part $n\kappa$ of the refractive index and the wavelength of light in vacuum λ_0 as shown above.

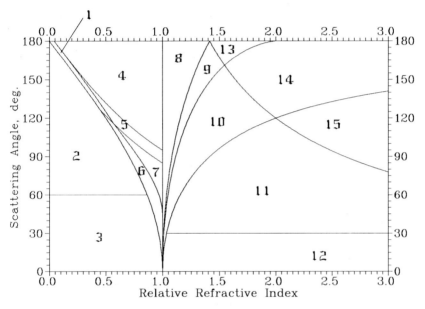

Figure 2: Scattering domains based on refractive index and scattering angle

Parameter a represents the fraction of the incident light absorbed in the largest particle during its transmission through the particle. A very large range of this parameter may be encountered in practice. Based on the numerical value of this parameter, the light absorbing character of particles may be classified as follows:

1. Very weak attenuation: $a \leq 0.005$
2. Weak attenuation: $0.005 < a \leq 0.05$
3. Moderate attenuation: $0.05 < a \leq 500$
4. Strong attenuation: $500 < a \leq 4000$
5. Very strong attenuation: $a > 4000$

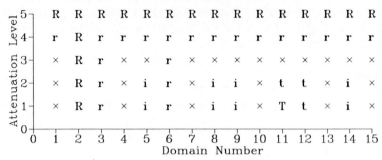

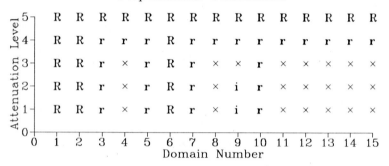

Figure 3: Scattering mode charts

In the case of weak attenuation, scattering mechanism is not significantly affected by light absorption. However, some caution is required unless attenuation is qualified as 'very weak'. Similarly, strong attenuation means that only reflected light will be dominant at all the scattering angles. However, the designer should proceed cautiously, unless attenuation is designated as 'very strong', which typically pertains to metallic particles. In the case of moderately absorbing particles, none of the geometrical scattering mechanism is generally applicable. Hence, the user needs to select the scattering domain and the state of polarization, so as to ensure that reflected light dominates.

Based on the above considerations, *scattering mode charts* of Fig. 3 are generated. For the 15 domains and 5 levels of attenuation, dominant scattering modes are indicated. Separate charts are given for parallel and perpendicular polarization. The differences between the two charts arise from the fact that various Brewster conditions are met only in the case of parallel polarization. The dominant scattering modes are represented by R, T and I that depict reflection, transmission (refraction) and internal-reflection respectively. The uppercase and lowercase symbols signify whether the level of confidence in the corresponding geometrical mechanism is high or low respectively. The symbol × is used to indicate that none of the three geometrical scattering modes is expected to dominate.

If the level of confidence is high, a linear relationship between phase and diameter is expected and geometrical scattering theory can be used to determine the phase-diameter conversion factor with an accuracy of 5% or better. If the confidence level is low, Mie scattering theory should be used to obtain the phase-diameter relationship. Nevertheless, a low level of confidence should not be interpreted as likelihood of poorer measurements. It simply means that a single geometrical mode of scattering is not fully dominant. However, accurate phase Doppler measurements may be possible.

A rigorous procedure for treatment of non-monotonic response, known as *joint probability method*, is described in Ref. 3. It involves two or more phase shift measurements, in order to eliminate the ambiguities associated with individual phase measurements. In many applications with well-optimized optical layouts, the non-monotonicity may be reduced to an insignificant level, so that a rigorous procedure, such as the joint probability method is not warranted. Nevertheless, the exact response curve needs to be "monotonized", in order to generate a phase-to-diameter conversion table. Two monotonizing methods are discussed here, i.e. *frequency-domain smoothing* and *linear segmenting*.

In the first method, the linear trend is subtracted from the actual response curve and the remaining part is high-pass filtered in the frequency domain. Initially, the cut-off filter frequency is high. It is gradually reduced until a monotonic response curve is obtained. This method is usually suitable if the largest particle is 100 times or larger in diameter than the wavelength of light, so that the response curve may consist of some low amplitude high frequency oscillations, besides a low frequency non-linearity. The difference between the frequencies of these two components is usually large, and hence, the high-frequency fluctuations can be removed by frequency-domain low-pass filtering without affecting the nonlinear trend.

For small particles, the frequency ranges of non-monotonicity and non-linearity begin to overlap, so that frequency-domain smoothing results in a distortion of the nonlinear behavior. In such cases, linear segmenting is an effective means to obtain a monotonic function that closely follows the actual response curve. In this method, the segments of non-monotonic response are identified and replaced with straight lines. The smoothness of the response curve may be preserved by replacing the non-monotonic segments with third or fourth order polynomials, instead of straight lines. However, measurement accuracy with linear segmenting is as good as with polynomial segmenting. Examples of both the monotonizing method are given later in Sec. 5.

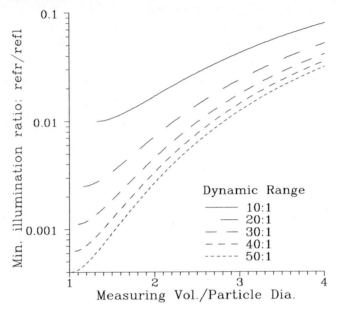

Figure 4: Non-uniformity of particle illumination

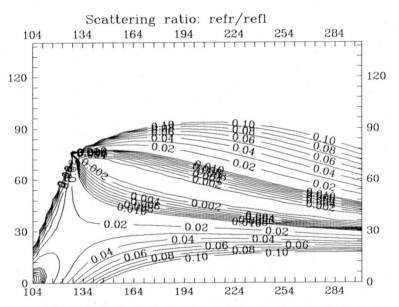

Figure 5: Reflection/refraction ratio in the scattered field.

3. A Consideration of Trajectory Effects

The measuring volume diameter is set by default to be comparable to the largest particle diameter in order to suppress the adverse effects of the Gaussian intensity profile. In the case of transparent particles with refractive indices larger than 1, the charts of Fig. 4 & 5 are consulted for further optimizing the scattering geometry in respect of the Gaussian profile effect. Figure 4 receives two inputs: (i) the measuring volume to particle diameter ratio, and (ii) the particle dynamic range.

A large dynamic range requires a low enough trigger level, so as to register the weak signals from small particles. This also means that a large particle will generate signals from a large volume, including the undesirable regions of the illuminating field. Figure 4 provides a measure of the non-uniformity of particle illumination. The light rays reaching a detector after reflection and refraction enter the particle at different locations. The separation between these two incidence locations is conservatively estimated as particle diameter itself. The output of Fig. 4, i.e. the ordinate (maximum incidence ratio: refraction to reflection), indicates how much the intensity of incident light at the incidence point of refraction — which is presumably the dominant scattering mechanism — could be smaller than that at the incidence point of reflection, under the worst conditions.

Figure 5 shows the ratio of reflected to refracted light in the scattered field (scattering ratio: reflection to refraction) for uniform illumination. The abscissa and ordinate of Fig. 5 represent 100 × refractive index and the scattering angle respectively. The latter should be chosen appropriately, so that the output of the second chart is smaller than that of the first chart. This condition ensures that the refracted light signal dominates the reflected light signal. This condition suffices to ensure that correct size measurements are obtained, at least in the neighborhood of the *critical particle diameter* introduced in Ref. 2.

Figures 4 & 5 are based on geometrical optics but are validated using generalized Lorenz-Mie theory [6] that fully takes into account the Gaussian structure of the laser beam.

4. Type of Layout and Receiver Geometry

For a given particle size range Δd_p , some key design parameters of the system can be selected using the scattering angle θ and the corresponding geometrical scattering mode determined through the procedures of Sec. 2 & 3. These parameters ascertain the elevation angle and the beam angle in the case of a standard phase Doppler system. As shown in Refs. 1 & 4, the measuring sensitivity (degrees/micron) is approximately proportional to $\sin\alpha\sin\psi$, which in turn depends primarily upon the beam spacing s_b , spacing between centroids of the receivers s_r and the focal lengths of the transmitting and the receiving units denoted by f_t and f_r respectively. These parameters are illustrated in Fig. 6 for a typical APV layout and are related as below:

$$\frac{f_t}{s_b}\frac{f_r}{s_r} \le \frac{\Delta d_p}{\lambda F_s(\theta,m)}.$$ (6)

The function F_s takes the following form for a reflecting particle,

$$F_s = 4\sin(\theta/2).$$ (7)

For a refracting particle,

$$F_s = \frac{4\cos(\theta/2)\sqrt{1+m^2-2m\cos(\theta/2)}}{m}.$$ (8)

The above equations follow from Eqs. (60) & (63) of Ref. 1. For particles with internal reflection as the dominant scattering mechanism, an explicit analytical expression for F_s does not exist and the function is computed using an iterative method.

The right-hand side of Eq. (6) will be referred to as the *range function*, as it is a non-dimensional measure of the particle size range under consideration.

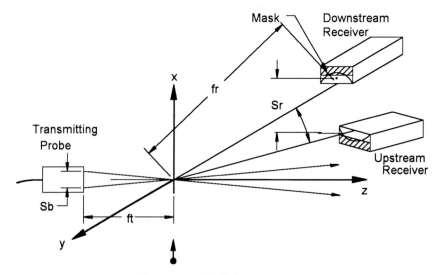

Figure 6: APV design parameters

Following Eq. (6), the optical parameters can be selected to cover a size range $\Delta d_p = d_{p\max} - d_{p\min}$, where $d_{p\min}$ may have a non-zero value. It may be reminded here that conventional approaches to phase Doppler design are based on the presumption that the smallest particle diameter is always nearly zero, which results in an unnecessary loss of sensitivity in the case of narrow size distributions with $\Delta d_p / d_{p\,max} \ll 1$. APV hardware obviates this restriction and provides maximum sensitivity in the case of narrow size distributions by using separate receivers, whose angular spacing can be increased to match a narrow size range. This is not possible with conventional phase Doppler receivers, where detector spacing is severely limited by the numerical aperture of the receiver lens.

It is also obvious from Eq. (6) that for very small particles or very small size ranges, the right-hand side of this equation would become so small that it would not be possible to obtain adequate sensitivity even if the largest practical value of receiver spacing is employed. For such applications, planar layouts are useful.

In a planar arrangement, scattering angle is represented effectively by the elevation angle ψ, whereas off-axis angle is zero. Invoking these conditions in Eqs. (60) & (63) of Ref. 1, the appropriate hardware settings of a planar layout are determined through

$$\frac{f_t}{s_b} \le \frac{\Delta d_p}{\lambda F_p(\theta, m)}, \tag{9}$$

where

$$F_p = \frac{1}{\cos(\theta/2)} \tag{10}$$

for reflecting particles and

$$F_p = \frac{\sqrt{1 + m^2 - 2m\cos(\theta/2)}}{m\sin(\theta/2)} \tag{11}$$

for refracting particles. Usually planar layout provides adequate sensitivity to measure particle diameters as small as a fraction of the light wavelength.

5. Some Examples

The above procedure for design and response determination of a phase Doppler system is illustrated below using examples of practical applications, where APV can be used effectively for particle sizing.

1. Drop sizing in the spray of a pressure atomizer

Considering water drops with a size range of 200 μm, Figs. 2 & 3 indicate that a scattering angle in the range 30°–80° combined with parallel polarization would ensure transmission as the dominant scattering mechanism. According to Figs. 4 & 5, a measuring volume diameter of 300 μm combined with an off-axis angle of 70° would allow the user to cover a dynamic range of 30:1 without introducing significant trajectory related errors.

350

Using the above specifications, the range function according to Eqs. (6) & (8), is about 205≅15×15 for a wavelength of 0.5145 μm (This value of wavelength is used in all the examples). Hence, the two terms on the left-hand side of Eq. (6) should be about 15 each. If one term is larger than 15, the other one should be smaller than 15.

The specifications of APV hardware meeting the above requirements are given in Fig. 7, which also includes the response curves based on Mie scattering theory, as well as geometrical optics. It can be seen that geometrical prediction is adequate, as indicated in Fig. 3.

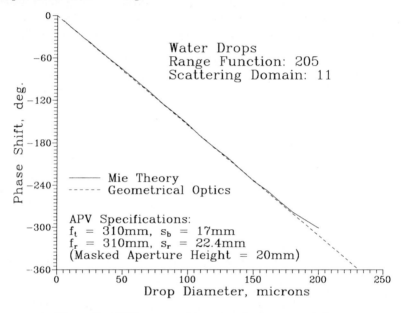

Figure 7: APV layout and response for large water drops

2. Partially opaque paint drops

Considering a refractive index of 1.33, attenuation coefficient of 1 mm^{-1} and size range of 100 μm, the following information is obtained using Figs. 2 & 3:

Since attenuation level is 3, operation in Domain 10 is appropriate. Reflection appears to be the scattering mechanism for an off-axis angle of 110° and perpendicular polarization. However, the level of confidence in this geometrical mode in low.

The range function for a standard layout is 7.7×7.7, which can be implemented using standard APV transmitting and receiving hardware. The appropriate arrangement is described in Fig. 8, which also includes the response curves based on Mie scattering theory and geometrical optics as well as the monotonized version of Mie response. It can be seen that the monotonized curve, which is based on frequency-domain smoothing, is a better representation of the actual response than the geometrical prediction.

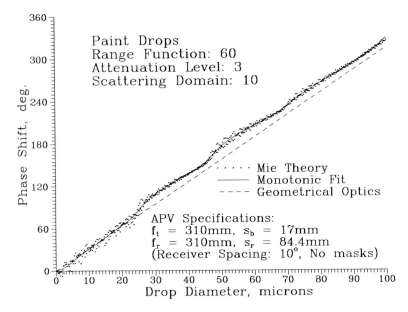

Figure 8: APV layout and response curve for partially opaque drops

3. Oil drops produced by a condensation aerosol generator

In this example, transparent oil drops smaller than 2 μm having a refractive index of 1.454 are considered. As in the first example, refraction would be the appropriate scattering mechanism for sizing. For a scattering angle of 70°, the range function is about 2. Obviously, a standard layout cannot be configured for such a small range function.

Considering a planar layout, the corresponding range function, according to Eqs. (9) & (11), is 3.8, i.e. the focal length to diameter ratio for the transmitting optics should be slightly larger than 3.8. This requirement can be easily met by an LDV transmitting probe, which uses a 50 mm beam spacing and the focal length can be as small as 120 mm.

Specifications of the system for the above application are given in Fig. 9. An asymmetric layout of the receivers is employed in order to suppress the oscillations in the response curve. Monotonized response curves, based on both the frequency-domain smoothing and linear segmenting are included in Fig. 9. As mentioned in Sec. 2, linear segmenting is a better choice for such small particles.

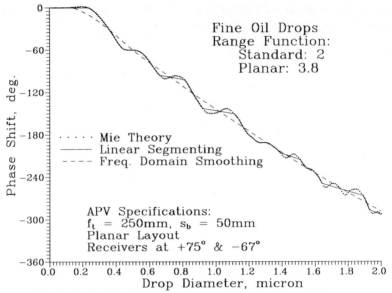

Figure 9: APV layout and response curve for fine oil drops

6 Conclusion

A tool box, consisting of algebraic equations, graphical aids and a numerical code, is introduced for the users of the phase Doppler technique. Once the information regarding particle size range, refractive index and attenuation coefficient is available, these tools permit a selection of the scattering mechanism, scattering angle and the state of polarization of the transmitted beams. Subsequently, a range function can be calculated that determines whether a standard or a planar layout should be employed. Furthermore, the range function is used for selecting the focal length to beam spacing ratio for the transmitting optics and focal length to detector spacing ratio for the receiving optics.

Finally, a Mie scattering code allows to optimize the layout and provides a monotonized response curve that is used for converting the measured phase shifts to particle diameters.

Graphical tools are also available for selecting optical parameters, so as to minimize the undesirable effect of particle trajectory on the system response.

We believe that these tools will allow the phase Doppler users to optimize the optical layouts for their specific applications, without necessarily acquiring an in-depth knowledge of the underlying physical principles. These developments are aimed at eventually contriving an expert system for layout and response determination of phase Doppler systems.

References

1. A. Naqwi & F. Durst, "Light scattering applied to LDA and PDA measurements, Part 1: Theory and numerical treatments", Part. Part. Syst. Charact. 8 (1991) 245–258.

2. A. Naqwi, "Innovative phase Doppler systems and their applications", Part. Part. Syst. Charact. 11 (1994) 7–21.

3. A. Naqwi & M. Ziema, "Extended phase Doppler anemometer for sizing particles smaller than 10 μm", J. Aerosol Sci. 23 (1992) 613–621.

4. A. Naqwi & F. Durst, "Light scattering applied to LDA and PDA measurements, Part 2: Computational results and their discussion", Part. Part. Syst. Charact. 9 (1992) 66–88.

5. H.C. Van de Hulst, "Light Scattering by Small Particles", Dover Publications, New York (1981).

6. G. Gréhan, G. Gouesbet, A. Naqwi and F. Durst, "Trajectory ambiguities in phase Doppler systems: use of polarizers and additional detectors to suppress the effect", Proc. 6th Intl. Symp. on Applications of Laser Techniques to Fluid Mechanics, Lisbon (1992) Paper # 12.1.

Measurement of Size and Velocity of Arbitrarily Shaped Particles by LDA Based Shadow Image Technique

Hiroshi Morikita, Koichi Hishida and Masanobu Maeda

Department of Mechanical Engineering,
Keio University
3-14-1 Hiyoshi, Kohoku-ku,
Yokohama 223, Japan.

Abstract. An instrument based on an imaging technique and laser Doppler velocimetry (LDV) was developed and evaluated for sizing particles of irregular shape. The signals from particles were detected by a 32 channel linear photodiode array focused on the measuring volume and the image was reconstructed from 'sliced' images of the particle provided by temporal outputs from each channel. The present instrument consists of a multi-channel transient recorder and independent, single-channel LDV processor, and provided size, velocity, shape and trajectory information of particles ranging from 30 to 250 μm with velocity up to 3 m/s; the particle size was obtained from the area of the shadow image and the trajectory was inferred for defocused images by cross-correlation between two images projected by two incident beams of the conventional LDV. For determination of particle size distribution, the biasing effect, caused by the size dependence of the sampling space, were eliminated by a correction scheme. The accuracy of the average size for spheres was better than 3.0 % referred to that measured by a microscope. The size distribution of non-spherical particles was similar to microscopic measurement and the maximum difference in arithmetic average diameter was approximately 10 %.

Keywords. Optical Particle Sizing, Non-spherical Particle, Laser Doppler Anemometry, Image Processing, Linear Diode Array

1. Introduction

The sizing of arbitrarily shaped particles is of technical interest and accurate determination of their equivalent diameter is important, because most particles in industrial processes are non-spherical. For sizing spheres, many laser-based techniques have been developed such as the phase Doppler anemometer [e.g. Durst & Zaré, 1975; Sanker et al., 1991; Kobashi et al., 1992] and an imaging system [e.g. Hovenac et al., 1985]. Although these instruments are accurate and robust, they can not be applied for sizing non-spherical particles since the methods rely on the light scattering from a sphere. The diffraction amplitude method [Orfanoudakis & Taylor, 1993, Morikita et al., 1994] or the other methods based on light scattering [e.g. Bottlinger and Umhauer, 1988] are suitable instruments for this purpose, however extreme care has to be taken to align the optical unit to avoid uncertainty caused by the non-uniform irradiation distribution of the incident laser beam. A recent method proposed by Kaye et al. [1991] results in higher accuracy than the single detector system, but requires large optical benches surrounding the test volume.

Imaging techniques [Knollenberg, 1970 and 1976, Bertollini et al., 1985, Sanker et al., 1994] have an advantage for sizing non-spherical particles because the two-dimensional images can eliminate problems arising from the non-sphericity of particles, although most techniques were developed for the measurement of spheres. The shadow Doppler technique, which is an instrument for sizing irregularly shaped particles developed by the authors [Hardalupas et al., 1994], is also based on an imaging method. In this instrument, the images were projected from illumination by a pair of beams of a conventional LDV, and the recorded image provided area-equivalent size. The maximum sizing error for spheres evaluated in our previous work was approximately 10% using 16 diode elements. The primary advantage of the method is that less care in alignment is necessary compared with the intensity-based methods and it can provide velocity and size information simultaneously.

The purpose of this study is to measure non-spherical particles in flows for evaluation of performance of the instrument and to establish a suitable signal processing technique. The measurements were implemented with free-falling and accelerating flows which provided independent estimations of diameters as described below. The signal processing, which provided velocity, size, shape and trajectory of the particle, is also introduced and assessed. In the following section, the instrument and method are described and experimental results of sizing can be found in section 3. Section 4 presents a method to determine the size distribution which eliminates the biasing effect caused by size-dependence of the area of sampling space.

2. Optical Setup and Signal Processor

2.1 Principle of Shadow Doppler Method

The optical arrangement, which includes a conventional LDV system, is illustrated in figure 1. The transmitting optics consists of beam expander, beam-splitter, a pair of acousto-optic modulator (AOM) cells and focusing lens. Laser beams of 514.5 nm wavelength (green) formed a measuring volume with an intersection angle of 4.936 degrees: the incident beams were reformed by a pair of cylindrical lenses which provided an elliptical cross-section to increase the effective energy density in the area detected by the diode array. The measured diameters of the ellipse were approximately 500 and 250 μm with the direction of the major axis coinciding with that of the diode array. Shift frequencies provided by the AOM component of were set at 50 to 500 kHz which did not cause large fluctuations on the sizing signals due to interference fringes.

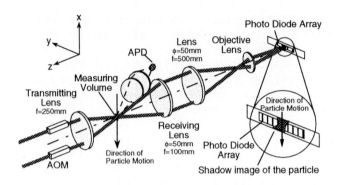

Figure 1 Principle of the method and arrangement of receiving optics. The direction of the particle motion was perpendicular to the axis of the photodiode array.

Two independent receiving units for size and velocity measurement were used, as shown in figure 1, for ease of alignment procedure. The image of a particle passing across the measuring volume was focused on the diode array by the relay lenses, with a magnification ratio, G, between 75 and 120, and the Doppler signal from the same scatterer was detected, via focusing lenses and a spatial filter (pinhole), by an avalanche photo diode (APD) (RCA, C30902E).

Signal detection of the particle size and shape was achieved by imaging the measuring volume on a 35 channel linear diode (Hamamatsu S4114-35). As a particle passed through the volume, the magnified shadow image moved across the detector in the direction of particle motion and the output amplitudes of the channels varied. The signals at each sample of the transient recorder provided a one dimensional shadow image of the particle normal to the direction of the sweep, so a two dimensional image of the particle could be reconstructed from the recorded time-series output of the detector.

Since the shadow images are from each of the two beams, the appearance of the images depend on the trajectory of the particle. Figure 2 (a), (b) and (c) show various patterns of projected shadows on the detector plane by a particle without and with 'defocus', i.e. the distance between the particle and the focal plane of the measuring volume, D_{def}, defined in figure 2. When the particle trajectory was on the focal plane of the receiving lenses, the two shadow images projected by each of the two beams coincided precisely and the output signals had two levels of amplitude, as shown in figure 3(a); the marking a-a' indicates the line in figure 2 and the plateau at the lower voltage level corresponds to the central, dark part of the particle image. Those two images separate gradually with defocus distance and the signals became tri-state (figure 3(b)) because the light intensity of the region which lies in the shadow of the particle illuminated by only one beam is half as much as that of the region illuminated by both, unblocked beams. Finally, when the particle is out of the measuring volume, the overlapping area and the

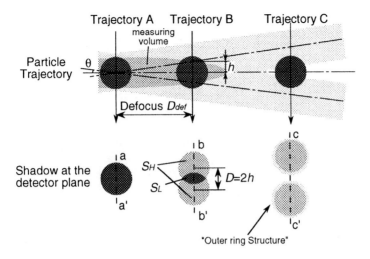

Figure 2   Shadow images of a particle focused on the photodiode array and trajectory dependence on their appearance (A: in-focus; B and C: out-of-focus).

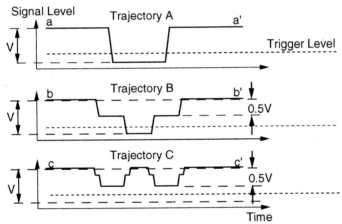

Figure 3   Idealised output signals from the photodiode array with trajectories defined in figure 2.

dark shadow disappear. It should be noted that the distance D (figure 2(b)) between the two images projected by each of the two beams is proportional to the defocus distance, D_{def}, and this was used to estimate the particle trajectory.

For the trajectories A and B, the equivalent size, d_a, is based on the average area of the projected shadow and is defined by the following equation:

$$d_a = \sqrt{\frac{4(2S_L + S_H)}{2\pi}}$$

(1)

where the S_L and S_H are areas of 'dark' and 'half-dark' shadow as illustrated in figure 2(b): S_H is the summation of the upper and lower parts of the 'half-dark' shadow.

The sizing accuracy of the presented method, therefore, relies on the quality of the image and deteriorates with large defocus distance because the image becomes less sharp. For this reason, particle images with trajectory C was not recorded by setting the trigger level between lower and middle signal levels as indicated in figure 3. The trigger level corresponded to about 20% of the maximum signal amplitude due to both beams of LDV. Since the system was triggered by the sizing signal, the sampling space of the APD was aligned to have a larger area than that for the sizing to detect Doppler signals for all the trigger events.

In contrast, when a particle was in the path of the laser beam but outside of the measuring volume, the recorder was not triggered, because the short depth-of-field of the collection lenses made such signals shallower and therefore untriggerable.

The novel features of the method are summarised as follows: the signal level is independent of size, material and refractive index of the particle, because the amplitude is determined by the presence of a shadow; hence, no biasing due to size dependence of the signal *amplitude* arises and transparent media, such as water or glass, can be measured without changing any optical configuration. The instrument can provide not only the size but also the trajectory of defocused particles. In addition, the required dynamic range of the detector is small and low intensity resolution is acceptable relative to intensity-based methods.

2.2 Signal Processor

For data acquisition of the output signals from the diode array, a computer-controlled transient-recorder system was employed. Figure 4 illustrates the block diagram of the processor: the system consists of a head module including the diode array and multi-channel transient recorder for sizing and an APD module and single channel recorder for velocity measurement. All the operating procedures were controlled by a personal computer (IBM-PC/AT 486-33 compatible).

The linear diode array had 35 diode segments and 32 of these were used as active channels to simplify the design of the transient recorder. Each diode had a 0.9 mm width x 4.4 mm height photo-sensitive surface and a slit aperture with 1 mm height was attached to the array to increase the spatial resolution; thus each diode had 1 x 1 mm active size, which corresponds to 10 x 10 μm of the measuring volume given by the magnification ratio, G, of 100 of the collection optics.

The multi-channel recorder comprised a master and three slave boards. Signals detected by the diode array were transferred to the recorder through buffer amplifiers and distributed to 8 analog multiplexers (Analog Devices, AD9300) which each had four inputs and one output. Two multiplexers on one board were connected to one A/D converter (Fujitsu MB40568) of 8-bit resolution and the converted results were stored in static RAM with 32 kByte on each board. The maximum conversion rate was 8 MHz from the system clock, which corresponded to 1 MHz total sampling rate. The trigger level was set at around 20 % of the maximum amplitude, as mentioned in the previous subsection; therefore, defocused particles required a pre-triggering function to obtain the whole signal corresponding to the leading negative-edge (figure 3(b)). For more detailed information of the hardware, see Hardalupas et al. (1994) and Kobashi et al. (1992).

For velocity measurement, a single channel recorder with 128 kByte storage memory was also provided. The clock of this board was synchronised with system clock of the multi-channel recorder with a sampling frequency up to 20 MHz. No trigger circuitry was incorporated on the board because this function was operated by the master board.

Stored signals were transferred to the host computer through an interface module connected on the ISA-bus and processed. The data processing and validation functions are described in the next subsection.

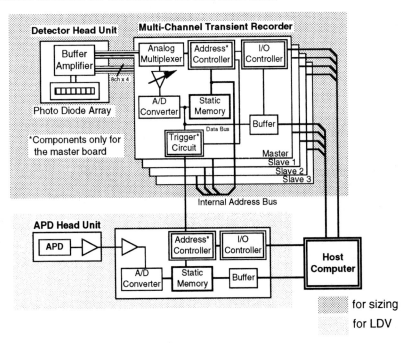

Figure 4 Signal processor including two head-units of the detectors for sizing and velocimeter.

360

2.3 Software Processing

The digitised data, stored in the static RAM, were transferred to the host computer and processed to reconstruct a 'pseudo-image' of the particle. In this subsection, the procedure to infer velocity, area-equivalent size, defocus distance and trajectory direction is described.

2.3.1 Velocity Calculation

First of all, the Doppler burst signal detected by the APD was analysed to obtain velocity information from a fast Fourier transform (FFT) using an adjusted Gaussian interpolation scheme in frequency space [Kobashi et al, 1990]. The applied validation routine was described in Kobashi et al. and the maximum error was estimated at 0.2% of the signal frequency. If the triggered signal was recognised as erroneous, it was rejected and another measurement was attempted.

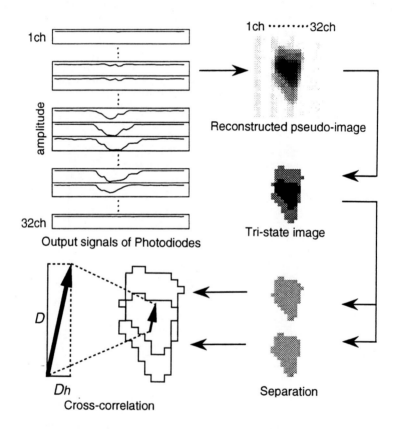

Figure 5 Procedure of the reconstruction of a 'pseudo-image', validation of the signal and determination of the defocus distance.

incident laser illumination in the absence of particles. The normalised amplitude of diode channel i at time t, $r_{norm}(i, t)$, was, hence, defined by the following equation,

$$r_{norm}(i,t) = \frac{r(i,t) - r_{min}(i)}{r_{max}(i) - r_{min}(i)} \tag{2}$$

Figure 5 shows an example of the normalised signals.

2.3.2 Normalization

The maximum amplitude of the signals from the diode array varies along the detector due to the Gaussian intensity distribution of the incident laser beam. Hence, amplitude normalization was required to adopt fixed level thresholding for all diodes. Since the amplitude was dependent on the presence or the absence of the particle image, the minimum level of channel i, $r_{min}(i)$, was equal to the ground level and the maximum level, $r_{max}(i)$, was equal to the amplitude of the

2.3.3 Temporal Windowing

Since the recorded data are temporal signals, time averaging of the temporal signals was necessary to convert these to a pixel image for a two-dimensional pseudo-image of a particle. The distance over which a particle, with velocity normal to the axis of diode array U [m/s], moved within one sampling period of the recorder, $1/f$ [s], was equal to U/f; and the pitch distance between each segment of the array was 1.0 mm, which corresponded to $10^{-3}/G$ [m] in the measuring volume of the LDV. Thus, the number of the samples, N, which corresponds to a distance moved by the particle equal to the channel pitch of the detector is

$$N = \frac{10^{-3}}{GU} \tag{3}$$

Consequently, sampled signals must be averaged over windows of N samples. The present software implemented using N integer where $N \geq 2.0$, and linear interpolation was applied where $N < 2.0$. It should be noted that the effect of this averaging resulted in negligibly small error compared with that caused by other factors.

2.3.4 Thresholding

Thresholding with two fixed levels were adopted to determine the boundary of the particle. Figure 5 shows the example of a tri-state image. The threshold levels were empirically set at normalised amplitudes of 0.4 and 0.75 as recommended in our previous work [Hardalupas et al., 1994]. The projected area was evaluated by a 'painting' routine to remove the shadow image of any other particles existing in the sampling space at the same time. The resultant projected area was substituted into equation (1) to calculate the equivalent diameter.

2.3.5 Image Separation and Cross-correlation

There are three purposes in this process: first, to determine the defocus distance of the particle; second, to estimate trajectory direction; third, to eliminate multiple scattering which can cause serious error in sizing. The tri-state image was separated as defined in figure 5, by a pattern recognition routine, into two binary images, which corresponded to shadows from each of the two incident beams. These separated images were used in a cross-correlation calculation to determine the displacement between two images. As shown in figures 2 and 5, the vertical component of the displacement, D, is proportional to the defocus distance. Thus, the defocus distance, D_{def}, can be obtained by

$$ D_{def} = \frac{D}{2\tan(\theta)} \tag{4} $$

where θ is the beam intersection half angle. When the defocus distance is comparable to, or larger than, the focal depth of the receiving optics, the quality of the image becomes poor. To eliminate this effect, particles with defocus larger than between about 300 to 800 μm were rejected.

Similarly, the horizontal displacement is proportional to the particle movement in horizontal direction D_h; hence, the trajectory direction, φ, of the particle in the plane normal to the axis of the anemometer can be described by the following equation.

$$ \varphi = \tan^{-1}\left(\frac{D_h}{D}\right) \tag{5} $$

This equation allows us to acquire two dimensional velocity measurement by use of a single channel LDV, although the accuracy decreases with decreasing defocus distance. This angle was also used for determination of the correct size distribution described in section 4.

When the two binary images (see above) have the same shape and area, the number of the correlated pixels agrees with the area of the projected shadows; therefore, the maximum value of the correlation function directly quantified the correspondence between two images and was used in the rejection of the image of multiple scattering. The tolerance of the correspondence was set at 30%, a number which was decided empirically. This function, for binary data, was performed with logical, rather than arithmetic, multiplication because it is economical of computation time.

3. Measurements of Velocity, Size and Defocus Distance

3.1 Spheres in an Accelerating Flow

Figure 6 shows the size-velocity correlation of polyethylene spheres (Flowbeads, Sumitomo Seika, coloured black) which was measured in an accelerating flow, produced by the suction flow induced by an inlet nozzle of 30 mm diameter set vertical with the maximum velocity approximately 4.0 m/s. The particles were released 100 mm above the nozzle entrance and were measured at 50 mm above the nozzle on the axis. The sampling frequency and magnification ratio, G, were 312.5 kHz and 116, respectively.

At the measurement point, the particle velocity varied with the size because of inertia. The velocity was clearly correlated to the measured size and the maximum RMS value of the size in each velocity class with 0.1 m/s bin width was about 10%. The flow was not carefully controlled, so that the scatter of the plots seemed to be caused by fluctuation of the flow. This result, therefore, confirms the precision of the size measurement and proves that the sizing accuracy was not affected by the allowed amount of defocus nor velocity of the particle.

3.2 Free-fall Particles of Arbitrary Shape

In order to evaluate the accuracy of the measurement, the velocity and size of particles with various shape were measured with particles in free-fall condition to provide direct assessment of sizing accuracy and comparison with the aerodynamic diameter. The experiments were performed in a chamber of 100 x 100 mm cross-section which was equipped with a particle feeder driven by an

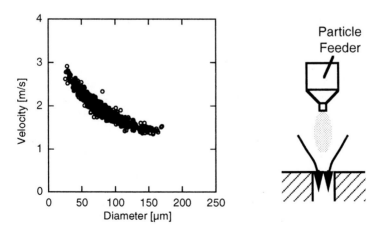

Figure 6 Size-velocity correlation of polyethylene spheres accelerated in the suction flow. Defocus distance is limited at 800 μm.

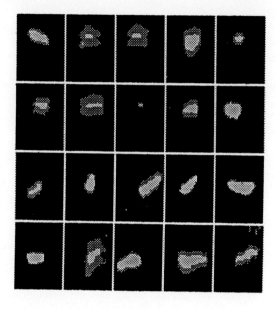

200 [μm]

Figure 7 Example of tri-state, two-dimensional reconstructed images of irregular stainless particles recorded by shadow Doppler velocimeter.

acoustic vibrator and the measuring volume was positioned 1.0 m downstream of the feeder at the bottom of the chamber. The sampling frequency and magnification ratio were set at 250 kHz and 105, respectively.

Figure 7 shows two-dimensional, tri-state images of irregular stainless-steel particles captured by the instrument. The images consisted of two overlapping parts with the same appearance, which were projected by each of two beams; hence, it is obvious that both overlapped images were from the single particle.

Scatter plots in figure 8 (a) and 9 (a) show resultant size-velocity correlation of spherical and non-spherical copper particles with defocus limit set at 800 μm. For comparison, the calculated terminal velocity of a sphere is shown by solid lines. Particles ranging up to 150 μm were estimated to be able to reach their terminal velocities within the allowed 1.0 m of free-fall.

The terminal velocity defines the aerodynamic diameter (Stokes diameter [Clift et al. 1978]), and figure 8(a) shows that the area-equivalent diameter of *spheres* coincides with the Stokes diameter. Since the velocity measurement was reliable and accurate, the result for the spheres in figure 8(a) allows us to evaluate

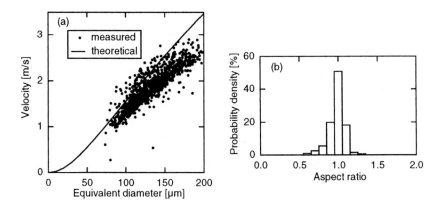

Figure 8 Size-velocity correlation (a) and distribution of aspect-ratios (b) of copper spheres in free-fall. Defocus distance was limited at 800 μm.

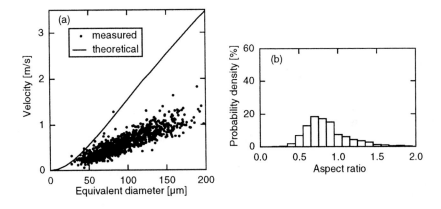

Figure 9 Size-velocity correlation (a) and distribution of aspect-ratios (b) of irregularly shaped copper particles in free-fall. Defocus distance was limited at 800 μm.

sizing accuracy of the method. The velocity of a particle was linearly related to its diameter for the whole size range and the rms value of the size distribution at 2.0 m/s was 10 μm approximately. The size tended to be overestimated by about 15% in contrast with the results of our previous work [Hardalupas et al., 1994] which resulted in 10% underestimation. The overestimation is likely to have been caused by either error in determination of the magnification ratio or error of the theoretical terminal velocity, since it is obvious that the error was systematic and the aspect ratio gave correct results as described below. The plots show wider scatter than that of polyethylene in figure 6, because the air flow was induced by the particles and the particles were not precisely spherical.

The image of a particle also provides useful information, for instance the aspect ratio of the projected shape which represents the particle orientation. The aspect ratio was defined by the ratio of the maximum sizes (Ferret diameters) relative to normal and parallel axes of the photodiode array as defined in figure 10. Figure 8(b) is the pdf of the aspect ratio from copper spheres which shows a narrow distribution around unity aspect ratio. Thus, this gives us confidence that the temporal windowing was accomplished correctly.

The results of non-spherical particles in figure 9(a), under the same experimental conditions, showed smaller velocities as compared with those of the spheres. In addition, most of the particles in Figure 9(b) show smaller aspect ratios than unity and it denotes the particles were 'aligned' by aerodynamic effect. The motion of the particles depends on their orientation because the drag force from the surrounding air can vary with the shape and orientation. At relatively low Reynolds number, the orientation especially tends to be such that the maximum cross section is aligned *normal* to the direction of the motion [Clift et al. 1978]. This explains the small values of aspect ratio in figure 9(b) and it denotes that the 'alignment' had the dominant contribution to the large deference on the terminal velocities. Since the plots showed monotonic increase of size with velocity and all the other irregular particles tested in this study showed the same tendency, this result confirmed that qualitative measurement is at least possible and suggests that the projected area is statistically related to the aerodynamic diameter which is defined by the terminal velocity.

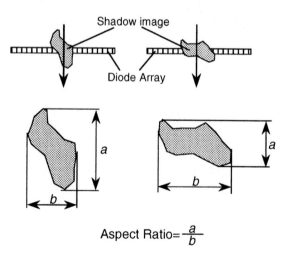

Figure 10 Definition of aspect ratio used in figure 8 and 9 and dependence on particle orientation.

4 Correction of Bias Errors for Determination of Accurate Size Distribution

For practical applications involving a dispersed phase in a flow, it is often necessary to determine the size distribution at each measuring point with high spatial resolution. As with many optical sizing instruments, sampling bias was caused by the dependence of the effective sampling space on particle size, defocus and shape; in order to eliminate these effects, data correction on the basis of particle trajectory was attempted as described below.

4.1 Area of Sampling Space

Figure 11 presents the size dependence on the effective area of the sampling space and its width (area) over which the instrument can be triggered. For simplicity, in-focus spheres only are shown. When the particle diameter is smaller than the size of the diode array, the width of the sampling space, d_w, is equal to the particle diameter because only the 17th channel of the 32 available was used for triggering; therefore, the triggerable width increased with particle diameter, as illustrated in figure 11(a).

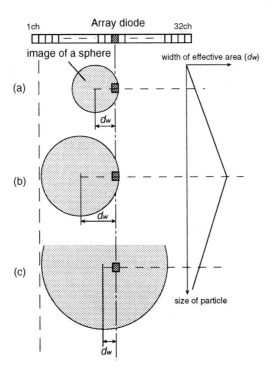

Figure 11 Effective width of the sampling space depending on the particle size: d_W denotes the valid width of sampling space for spheres of in-focus condition.

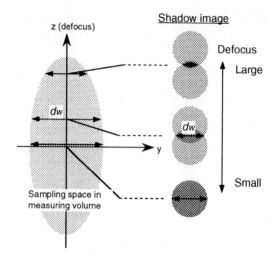

Figure 12 Relationship between defocus distance and effective area of the sampling space for spheres.

For determination of the area-equivalent size, the whole projected image must be recorded and the last channels at both edges of the diode array, i.e. channels 1 and 32, were used to detect the condition at the edge of the image. If the diameter of the particle image exceeds the half width of the array (figure 11(c)), the image was captured but was not validated and hence the width d_w decreases with increasing particle size. Thus, a particle of which the diameter is equal to the width of 15 segments of the diodes has the largest sampling space (figure 11(b)) and the maximum measurable diameter of the particle image was equal to the width of 30 segments of the diodes.

At the same time, the area of the sampling space reduces with defocus distance due to reduction of the overlapped area of the image as shown in figure 12. The sampling space in the measuring volume remains elliptical for all particle sizes but its width, d_w, is dependent on the particle size[1]. As a consequence, width of sapling space, $d_w(d_a,z)$, at distance z (figure 12) for particle diameter of d_a can be described as the following equations:

$$d_w(d_a,z) = \sqrt{d_a^2 - 4z^2 \tan^2(\theta)} \qquad\qquad (d_w<W)$$

$$d_w(d_a,z) = W - d_a \qquad\qquad (d_w>W) \qquad (6)$$

---

[1]If a particle passed through the measuring volume but outside sampling area, the system was not triggered because the overlapping area could not be on the trigger channel.

where W is the width of 30 segments of the diode array and area of sampling space for particle size of d_a, $A(d_a)$, with defocus limitation at l is

$$A(d_a) = \int_{-l}^{+l} d_w(d_a, z)dz \qquad (7)$$

Figure 13 shows an example of the size of the sampling space for spheres with magnification ratio of approximately 75, and the blocked and open symbols correspond to the area without defocus limit and with limit at 400 μm, respectively. In principle, the effective area was proportional to square of the particle diameter, however, the right side of the graph is limited by the effects of the image exceeding the edge of the detector. This function was used as a weight function to eliminate biasing effect caused by size dependence of the sampling space.

The overlap area can vary with particle shape, so that the sampling space is also a function of the shape. If the particle shape is so complex that the area of the sampling space is not a monotonic function of the defocus distance, correct measurement would require some suitable algorithm to estimate the effective area. In the present study, an equivalent ellipse, defined to have the same projected-area and same aspect-ratio of 'fixed axis' Feret diameter as the particle to be measured, was used to represent the shape in order to simplify the calculations. The reconstructed image of the particle was converted to this ellipse, which was the crude represents of the particle shape, and the size bias correction was implemented using the axis of the ellipse parallel to that of the array.

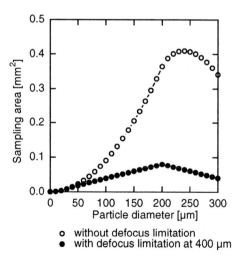

o without defocus limitation
● with defocus limitation at 400 μm

Figure 13 Example of effective size of the sampling space for spheres as a function of particle size with and without defocus limitation.

Figure 14 shows the distribution of particle passage positions for three size classes in the bisector plane of the incident beams which was inferred from the procedure described in section 2. The particles passed normal to the plane of the graph and each symbol corresponds to one particle. Note that positive and negative values of defocus are not distinguished; i.e., particles which are at a defocus distances of +200 and -200 μm were plotted at the same ordinate, +200 μm.

All the distributions were elliptical, which corresponded to the shape of the sampling space, and the position of the scatter plots for each size class remained within the sampling space determined by equation 6 as indicated with solid lines. The plots of the small size range in figure 14(a) was distributed uniformly, but the large particles with large defocus in figure 14(c) were not sampled as many as those with small defocus. This was due to the 'shallow' images which could not be triggered as stated in chapter 2 and this effect was removed by providing the defocus limitation. In figure 14, particles with defocus up to 400 μm were not affected by the shallow images and were acquired equally; thus, the particle size distribution could be corrected with sampling area of the spheres in equation 7 with defocus limitation at 400 μm in this configuration.

The empty region between defocus distances of 0 and 100 μm was caused by finite responsiveness of the photodetector, and particles there were estimated to be around D_{def}=100 μm which corresponded to one pixel displacement of the two projected images. The data, which has inconsistently large defocus distance referring to the diameter, was also rejected.

4.2 Correction of Size Distribution

Figure 15 compares the size distribution of polyethylene spheres with area-equivalent diameter obtained from microscope measurement with a CCD camera and frame grabber system. The distribution was calculated from the same data shown in figure 6 and was corrected by equation (7) and the microscopic data were provided with the area-equivalent size similar to the size from shadow image.

The size distribution corrected by the scheme based on the sampling area coincides with those from a microscope and it proves that the bias error has been effectively removed. The results showed higher reproducibility, as compared with amplitude-based method, because the fluctuation of the laser power, and the temperature drift of the detector, did not affect to the accuracy. The maximum error of average diameter was about 3.0 % and this result gives confidence that the correction scheme was appropriate for the method.

The corrected size distributions of non-spherical copper and stainless-steel particles are shown in figure 16 and 17, respectively, and compared with microscopic and aerodynamic diameter obtained from the terminal velocities. The results were from the free-fall particles presented in figure 9 and the biasing effect was corrected with the scheme described in the previous subsection. The

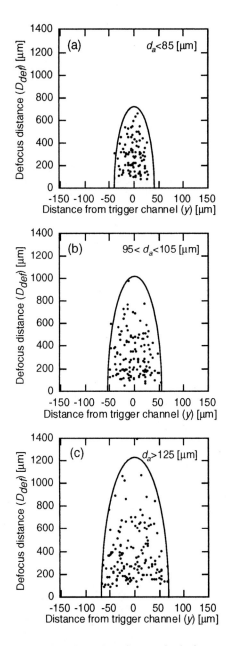

Figure 14 Position of particle trajectories of non-spherical copper particles for three size classes: (a) d_a<85 [μm]; (b) 95 [μm]<d_a<105 [μm]; (c) 125 [μm]<d_a. Particles motion is perpendicular to the graph.

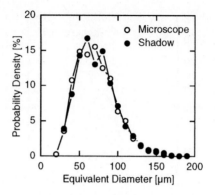

Figure 15 Size distribution of polyethylene spheres measured by a microscope (open) and shadow Doppler velocimeter (blocked).

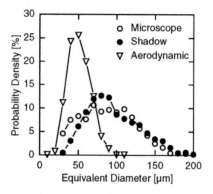

Figure 16 Size distribution of non-spherical copper particles measured by microscope and shadow Doppler velocimeter.

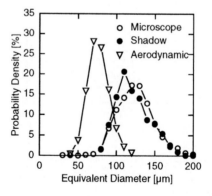

Figure 17 Size distribution of non-spherical stainless-steel particles measured by a microscope and shadow Doppler velocimeter.

distributions agree well with the microscopic, rather than the aerodynamic, diameters, calculated from their terminal velocities, and with better accuracy than the intensity-based methods which occasionally measure particles with large error. The large difference from aerodynamic diameter denotes that the particle has smaller mass than a sphere of the same projected area, and this tendency seemed to increase as increasing the irregularity of the particle shape. Unfortunately, there were not enough data to quantify the relationship, so further experiments are required in order to determine aerodynamic diameters by the instrument.

The maximum difference of average diameter from microscopic measurement was approximately 10 %, and the distribution measured by shadow Doppler tended to be biased towards smaller diameter as we expected: particles were likely to present their largest area to the observer when measuring by microscope and, hence, the distribution was biased to larger diameter than that from the present instrument. This result suggests that the diameter of an irregular particle measured by a microscope differs from the aerodynamic diameter, especially when Feret diameter, instead of area-equivalent, is used to define the particle size; so that the difference should be taken into account to estimate particle motion caused by the aerodynamic effect in applications relevant to two-phase flows.

5. Conclusion

In this study, the performance of shadow Doppler instrument was assessed with irregularly shaped particles in fundamental flows. Novel results are summarised as follows:

1. A signal processing method for shadow Doppler velocimetry has been demonstrated using computer controlled hardware. The procedure, which comprised temporal windowing, pattern recognition and correlation of the image, successfully provided information of area-equivalent size, shape and trajectory direction of the particles.

2. Simultaneous measurement of size and velocity of irregularly shaped particles has been achieved in an accelerating flow and quiescent air. Measurement of defocus distance allowed us to identify the erroneous data which was caused by poor images of particles with large defocus distance. Polyethylene particles in the suction flow resulted in about 10 % precision, which including a contribution from the fluctuation of the flow.

3. The size distribution of the particles was obtained using a correction scheme based on the effective size of the measuring volume. The resultant size distribution of polyethylene spheres agreed with the microscopic measurement within 3.0 μm error of arithmetic average. For irregularly shaped particles, the size information inferred from the two-dimensional projected image showed greater similarity with the distributions from the microscopic measurement, rather than that of aerodynamic diameter, and the maximum difference of arithmetic average diameter from microscopic measurement was about 10%.

374

6. Acknowledgement

The authors are grateful to Prof. J. H. Whitelaw, Dr. A. M. K. P. Taylor and Dr. Y. Hardalupas for their useful advice, and Mr. K. Kobashi for his help on construction of the hardware. We also thank Mr. Y Takubo and Mr. K Ishida for their assistance with the experiments.

7. Nomenclature

A	area of sampling space [m^2]
D	distance between two shadow images normal to the array axis [m]
d_a	area-equivalent diameter [m]
D_{def}	defocus distance [m]
D_h	distance between two shadow images parallel to the array axis [m]
d_w	width of sampling space [m]
f	sampling frequency [Hz]
G	magnification ratio of receiving optics[-]
φ	particle trajectory angle [deg]
l	defocus limitation [m]
N	ratio of sample number to the channel pitch of the array [-]
θ	beam intersection half angle [deg]
r	absolute amplitude of the signal [V]
r_{max}, r_{min}	maximum and minimum absolute amplitudes of the signal [V]
r_{norm}	normalised amplitude of he signal [-]
S_H, S_L	area of the shadow image below higher/lower threshold levels [m^2]
U	particle velocity [m/s]
W	width of active area of diode array [m]

8. References

Bertollini, G. P., Oberdier, L. M. and Lee, Y. H.: Image processing system to analyze droplet distributions in sprays, Opt. Eng. , Vol. 24, No. 3, 1985.

Bottlinger, M. and Umhauer, H.: Scattered light particle size counting analysis: influence of shape and structure, Optical Particle Sizing: Theory and Practice, G. Gouesbet and G. Gréhan Eds., Plenum Press, New York, 1988, pp. 363-369.

Clift, R., Grace. J. R. and Weber, M. E.: Bubbles, Drops, and Particles, Academic Press, 1978.

Durst, F., Zaré, M.: Laser Doppler Measurements in Two-Phase Flow, Proceedings of the LDA Symposium Copenhagen, 1975, pp. 403-429.

Hardalupas, Y., Hishida, K., Maeda, M., Morikita, H., Taylor, A. M. K. P. and Whitelaw, J. H.: A Shadow Doppler technique for sizing particles of arbitrary shape, Appl. Opt., 1994. In the press.

Hovenac, E. A., Hirleman, E. D. and Ide, R. F.: Calibration and sample volume characterization of PMS optical array probes, proc. of ICLASS, London, U.K, 1985.

Kaye, P. H., Eyles, N. A., Ludlow, I. K. and Clark, J. M.: An instrument for the classification of airborne particles on the basis of size, shape, and count frequency, Atmospheric Environment, Vol. 25A, No. 3/4, pp. 645-654, 1991.

Knollenberg, R. G.: The Optical Array: An Alternative to Scattering or Extension for Airborne Particle Size Determination, J. of Appl. Meteorology, vol. 9, pp. 86-103, 1970.

Knollenberg, R. G.: The use of low power lasers in particle size spectrometry, SPIE, Vol. 92, Practical applications of low power lasers, 1976, pp. 137-152.

Kobashi, K., Hishida, K. and Maeda, M.: Measurement of Fuel Injector Spray Flow of I.C. Engine by FFT Based Phase Doppler Anemometer, in *R. J. Adrian, D. F. G. Durao, F. Durst, M. Maeda, J. H. Whitelaw* (Eds.) Applications of Laser Technique to Fluid Mechanics, Springer-Verlag, 1990, pp. 268-287.

Kobashi, K., Hishida, K. and Maeda, M.: Multi-Purpose High Speed Signal Processor for LDA/PDA Using DSP Array, Sixth International Symposium Applications of Laser Techniques to Fluid Mechanics, 21.6.1-21.6.6, 1992.

Morikita, H., Hishida, K. and Maeda, M.: Simultaneous Measurement of Velocity and Equivalent Diameter of Non-spherical Particles, Part. Part. Syst. Charact. Vol. 11, No.3, pp. 227-234, 1994.

Orfanoudakis, N G, Taylor, A M K P.: Evaluation of a sizing anemometer and application to a small scale swirl stabilized coal burner, *Second International Conference on Combustion Technologies for a Clean Environment*, 19 - 22 July 1993, Lisbon.

Sanker, S. V., Buermann, D. H., Inenaga, A. S., Ibrahim, K. M. and Bachalo, W. D.: Coherent scattering in phase Doppler interferometry: response of frequency domain processors, 7th Int. Symp. on Appl. of Laser Techniques to Fluid Mechanics, Lisbon, paper 24. 4, 1994.

Sanker, S. V., Weber, B. J., Kamemoto, D. Y. and Bachalo, W. D.: Sizing fine particles with the phase Doppler interferometric technique, Appl. Opt., Vol. 30, No. 33, 1991.

Part IV
Instrumentation for Whole-Field Velocity

Interrogation and Validation of Three-Dimensional Vector Fields

C.D. Meinhart, D.H. Barnhart and R.J. Adrian

Department of Theoretical and Applied Mechanics, University of Illinois, Urbana, IL 61801, USA

Abstract. A holographic particle image velocimetry (HPIV) system has been developed to measure instantaneously all three components of velocity at order 10^6 points in a three-dimensional volume. A high speed interrogation system and an automated vector validation procedure is used to make this measurement technique applicable to turbulence research. The procedure by which holographic particle images are analyzed to obtain the three-dimensional velocity measurements from a single hologram, and how these measurements are validated using automated statistical algorithms is described in this paper.

Keywords. Particle Image Velocimetry, Holography, PIV Validation.

1 Introduction

The physics of turbulence can be better understood when its instantaneous three-dimensional structure is captured by measuring all three components of velocity in a three-dimensional volume with high spatial and velocity resolution. Over the past several years, two-dimensional PIV techniques have been used to measure two-dimensional slices of coherent structures in turbulent flow fields at relatively high Reynolds numbers. Holographic PIV is an extension of the standard two-dimensional PIV technique to three dimensions, enabling the three-dimensional structure of high Reynolds number turbulent flows to be measured.

The analysis of large three-dimensional holographic image fields using correlation techniques requires large numbers of computations. For example, the analysis of a $100 \times 100 \times 100$ mm³ volume, with 1 mm grid spacing, requires 10^6 two-dimensional correlations. If a conventional interrogation system can process 1-4 vectors per second using a pixel resolution of 128×128, it would take between 70 and 300 hours of processor time to analyze a 10^6 vector field. This is an unacceptably long time, especially if one wishes to interrogate many such fields. Therefore, an eight array parallel processor system has been developed which is capable of analyzing 100 vectors per second using 128×128 single-frame cross-correlations (see, Meinhart *et al.* 1993). This corresponds to analyzing 10^6 vectors in less than 3 hours. For comparison purposes, the peak theoretical performance of the eight array processors operating optimally in parallel is about 2.62 Mpixels per second.

After the image fields have been interrogated, the resulting vector maps contain spurious vectors. These spurious vectors are the result of noise in the correlation functions of the image fields during interrogation and they must be removed before the velocity fields can be considered valid measurements. It is both time consuming and tedious to sort through 10^6 vectors and remove the erroneous vectors manually. Therefore, an automated vector validation procedure has been developed, which uses algorithms to analyze statistically the reliability of each velocity measurement. When a vector is deemed erroneous, it is tagged and then removed automatically. The result is a reliable three-dimensional velocity field consisting of order 10^6 three-dimensional velocity vectors.

2 Analysis of Holographic Image Fields

2.1 Holographic Image Recording

The holographic imaging system used to record particle images is described in detail by Barnhart *et al.* (1994). This system records four distinct holographic images of every particle in the field of view onto a single holographic plate. Four particle images result by recording each particle from two different stereo views at the two different times, t_1 and t_2. The two stereo views are used to reconstruct the image field from two different perspectives, providing displacement information in all three directions. Exposures from times t_1 and t_2 are recorded with separate reference beams. This enables the exposures to be reconstructed individually and cross-correlated.

2.2 Holographic Image Acquisition

Fig. 1 shows the optics and the computer system used for reconstruction, acquisition and interrogation of the holographic image fields. The interrogation involves four steps: (1) measuring the two-dimensional displacements of the images that are recorded and reconstructed through L_{12}-P_1-L_{11}, (2) measuring the two-dimensional displacements of the images that are recorded and reconstructed through L_{22}-P_2-L_{21}, (3) validating the two-dimensional fields to remove erroneous velocity measurements, and (4) combining the two sets of two-dimensional measurements to obtain the three-dimensional displacements. The procedure by which two stereo-paired displacement fields are combined to obtain one three-component displacement field is similar to the stereo PIV technique described by Prasad *et al.* 1993.

During the first interrogation, the optical path through L_{12}-P_1-L_{11} is blocked so that only the particles images recorded through L_{22}-P_2-L_{21} are reconstructed. The images from t_1 and t_2 are reconstructed and digitized separately, using computer controlled switching between reference beams $R_1{}^*$ and $R_2{}^*$, which is synchronized with the image capture of the frame grabber. These image fields are then double-

(a)

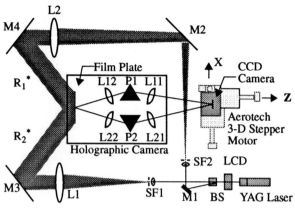

(b)

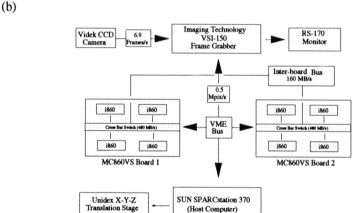

Fig. 1. Schematic of reconstruction and interrogation system for holographic images: (a) Optics configuration (top view), (b) Computer configuration (reproduced with permission from Meinhart *et al.* 1993).

frame cross-correlated, using the system described by Meinhart *et al.* (1993), to determine the two-dimensional displacement field of the images as viewed through L_{22}-P_2-L_{21}, see Fig. 1. This procedure is repeatedly applied to the entire reconstructed image volume by scanning the CCD camera over many X-Y planes.

The entire volume is then re-analyzed, except that the optical path through L_{22}-P_2-L_{21} is blocked so that only particle images recorded through L_{12}-P_1-L_{11} are reconstructed, digitized and interrogated. The resulting two-dimensional displacement measurements are validated using the automated routines, which remove erroneous measurements. The two-dimensional displacements fields are then combined to obtain all three displacement components on a three-dimensional grid. The process by which two stereo-paired velocity fields are combined is similar to the process used in two-dimensional stereo PIV by Prasad *et al.* (1993). This procedure is referred to as *stereo-stereo holography*, because it involves recording the images stereoscopically and reconstructing them stereoscopically. It is not the same as reconstructing a single holographic image and viewing it stereoscopically.

The displacement of the holographic images at each interrogation spot is found by double-frame cross-correlating the images of the first exposure I_1 and the second exposure I_2, where the cross-correlation function is defined as

$$C(s) = \int I_1(X) I_2(X+s)\, dX. \tag{1}$$

A comparison between autocorrelation and cross-correlation is illustrated in Fig. 2, where the autocorrelation function $R(s)$ of two superposed image fields $I_1 + I_2$, and the cross-correlation $C(s)$ between the image fields I_1 and I_2 are plotted. Clearly, the height of the signal peak (labeled C_D and R_D^+ for cross-correlation and autocorrelation, respectively) compared to height of the surrounding noise peaks is much higher for cross-correlation than it is for autocorrelation. The use of cross-correlation instead of autocorrelation is an important part of the holographic interrogation system, because in demanding situations cross-correlation provides more reliable velocity measurements than autocorrelation. In addition, double-frame cross-correlation eliminates the need for artificial image shifting, when the displacement vector is near zero or reverses direction. See Keane and Adrian (1992) for a detailed discussion of the theory of cross-correlation of PIV image fields.

3 Vector Validation

3.1 Introduction

The process used to validate the large numbers of vectors generated by holographic image fields is discussed below. Algorithms are used that analyze statistically the reliability of each velocity measurement, remove erroneous measurements and replace empty grid points with alternatively measured vectors.

A two-stage strategy is used to validate the velocity fields. During the first stage, all or nearly all the erroneous velocity vectors are identified and removed. During the second stage, alternatively measured velocity vectors are sought to fill the empty grid points, by determining the reliability of the alternatively measured vectors.

An interpolation scheme is then used to fill in empty grid points that are absent of velocity data with information from the neighboring velocity field. The vector fields are then smoothed by convolving the velocity vectors with a Gaussian smoothing kernel.

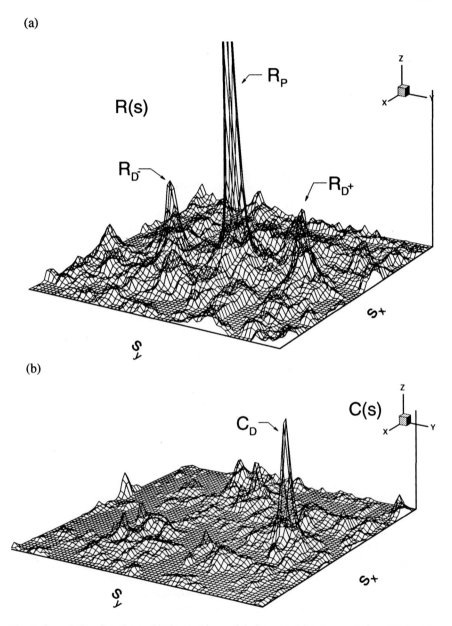

Fig. 2. Correlation functions of holographic particle images: (a) autocorrelation, (b) double-frame cross-correlation.

3.2 Removal of Erroneous Measurements

When removing erroneous vectors, it is important to remove all the outliers, even at the expense of sacrificing a few valid vectors. If the opposite viewpoint is adopted, where one tries to retain all the valid vectors with the possibility of retaining a few outliers, the outliers which are large in magnitude will likely corrupt the flow statistics.

We use three automated techniques, which are based upon the spatial relationships between in-plane velocity vectors, to remove erroneous measurements in PIV vector fields: (1) *tolerance removal*, (2) *magnitude difference removal* and (3) *quality removal*. These three in-plane techniques are supplemented by a *stereo matching* technique which matches two stereo-paired PIV velocity fields to remove erroneous vectors from stereo PIV image fields.

3.2.1 Tolerance Removal

Tolerance removal is considered a global technique because its criteria is based upon statistical quantities that are spatially averaged over the entire vector field, as opposed to statistics averaged locally in space. A tolerance is placed by requiring the vectors to be within a specified number of standard deviations from the mean, where the standard deviation (rms) and mean are estimated by averaging either in the x direction, the y direction, or by averaging in both the x and y directions. Global averaging decreases the effect that nearby erroneous vectors have on the estimated mean and standard deviation. In extreme situations, numerous outliers can adversely affect the statistics, causing unreasonably large values in the estimated rms velocities, which limits the effectiveness of the tolerance removal technique. In these situations, it becomes necessary to impose reasonable limits on the rms velocities, which enhances the effectiveness of the tolerance removal technique.

3.2.2 Magnitude Difference Removal

Magnitude difference removal is a local technique that mathematically defines how well a particular *vector in question* fits in with its surrounding neighborhood. This is considered a local technique because it is primarily dependent upon the vectors in the immediate neighborhood. Fig. 3 shows the spatial relationship of a vector in question and its surrounding 3×3 neighborhood. The velocity vector of the neighborhood is estimated by calculating the median of the neighborhood for each velocity component[1]. The median is a more robust average than the conventional mean average (Westerweel 1994). If a neighborhood contains several erroneous vectors that are large in size, the mean average of the neighborhood will be an inaccurate estimate of the actual neighborhood velocity, while the median average may essentially be unaffected by the neighboring outliers.

We define the magnitude difference Md as

---

1. The median technique was suggested by Jerry Westerweel.

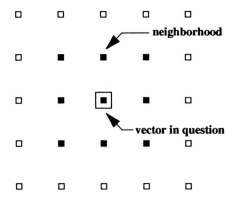

Fig. 3. Velocity vector grid showing the spatial relationship between a 3 × 3 neighborhood and its vector in question. This concept is used by the magnitude difference vector removal technique.

$$Md = \sqrt{\frac{(u - u_{med})^2}{u_{rms}^2} + \frac{(v - v_{med})^2}{v_{rms}^2}}, \qquad (2)$$

where u, v are the horizontal and vertical components of the vector in question, u_{med} and v_{med} are the horizontal and vertical components of the neighborhood median velocity. The velocity difference between the horizontal and vertical components of velocity between the vector in question and its neighborhood are normalized by their respective components of globally averaged rms velocity, u_{rms} and v_{rms}. The magnitude difference is the Euclidean norm of the normalized velocity components and the magnitude difference defines a functional that is an estimate of how well a vector fits in with its neighborhood.

By specifying a high value and a low value of magnitude difference, all the velocity measurements can be divided into three categories: erroneous vectors, questionable vectors and reliable vectors, according to Fig. 4. When the magnitude difference routine is executed, the erroneous vectors are removed, the reliable vectors are retained, and the questionable vectors are sent to the *vector-checker*.

The vector checker is the PIV analog of the spelling checker used in word processors. The idea is rooted in the fact that there may not exist a clear-cut mathematical line separating good vectors from bad vectors, but instead, a gray area between good and bad vectors, where vectors are considered questionable. While it may be difficult to develop an automatic algorithm that can determine whether a borderline vector is reliable or erroneous, it is much easier for an algorithm to seek out and identify questionable vectors, display them at the center of the computer screen, and prompt the user for a keep or remove decision. This

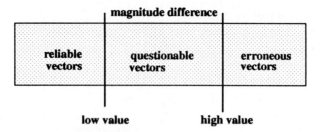

Fig. 4. Division of vectors into the three categories of reliable, erroneous and questionable, based upon the magnitude difference criterion.

utility makes it possible to cast judgement on many vectors in just a few seconds. The vector checker utility is only used as a last resort in difficult situations, when it is not possible to remove all the bad vectors with automatic routines.

3.2.3 Quality Removal

The quality of a vector is measured in the correlation plane during interrogation, and is defined as the ratio of the height of the highest displacement correlation peak to the second highest displacement correlation peak. Since the quality is a measure of the signal-to-noise ratio, it should be a good criterion for estimating the reliability of the measurement.

Experience shows, however, that quality is in fact a poor measure of vector reliability (Westerweel 1994). While it is true that nearly all bad vectors have quality values only slightly larger than unity, it turns out that reliable measurements can often have qualities with similar values to that of erroneous measurements. Therefore, we usually do not use quality to validate PIV measurements.

3.2.4 Stereo Matching

In stereo PIV, all three components of velocity are calculated by first calculating in-plane components of velocity in two stereo-paired velocity vector fields, and then combining the stereo-paired velocity vector fields to obtain all three components of velocity (see Prasad *et al.* 1993). The system of equations used to combine the two perspectives is over constrained, thereby over specifying one component of velocity. Since one component of velocity is measured independently from both the right and left perspectives of the hologram, the two measurements for each grid point should agree to within the accuracy of the HPIV measurements. With stereo matching, erroneous measurements are identified when the difference between the two vertical components of velocity at a given point is greater than a specified value.

3.3 Replacement with Alternatively Measured Vectors

The second part of the validation strategy is to replace the empty grid points with alternatively measured good vectors. The alternative replacement algorithm fills empty grid points with alternatively measured velocity vectors, if the alternatively measured vectors satisfy both a global criterion and a local criterion. The global criterion requires that the alternatively measured vectors (the Choice 2 or Choice 3 vectors) reside within a user specified tolerance. The local criterion requires the successful replacement vector to have a lower magnitude difference value than all the other alternatives. In other words, an empty grid point is filled with an alternatively measured velocity vector, if the alternatively measured vector is within a specified tolerance and fits in well with its neighborhood.

3.4 Interpolation

Isolated empty grid points that have reliable and well represented surrounding neighborhoods can be assigned an estimated velocity, u', based upon a weighted spatial average of the neighborhood

$$u'(x') = \int w(x - x') u(x) dx, \tag{3}$$

where the weighting function, $w(x - x')$, is a Gaussian or a similar-type function. A convolution-type average is typically less sensitive to noise than the median average and utilizes more information from neighboring velocity vectors than a simple linear interpolation. The convolution integral is evaluated in fully three-dimensional space for holographic vector fields and in two-dimensional space for standard two-dimensional PIV vector fields.

If empty grid points exist within the neighborhood, those points are excluded from the calculation. The reliability of the neighborhood estimate is ensured by requiring that a specified percent of the neighborhood velocity vectors are present before the empty grid point is assigned the estimate u'.

3.5 Low-Pass Spatial Filtering

The previous sections have discussed nonlinear filtering techniques used to remove erroneous measurements and to replace alternatively measured velocity vectors. A linear low-pass filter can also be applied to the velocity field by convolving it with a Gaussian kernel

$$G(x - x') = \exp\left(-\frac{(x - x')^2}{x_k^2}\right), \tag{4}$$

where x and x' are the spatial coordinate vectors, x_k is size of the kernel. This convolution is evaluated in three-dimensional space for holographic vector fields. The Gaussian filter removes high frequency noise from the velocity measurements which can result from random white noise generated by film grain noise and pixel digitization during interrogation.

3.6 Batch Mode Execution

Once a sequence of vector validation routines and their operating parameters are found to work well for several two-dimensional planes of a three-dimensional HPIV vector field, they are stored in a macro file. An entire holographic vector field consisting of $\sim 10^6$ vectors can then be validated automatically in about 3 hours by executing the validation program in batch mode.

4 Vector Validation Example

Typically, the mode of operation for validating three-dimensional holographic vectors is to: (1) use stereo-matching to remove vectors whose vertical components of velocity do not agree, (2) apply the tolerance removal technique to remove erroneous vectors that have very large magnitudes, (3) apply the magnitude difference technique to remove erroneous vectors that do not fit in well with the neighboring vectors, (4) replace the empty grid points with alternatively measured vectors (where applicable), (5) reapply tolerance and magnitude difference to remove any erroneous vectors not previously removed or incorrectly added during Step 4, and (6) three-dimensionally interpolate and smooth velocity fields (where applicable).

The performance of this sequence of operations was evaluated by applying the automated routines to a typical two-dimensional PIV vector field using steps (2)-(5) and then comparing those results to a manual validation of the same vector field. Fig. 5 shows the velocity field before and after the sequence of automated operations. The performance of each operation and the convergence to the manually validated field is displayed in Fig. 6. Fifty-five percent of the erroneous vectors were initially removed during interrogation, because there was not an adequate signal peak detected in the correlation function. After applying tolerance removal (Step 2) and then magnitude difference removal (Step 3), 93% and 97% of the erroneous vectors were removed, respectively. The application of the alternatively measured vector routine (Step 4) resulted in the replacement of 95% of the reliable Choice 2 and Choice 3 vectors. Magnitude difference was then applied for the second time (Step 4). The second application of magnitude difference removed 100% of the erroneous measurements.

The vector field used in this example contains roughly 10,000 grid points, out of which the automated removal routines correctly removed 100% (387 vectors) of the erroneous vectors, and the automated replacement routine correctly replaced 95% (93 vectors) of the reliable Choice 2 and Choice 3 vectors. The only error occurred when the automated routines failed to replace two Choice 2 vectors and two Choice 3 vectors. The performance of the automated routines displayed in Fig. 6 is not always achieved, but it is characteristic of most PIV vector fields, when data dropout is not too severe.

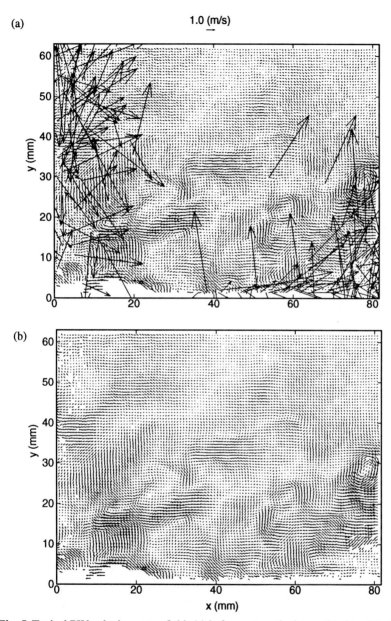

Fig. 5. Typical PIV velocity vector field: (a) before automatic data validation, (b) after data validation.

390

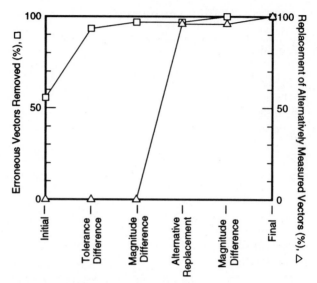

Fig. 6. Performance of a typical vector validation sequence. This graph shows the effect of the various automated validation routines on the percent of erroneous vectors removed and the percent of alternatively measured vectors. Convergence to 100% is achieved when the automated routines agree identically with manual decisions.

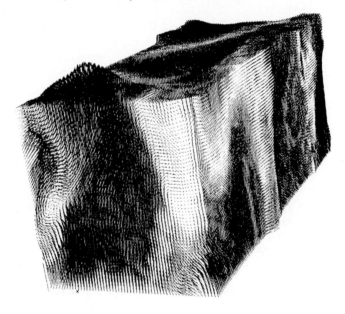

Fig. 7. Three-dimensional holographic velocity vector field (reproduced with permission from Barnhart *et al.* 1994).

5 HPIV Measurements of a Jet Flow

Fig. 7 shows an HPIV vector field, covering a volume of $24.5 \times 24.5 \times 60$ mm^3, of air flow issuing from a 76 mm turbulent pipe at Reynolds number Re = $U_B D/v$ = 6000. The interrogation spots are 0.87 mm^3 and overlapped by 50% in all three directions, giving nearly 400,000 three-dimensional velocity measurements. This vector field resolves in three dimensions nearly all the turbulent motion in the flow. The interrogation and vector validation procedures described in this paper allow three-dimensional velocity measurements such as the ones presented in Fig. 7 to be obtained with just a few hours of computer time.

6 Conclusions

A fully-automated high-speed holographic particle image velocimetry interrogation and vector validation procedure has been developed which can be used to measure over 400,000 three-dimensional velocity vectors on a three-dimensional grid. This technique is made feasible by incorporating a high-speed parallel processing system to analyzed over 100 vectors per second, using 128×128 double-frame cross-correlations. By using double-frame cross-correlation instead of autocorrelation, a larger percentage of reliable vectors is obtained, because cross-correlation produces a greater signal-to-noise ratio.

Automated vector validation routines have been developed that remove efficiently erroneous velocity measurements, which result from noise in the correlation functions during interrogation. These automated validation routines make it possible to validate millions of PIV velocity vectors in just a few hours. Test results show that, for a typical two-dimensional vector field, decisions made by the automated validation routines are nearly identical to manually made decisions.

Acknowledgments

This research was supported by the Office of Naval Research Grant N00014-90-J-1415 and a grant from TSI, Inc.

References

Adrian, R. J. 1991. Particle-imaging techniques for experimental fluid mechanics. *Annu. Rev. Fluid Mech.*, **Vol. 23**, pp. 261-304.

Barnhart, D. H., Adrian, R. J. and Papen, G. C. 1994. Phase-conjugate holographic system for high resolution particle image velocimetry. *Appl. Optics*, **Vol. 33**, No. 30, pp. 7159-7170.

Keane, R. D. and Adrian, R. J. 1992. Theory of cross-correlation analysis of PIV images, *Appl. Sci. Res.*, **Vol. 49**, pp. 191-215.

Meinhart, C. D., Prasad, A. K., Adrian, R. J. 1993. A parallel digital processor system for particle image velocimetry. *Meas. Sci. Technol.*, **Vol. 4**, pp. 619-626.

Prasad, A. K. and Adrian, R. J. 1993. Stereoscopic particle image velocimetry applied to liquid flow, *Exp. Fluids*, **Vol. 15**, pp. 49-60.

Westerweel, J. 1994. Efficient detection of spurious vectors in particle image velocimetry data. *Exp. Fluids*, **Vol. 16**, pp. 236-247.

HIGH-SPEED 3-D SCANNING PARTICLE IMAGE VELOCIMETRY (3-D SPIV) TECHNIQUE

Yann G. Guezennec, Yang Zhao and Thomas J. Gieseke

Mechanical Engineering Department, The Ohio State University
206 West 18th Avenue, Columbus, Ohio 43210, USA

Abstract. A 3-D Scanning Particle Image Velocimetry (3-D SPIV) technique has been developed to overcome the resolution problem associated with conventional 3-D Particle Tracking Velocimetry (3-D PTV) without requiring the complexity of 3-D holography. The technique employs a high repetition rate Copper Vapor Laser synchronized with a scanning mirror drum and sheet forming optics to scan the volume of interest. 25 or 50 individual mirrors are adjusted to form an array relatively thick laser sheets parallel to each another, hence effectively covering the entire volume of interest (typically 100x100x100 mm^3). This device is capable of recording the time evolution of a flow field in 3-D with a temporal resolution of 200 to 400 Hz, which is more than sufficient for most water flows. The digitized 16mm movies are then processed by a combination of techniques: conventional 2-D cross-correlation and 2-D PTV within each plane. Due to the 2-D nature of the recorded information, relatively high seeding densities can be used to yield 20,000 to 50,000 vectors per scan, but yet provide the temporal information associated with successive scans at a rapid rate. The physical implementation of this technique, the overview of the processing and some sample images are presented in this paper.

Keywords. scanning, stereo, PIV, PTV, imaging processing

1 Introduction

Over the last decade, the 2-D Particle Image Velocimetry (2-D PIV) has been well developed to measure two components of the flow velocity over a plane (Adrian 1986, 1991; Grant and Smith 1988, and many others).   This technique has significant advantages over other measuring techniques such as being non-intrusive and capable of providing instantaneous, multi-point measurements in a plane with a high spatial resolution. However, the conventional 2-D PIV technique, which relies on image auto- or cross-correlation is inherently restricted to two-dimensional measurements in a plane, as neighboring particles images are also neighboring in space and hence have relatively similar motion.   Recently, many attempts have been made to extend 2-D PIV to three-dimensional measurements.   Weinstein et al. (1985) employed a dual view, high speed, holographic movie technique for studying turbulent flow physics. Among others, Meng and Hussain (1991) applied a holographic particle velocimetry for 3-D measurements of vortex interactions and coherent structures. However, to minimize speckle noise, a very low seeding density must be used with in-line holography. In the long run, holography probably has the best potential for 3-D measurements, but the complexity of the experiments and of the hologram interrogation and processing makes this technique only usable by a few researchers in very controlled environments. In particular, the problem of optical distortion which is omnipresent in virtually any realistic experimental setup represents a serious challenge for holographic reconstruction, unless the same physical setup is used for hologram recording and for interrogation. In addition, the processing time for high density holography is very large and usually requires the implementation of the processing on parallel array processors as described by Adrian and his coworkers.

The other approach which has been used for 3-D flow measurement is 3-D Particle Tracking Velocimetry (3-D PTV) and its earlier implementation 3-D Particle Streak Velocimetry. Using long time exposures, the use of stereoscopic Particle Streak Velocimetry for estimating 3-D velocity field has been described by Sinha and Kuhlman (1992). Racca and Dewey (1988) carried out a study of 3-D flow fields where experiments were performed using an orthogonal-view cinematographic method. Adamczyk and Rimai (1988) have investigated the use of orthogonal views to reconstruct a 3-D flow field by two video cameras. More recently, Maas et al (1993) and Malik et al. (1993) reported the use of 3 or 4 synchronized CCD cameras for the determination of velocity vectors within an observation volume. Similarly, Guezennec and his co-workers (Guezennec et al. 1994, Trigui et al. 1994, Kent et al. 1993) have developed a very efficient and robust technique to implement 3-D Particle Tracking Velocimetry using only a pair of cameras. Their implementation uses single exposure sequence of five successive images taken by two cameras to reconstruct without any ambiguity or operator guidance the 3-D vectors. Furthermore, the use of an *in situ* calibration technique accounts very accurately and efficiently for all sources of distortion encountered in realistic experiments.

However, the chief disadvantage of all the stereoscopic 3-D techniques stems from the fact that the 3-D space is projected onto 2-D image planes. This contributes to the "crowding" of the images which inherently limits the number of tracer particles to several hundreds to a few thousands depending on the imaging resolution. All these techniques rely on the fact that distinct images of each flow tracer can be obtained with little or no image overlap, hence requiring relatively low seeding densities. On the other hand, they provide a rather convenient way of obtaining 3-D information without the difficulties associated with holography. In an attempt to extend the nature of the information obtainable from 2-D PIV, Brücker and Althaus (1992) applied PTV combined with volume scanning by a periodical sweep of a laser sheet to study the phenomenon of vortex breakdown. Their photographic images were digitized manually on a digitize table and they mainly obtained 2-D velocity distribution in the scanning volume. Prasad and Adrian (1993) recently provided an application of a twin-camera, stereoscopic system to obtain 3-D vectors on planar domains using conventional double-exposure laser sheet lighting. Owing to the planar nature of the illumination, again the seeding density can be quite high and this technique can easily yield several thousand 3-D vectors on a single plane.

To overcome the limitations of the various approaches, a 3-D scanning particle image velocimetry technique capable of measuring 3-D velocity vectors in a volume is presented in this paper. This approach is similar to the work of Brücker and Althaus, but differs in many respects. Through the use of a high speed movie camera coupled to a high repetition rate Copper Vapor Laser (CVL) and a specially designed scanning system, we attempt here to combine the advantages of the previous techniques to provide 3-D measurements with a relatively high seeding density. Sample images showing the feasibility of the physical implementation of the technique, as well as numerical tests aimed at developing and validating the processing algorithms on synthetic (yet realistic) images are presented in the next sections.

2 Experimental Apparatus

The experimental apparatus required to implement this technique consists of a Copper Vapor Laser (CVL) coupled with a high-speed movie camera and a rotating drum scanning device. The Copper Vapor Laser was chosen for its high repetition rate (up to 10 kHz for the one used here) which enables the rapid scanning of the volume with a scanning period commensurate with the time scales of the flow field of interest. The CVL produces very short (30 ns.) light pulses in the visible part of the spectrum and provide an effective way of freezing the motion of the particles. The output beam from the CVL is first collimated to provide relatively thick (1 to 2 mm.) laser beam. This laser beam is then sent to a rotating drum scanner which was designed and built specifically for this application.

2.1 Scanning Drum

Fig. 1 schematically depicts the configuration of the scanning device. It consists of a circular drum onto which 25 or 50 mirror holders are attached at equal intervals along the circumference of the drum. Each of the mirror holders is mounted in a key slot parallel to the axis of the drum which allows the positioning of the mirrors to be easily adjusted in the direction of the drum axis. For the remainder of this paper, the mirror holders were uniformly spaced along the z-axis, creating a regular helix pattern on the drum surface.

A small mirror is mounted on each holder at a 45° angle with respect to the axis of the drum. The incoming laser beam from the beam collimator is aligned with the drum axis and is aimed at the center of the mirrors. Upon striking each mirror, the laser beam is reflected perpendicular to the axis of the drum (in the meridianal plane) at the axial position of the specific mirror. By arranging the mirrors into an helix pattern around the periphery of the drum, the output laser beams forms a series of parallel laser beams, all mutually perpendicular to the axis of the drum and with a spacing set by the mirror positions. These beams are then expanded into parallel laser sheets using a long cylindrical lens mounted parallel to the drum axis. The drum is driven by a DC motor which rotates at constant (although adjustable) speed. An optical encoder also coupled to the drum shaft provides synchronizing pulses to trigger the Copper Vapor Laser. Interfacing electronics were designed to generate an adjustable number of pulses to the laser synchronized with each mirror.

The prototype of this scanning device used here is capable of forming 25 parallel laser sheets with a fully adjustable spacing. Different number of laser sheets can easily be obtained with a differently machined drum. Other scanning patterns can also be obtained with this same device by adjusting the position of each mirror holder in its key slot along the drum. Hence this device is inherently superior to single scanning mirror arrangements which do not generate parallel output beams and cannot follow arbitrary scanning patterns except at the slowest speeds. Since the drum of our device rotates at constant speed (and has significant inertia), a very stable scanning rate is easily achieved, which is in fact used as the "master" clock for the entire system. For the purpose of the technique described in this paper, the laser sheets are relatively thick (about 1 to 2 mm) and the spacing is adjusted such that the volume of interest is fully illuminated by adjacent or slightly overlapping laser sheets.

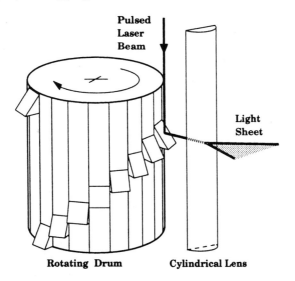

Fig. 1. Schematic Scanning Drum Device

2.2 Recording Device

A 16mm high-speed, rotating prism movie camera (Photec, 10,000 fps) is used to record the flow field. Since it is nearly impossible to have a truly constant camera speed, "streak mode" recording is used. Instead of exposing the film through the front-mounted lens and the rotating prism, the film is exposed directly through a lens mounted in place of the view finder without going through the rotating prism. Since our light source is pulsed, there is no blurring of the images on the film and the rotating prism is not necessary. This recording mode has the significant advantage of synchronizing the image capture to an external clock (here provided by the scanning drum and the laser) as long as the filming speed is high enough to avoid image overlap on the film. Timing lights are mounted inside the camera to expose registration marks on the edge of the film for later image registration and these timing lights are slaved to the external clock. Since the recording of the images is sequentially ordered on the film, we do not have the directional ambiguity associated with traditional double exposure PIV.

For this scanning technique to be practical, it is imperative to have sufficient depth of field to obtain sharp particle images at all positions along the scan. This requirement was addressed in several ways: One, by using relatively large particles (of the order of 50 microns) so that the particle images are not too small and a slight blurring near the edges is tolerable; Second, by using a good quality "long" macro lens used at a moderate F-number to maximize the depth of field. Our Photec 16mm camera was equipped with a Mamiya M-645, 80mm macro lens which is designed for use with 2"x2" cameras, and hence provides exceptional depth of field at an F-number of 8, since only the central most part of the lens is used for imaging on 16mm film. For the test experiment described here, the camera was placed approximately 80 cm from the test section and a 75 x 75 x 75 mm^3 volume was imaged. Good focus was achieved over the entire depth of the volume with this setup.

Since the energy per pulse is relatively low with our 10 W Copper Vapor Laser (approximately 1.5 mJ/pulse), a very sensitive B&W film was used to obtain good contrast images. Initially, we used a panchromatic Kodak recording film 2484, processed in D-19 for 8 minutes at 20°C. This approximately corresponds to a 3200 ASA film speed, at the price of a relatively coarse film grain. More recently, we have used a better Instrumentation grade film made by Kodak (SO-078) which has a finer grain at equivalent sensitivity and can be processed with standard TMAX processing. However, the size of the tracer particles and the magnification used here provided particle images which were at least 10 times larger than the film grain size. Hence the film graininess had little effect on the quality of the images for the intended purpose.

The system as described here provides an effective mean for recording high speed scans of the volume of interest onto a single film strip and where the images captured can be processed using conventional image cross-correlation to extract in-plane velocity component. Following the approach of Prasad and Adrian (1993), we are currently developing a split mirror arrangement to simultaneously record two views of the same volume onto the same strip of film. Such arrangement is depicted schematically in Fig. 2. Each view is recorded onto separate, side by side areas for each frame. This provides the necessary information for obtaining three components of velocity within each plane of illumination. This technique also eliminates the difficulties which are usually encountered when synchronizing two film cameras (which is impossible to do at the high filming speeds required by this technique). An *in situ* calibration similar to our previous work (Kent et al. 1993) allows to take into account all sources of distortion (index of refraction mismatch, perspective, lens aberrations, etc.) for the stereo-photogrammetric reconstruction of the 3-D vector field from the pairs of stereo images. This set-up allows the recording of the time evolution of a flow field in 3-D with a temporal resolution of 200 to 400 Hz (depending on the number of laser sheets per scan). This scanning rate is more than sufficient for most water flows and some low speed air flows.

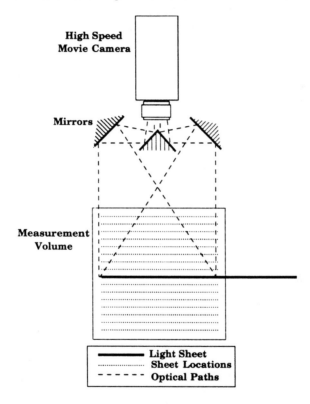

Fig. 2. Schematic of Split-Image Arrangement for Movie Camera

3 Data Processing

The resulting 16mm movies are digitized and processed by a combination of techniques. Since we were performing a proof-of-concept study, a low resolution was used to digitize the film. A 512x512 CCD camera was mounted on the digitization stand to provide images easy to manipulate on a PC. After digitization, the data base for each view was built by organizing the images in order of their sheet position and scan number, as suggested in Fig. 3. Since the

398

laser sheets are relatively thick, most of the particles do not move out of illuminated 'planes' between two successive scans.

Before carrying out any of the PIV- or PTV-specific processing, the quality of the images was enhanced by some simple image pre-processing. As it will be described later, we need to locate the individual particle in each image. To facilitate this step, it is best to enhance the quality of the image by removing the effect of lighting non-uniformity, as well as stationary features of the images (glares, boundaries, etc.). Two approaches are possible: One involves only a single image at a time, while the other involves multiple images. In the case of a single image, the enhancement is performed by subtracting from each pixel its

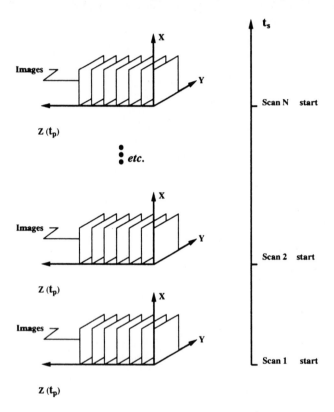

Fig. 3. Organization of Image Data Base
(for each side)

neighborhood average (computed by a Gaussian window filtering) and then stretching the contrast of the subtracted image. This effectively removes all the slow-varying components (in the spatial sense) of the image such as lighting non-uniformity. The resulting image is then easily binarized to clearly segment it into particles and background. An example of such processing will be shown in the Result Section for actual experimental images. The other approach which is more effective at removing not only slow-varying components of the image, but also all stationary features (regardless of their spatial spectral content), is to compute a running average of multiple successive scans (at the same z-position) and to subtract this running average from each individual picture. Our experience with 3-D PTV has shown that a running average of 5 to 10 pictures is more than adequate for this image enhancement step.

Once this image pre-processing is performed on all frames, the images from each view and from two successive scans can be processed by a conventional 2-D cross-correlation PIV technique. As mentioned earlier, since the images are singly exposed at each scan, there is no directional ambiguity as a true cross-correlation is used, as opposed to an auto-correlation for doubly-exposed PIV.

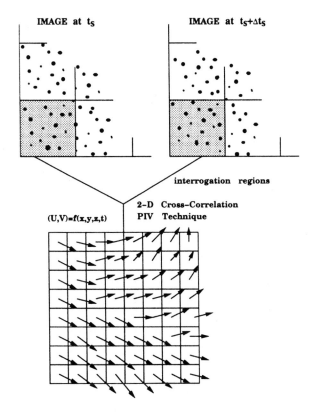

Fig. 4. Schematic of the 2-D Cross-Correlation Processing between Successive Scans

This process is schematically depicted in Fig. 4. Furthermore, if the cross-correlation levels are too low, possibly indicating that the out-of-plane displacements are larger than the thickness of laser sheet, space-time cross-correlations with adjacent laser sheets positions can also be computed and the location of the cross-correlation peak can be determined (albeit with a large quantization error) in the z-direction as well.

Unlike conventional 2-D PIV, we do not stop the processing at this stage and we use the velocity obtained from the cross-correlation to perform a "guided" tracking of each individual particle. This technique, which we pioneered several years ago (Guezennec and Kiritsis 1990) , has the advantage of increasing the data yield (number of velocity vectors obtained vs. the number of flow tracers imaged) by a factor of 10 or more as at least 15 to 20 particles per interrogation region are required for the correlation methods to provide a reliable neighborhood-averaged velocity. Furthermore, the higher yield of our technique allows to resolve velocity gradients more effectively and accurately. The computational overhead required to locate particles and track them (in a guided way with the correlation results) represents a negligible increase in processing time. This guided tracking is illustrated graphically in Figs. 5 and 6.

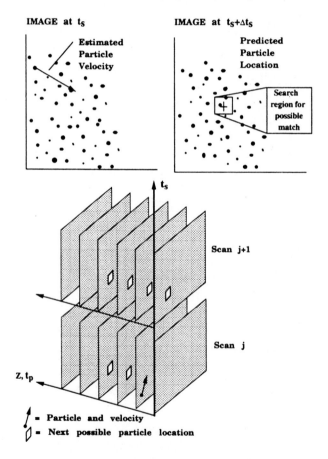

Fig. 5. Schematic Illustration of the 2-D "Guided" Tracking Between Successive Scans
(and Possibly Adjacent Laser Sheets)

Fig. 5 schematically depicts the procedure used for tracking between successive scans (and possibly z-planes). Based on the velocity estimated for a particle in the first frame, the corresponding particle position in the second frame is predicted. Then a simple outward spiral search (within some limits) is conducted to locate the closest actual particle in the second frame, and to effectively establish the match between particles. The velocity vector is then calculated directly from the particle positions and the scanning rate.

An illustration of this process with synthetic data is shown in Fig. 6. Fig. 6(a) depicts a section of a "true" velocity field (synthetically generated to test the algorithms, i.e. analytically prescribed and used for comparison and validation), while Fig. 6(b) shows the results which are obtained by 2-D cross-correlation over this region (divided into 8 x 8 sub-regions here). Naturally, the 2-D cross-correlation captures the essence of the flow field, but fails to adequately describe the local gradients present in this vortical flow field.

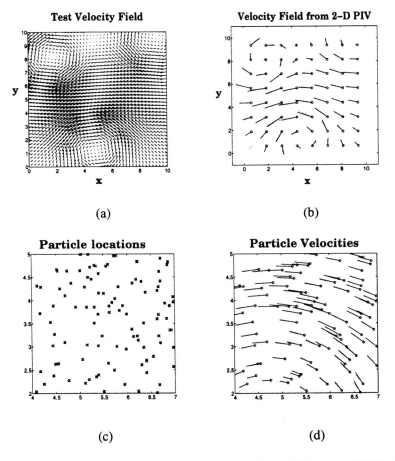

Fig. 6. Example of 2-D Cross-Correlation Processing and Subsequent 2-D Tracking

Furthermore, the data yield is low (64 vectors while approximately 1200 particles were imaged in this flow field). Naturally more vectors can be obtained from the correlation by processing with overlapping windows, but this is merely equivalent to interpolating or low-pass filtering the field. Fig. 6(c) shows the actual locations of particles which were present in a small sub-section of this flow field (around the top of the vortex located near the bottom center of the field). Based on the cross-correlation results of Fig. 6(b), the velocity field is first bi-linearly interpolated at the location of each of the particles. Then, using this interpolated velocity as an estimator, the particle is actually tracked between two successive frames as described in Fig. 5. This yields the results shown in Fig. 6(d) where the measured (solid lines) and "true" (dashed lines) velocity vectors are compared. As can be seen, this technique markedly improves the spatial resolution and accuracy of the cross-correlation processing alone. Typically, 1200 to 1500 individual velocity vectors can be extracted in this manner from a pair of 512x512 images.

This process can be performed for each view for all scans and sheets to obtain 2-D vectors over an entire volume and in time. At this point, the information from the two views can be recombined, using the *in situ* calibration information to compute 3-D vectors for each sheet. Two processing approaches are possible here: First, our initial idea was to actually track each individual particle in 3-D. This approach presents severe challenges as the third direction (z-axis) is highly quantized by the discrete laser sheets and, furthermore, particles may "drop out" between successive laser sheets due to poor or no illumination. The second approach is to take the results from the 2-D particle tracking described earlier performed for each view and smoothly interpolate these results on a regular grid for each side. Then, at each of the grid points, the two projections are recombined using the calibration information to calculate the three velocity components and the corresponding 3-D position. This approach is far superior to the first one described above, as local "drop-outs" in one view do not significantly affect the ability to stereo-match the data between the two sides. Based on some of our previous work in 3-D, we have found that an Adaptive Gaussian Window (AGW) filter is the best (and most robust) approach to perform reconstructive filtering on randomly spaced data in 2-D or 3-D. The width of the Gaussian interpolating filter is self-adjusting based on the average data density, hence avoiding too much or too little filtering.

Specifically, the 3-D reconstruction is implemented as follows: After the particles are tracked in 2-D on both sides, a regular grid is laid out on the data from the left side. The grid spacing in the laser sheet plane is self-adjusted to match the average data density in that plane, while the z-spacing is given by the physical laser sheet separation. Then, for each of these grid points (x_1,y_1) in the left view, the velocity projections (u_1,v_1) are computed at the grid points using the AGW filter. Using the calibration information, the image plane coordinates (x_1,y_1) are combined with the z-position (known by the laser sheet position) to first calculate the physical 3-D coordinates (x,y,z) and then associate the image coordinates in the right view (x_2,y_2) (in the corresponding laser sheet). Then an

AGW interpolation is perform locally around that point in the right view to evaluate the velocity projection in that plane (u_2, v_2). Finally, the 3-D velocity components are computed from the two matching projections using the calibration information to yield (u, v, w) at a known (x, y, z). This procedure has the advantage of being computationally explicit, take into account all sources of distortion without performing ray tracing, and avoid the "drop-out" problem which may not allow a particle-to-particle stereo match. Effectively, on the average it preserves the data yield obtained for the 2-D tracking through the stereo-matching step.

This new scanning technique has significant advantages over our previous 3-D PTV system, in that relatively high seeding densities can be used to obtain of the order of 20,000 to 50,000 velocity vectors per scan (due to the 2-D nature of the illumination). With the 3-D PTV, the entire 3-D field was illuminated and projected onto 2-D imaging planes, which resulted in "crowding" problems in the images for high particle densities. Furthermore, the tracking procedure is significantly simplified since the depth direction (along the camera's optical axis) is explicitly known by the position of scanning sheet.

4 Results

A test experiment was performed as a proof of concept, primarily designed to validate the approach and resolve many of the experimental and processing difficulties associated the technique as described in the previous sections. A small water tunnel with a square cross-section of 75 x 75 mm^2 was built. A small cylinder model with a diameter of 6 mm. was mounted across the center of the test section. The 3-D flow field in the near wake of the cylinder was imaged with the setup described here at a scanning rate of 35 Hz with 25 sheets per scan. The free-stream velocity in the test section was of the order of 5 cm/sec. These preliminary tests were performed to primarily verify that reasonable quality images could be captured on film and to test the complex synchronization between the Copper Vapor Laser, rotating drum scanning device and the high-speed camera.

Fig. 7 shows a typical subset of images taken in this wake flow. Only four images out of many thousands are shown here, corresponding to two successive scans in two adjacent laser sheets separated by 5 mm. It should be noted that the film used here is a negative film, hence the brightly illuminated particles appear as dark spots on the film. As mentioned earlier, the quality of the images is adequate, but significant lighting non-uniformity are present (mostly due to the intensity distribution within the incoming laser beam). However, even in these raw images digitized at 512x512 resolution, many particles are clearly visible. To enhance the quality of these pictures, the pre-processing described earlier was applied. Fig. 8 shows a typical example based on one of the raw images from Fig. 7. Fig. 8a displays the raw image, while the filtered image is shown in Fig. 8b. Finally, the binarized image is shown in Fig. 8c. For this case, the image pre-processing based on a single image was applied, which explains why the boundary reflections on each side are still visible in the final processed image. However at this stage, it is rather trivial to locate the particles using a simple thresholding technique coupled with our fast, single-scan particle locator software used in all our previous work (Guezennec and Kiritsis 1990, Guezennec et al. 1994). While this pre-processing is not required for the cross-correlation step of the PIV processing, the image enhancement tends to provide cleaner correlation maps due to the removal of stationary features and of most of the pixel noise, i.e. only the particle images are cross-correlated.

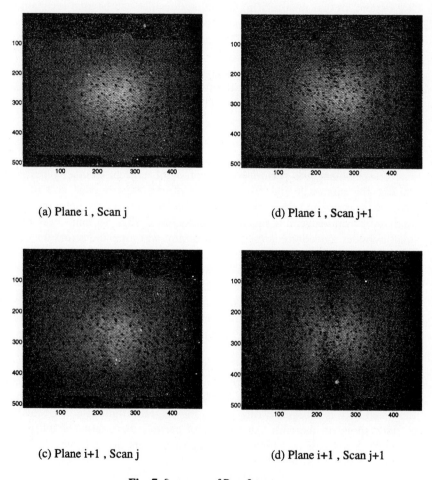

(a) Plane i , Scan j   (d) Plane i , Scan j+1

(c) Plane i+1 , Scan j   (d) Plane i+1 , Scan j+1

Fig. 7. Sequence of Raw Images

We have not yet carried out the entire PIV/PTV processing on these real experimental images, as we are carrying out further developments in image registration on the film. Since the PIV processing is performed by crosscorrelating two different images on the film (in this case separated by one entire scan of 25 images), it is imperative to have excellent registration of the film during digitization to ensure accurate results which are not biased by an overall image shift. The approach which we are currently working on is to expose four corner registration marks (dots) on the film, between the image area and the film perforations. In a preliminary step, these marks will be digitized and a cross-correlation of these relatively large dots will be performed to provide sub-pixel image registration information. This information will then be fed to our servo-controlled film transport system to precisely position the film before final digitization of the actual frame.

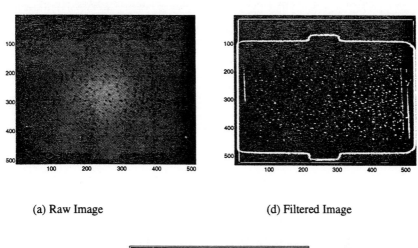

(a) Raw Image   (d) Filtered Image

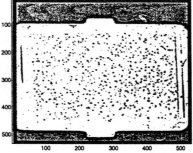

(c) Binaried Image

Fig. 8. Image Preprocessing

For the purpose of developing and testing much of the PIV/PTV processing, synthetic images corresponding to known (prescribed) flow fields were generated using typical experimental particle seeding densities, particle image size, calibration information, etc. These synthetic images were very similar in nature to the ones obtainable from the actual experimental setup after pre-processing (Fig. 8c). This approach, which we have used extensively in our previous 2-D and 3-D work (Guezennec and Kiritsis 1990, Guezennec et al. 1994), allows to systematically examine every step of the processing and understand the sources of errors, hence performing a validation which is not really possible with actual experimental data. Preliminary results from this study show that data yield of 1000 to 1200 vectors per laser sheet are easily achievable, even with the low digitization resolution used here (512x512). The typical accuracy for the 2-D vector projections is of the order of a few percent (as illustrated in Fig. 6). We have not yet combined the last 3-D reconstruction step into our software, but based on our previous 3-D PTV work (Kent et al. 1993), we know that calibration and 3-D reconstruction errors can be kept below 1%. Hence, we do not anticipate that the intrinsic accuracy of the entire processing will be dramatically degraded. As we had found before, the most controlling factor for the velocity error is the accuracy with which particles can be located in each 2-D image. Here, in addition, there is an inherent averaging (and quantizing) effect over the laser sheet thickness and position. This effectively diminishes the accuracy of the z-direction and the corresponding w-velocity by comparison with the in-plane components.

5 Summary and Conclusions

A 3-D, time-resolved, scanning measurement technique has been developed and tested. This approach is characterized by a relatively practical physical implementation, which is far simpler than an holocinematographic approach. This technique can provide 3-D velocity components in a 3-D volume with a relatively high yield (of the order of 20000 to 50000 vectors per scan, depending on the number of laser sheets used), and at the same time provide the temporal evolution of this flow field. Our current implementation of the processing software on a Pentium-based PC is capable of processing each image pair (pre-processing, zonal cross-correlation and 2-D tracking) in approximately 30 seconds. Hence, a complete scan of 25 sheets with two stereo images per sheet can yield 20000 to 50000 vectors in approximately 25 minutes of processing on a modern PC. This is much faster than hologram processing which is extremely computationally expensive for high yields and usually requires dedicated parallel array processors to be practical. In summary, this approach represent a practical alternative to holography when relatively high data density is desired in 3-D.

Acknowledgment

This work was supported by Ford Motor Company and FloCoTec, Inc. Their support is gratefully acknowledged.

References

Adamczyk, A. A. and Rimai, L. 1988: Reconstruction of 3-dimensional flow field from orthogonal views of seed track video images. Exp. Fluids **6**, 380-386.

Adrian, R. J. 1986: Multi-point of simultaneous vectors in unsteady flow - a review. Int. J. Heat Fluid Flow **7**, 127-145.

Adrian, R. J. 1991: Particle-imaging techniques for experiment fluid mechanics. Ann. Rev. Fluid Mech. **23**, 261-304.

Brücker, Ch. and Althaus, W. 1992: Study of vortex breakdown by particle tracking velocimetry (PTV). Part 1: Bubble-type vortex breakdown. Exp. Fluids **13**, 339-349.

Grant, I. and Smith, G. H. 1988: Modern developments in particle image velocimetry. Optics and Lasers in Engineering **9**, 245-264.

Guezennec, Y. G. and Kiritsis, N. 1990: Statistical investigation of errors in particle image velocimetry. Exp. Fluids **10**, 138-146.

Guezennec, Y.G., Brodkey, R. S., Trigui, N.T. and Kent, J. C. 1994: Algorithms for fully automated three-dimensional particle image velocimetry. Exp. Fluids **17**, 209-219.

Kent, J. C., Trigui, N. T., Choi, W.-C., Guezennec, Y. G. and Brodkey, R. S. 1993: Photogrammetric calibration for improved three-dimensional particle tracking velocimetry (3-D PTV). Proc. SPIE' 93 **2005**, 400-412.

Malik, N. A.; Dracos, Th. and Papantoniou, D. 1993: Particle tracking velocimetry in three-dimensional flows : Part II Particle tracking. Exp. Fluids **15**, 279-294.

Mass, H. G.; Gruen, A. and Papantoniou, D. 1993: Particle tracking velocimetry in three-dimensional flows : Part I Photogrammetric determination of particle coordinates. Exp. Fluids **15**, 133-146.

Meng, H. and Hussian, F. 1991: Holographic particle velocimetry: a 3-D measurement technique for vortex interactions, coherent structures and turbulence. Fluid Dyn. Res. **8**, 33-52.

Prasad, A. K. and Adrian, R. J. 1993: Stereoscopic particle image velocimetry applied to liquid flows. Exp. Fluids **15**, 49-60.

Racca, R. G. and Dewey, J. M. 1988: A Method for automatic particle tracking in a three-dimensional flow field. Exp. Fluids **6**, 25-32.

Sinha, S. K. and Kuhlman, P. S. 1992: Investigating the use of stereoscopic particle streak velocimetry for estimating the three-dimensional vorticity field. Exp. Fluids **12**, 377-384.

Trigui, N. T., Kent, J. C., Guezennec, Y. G. and Choi, W.-C. 1994: Characterization of intake-generated flow fields in IC engines using 3-d particle tracking velocimetry. SAE Paper No. 9402790.

Weinstein, L. M., Beeler, G. B. and Lindemann, A. M. 1985: High-speed holocinematographic velocimeter for studying turbulent flow control physics. AIAA-**85**-0526.

Multiple Light Sheet Particle Holography for 3-D Flow Velocimetry

H. Hinrichs and K.D. Hinsch

FB 8 - Physik, Carl von Ossietzky University,
Pf 25 03, D-26111 Oldenburg, Germany

Abstract. Holographic recording promises the ability to extend Particle Image Velocimetry (PIV) to a fully 3-D measuring method, where PIV is no longer limited to the instantaneous registration of 2-D velocity vectors within a single sheet. The recording of small particles in a 3-D volume introduces, however, additional noise in the interrogation step. This is investigated in more detail. First the holographic recording of a continuous deep volume is considered. Sampling the depth by spaced lightsheets allows a first improvement of the signal to noise ratio and provides a higher validation rate in the interrogation step. In addition, the light for illumination of the particles is used in a more economic way. If dense sampling is required the concept of simultaneous recording of all light sheets, but reconstruction of single sheets each can be used. This is possible by taking advantage of the limited coherence of the laser. A demonstration of this method is included. The validation characteristics of typical evaluation methods like autocorrelation and crosscorrelation techniques are estimated by Monte-Carlo-simulations. This gives the ability to find the method best suited for the fluid mechanical problem under investigation.

Keywords. Holographic Particle Image Velocimetry (HPIV), 3-D velocity fields, validation, light sheet, coherence, Kármán vortex street

1 Introduction

In recent years, the problems tackled by experimentalists and theoreticians in fluid dynamics have grown considerably. Sophisticated flow facilities like large size or high speed wind tunnels are the basis for ambitious experiments, and powerful computers allow modelling of very complex flow fields. Interest has turned to nonstationary flows, their spatial structures and their development in time. An ultimate aim is to measure and understand the evolution of turbulence.

The metrological tools for the experimental study of flow fields have also seen an impressive improvement. Optical probing of the flow through tracer particles has been preferred because it allows noncontact remote measurements with high spatial resolution. It relies, however, on the presence of small particles in the flow and assumes, that the particle motion faithfully resembles the fluid motion. In gaseous flows this requires very small particles and sets limits to the largest velocity gradients acceptable.

Fluid velocity, the characteristic parameter of the flow, is a three-dimensional quantity and many flow configurations of interest are of three-dimensional nature. Especially the structural character of a flow, a topic of current interest, is only revealed by its spatial features. When the flows are very voluminous or measuring time is expensive, the effort or the cost to scan the flow with a single point probe are enormous and may be even prohibitive. Even more, when the flows are nonstationary there is no time to move a single point probe through the flow field.

In view of these problems, much recent effort has been dedicated to the development of flow velocimetry in several dimensions. Particle image velocimetry (PIV) is presently the most successful technique for quantitative whole-field flow velocimetry (Hinsch (1993a)). It relies on the double exposure high-speed recording of tracer particles and provides the simultaneous registration of the in-plane velocity components within a plane from the flow volume. The field of observation is selected by the placement of the illuminating light sheet. Usually the particle pair separation is evaluated by processing small interrogation areas. Photographic and video techniques contend for the recording, digital and optical methods for the processing.

Strictly speaking, PIV is not a whole-field technique because velocity is measured at locations in a single plane only and without the depth component. The study of complex and nonstationary flows calls for extensions in all dimensions. Two basic approaches are pursued presently:

- Conventional imaging with more than one camera from different directions. This includes stereoscopic recording of a light sheet as well as video imaging of a large field with several cameras. Depth of focus is a limiting factor in these methods.
- Holographic recording of a deep volume. This solves the focus problem and allows almost unlimited depth of field. Thus, for true three-dimensional studies holography is the method of choice. The reconstructed particle fields can be processed by conventional techniques, stereoscopically, for example.

Holographic records cannot be evaluated by the reliable interrogation schemes developed for PIV. Often, costly three-dimensional particle tracking must be employed. We have therefore developed a concept that preserves favourable properties of ordinary PIV while offering the extension into the third dimension. Yet, the method is restricted in its application to moderately three-dimensional flows because the flow is still sampled by light sheets. For a deep-volume study the system features are:

- The light sheet is retained to establish the region of investigation.
- The deep volume is covered by a hologram of several light sheets at proper spacings in depth.
- Optimum use is made of the illuminating light by multiple pass through the volume.

- Crosstalk between light sheets can be avoided by coherence adjustments in the holographic recording.
- Light sheets are evaluated one by one by the wellknown methods of interrogating a PIV transparency.

2 Holographic Velocimetry of Particle Fields

2.1 The Principle of Holographic Particle Recording

Holography is the only method for true imaging of three-dimensional space. All particles of interest in the flow are illuminated with a properly taylored laser light wave. In a commonly employed configuration (Fig. 1), the complex light field scattered by the particles is stored by superposing a reference wave from the same laser source and recording the interference pattern on a photographic plate. Due to the oblique angle between reference wave and object light this is called off-axis holography. The coherence length of the laser must be adequate to guarantee interference over all path length differences encountered. The developed pattern (the hologram) is used to reconstruct an image of the object. For this purpose the hologram is illuminated with the original reference wave that reconstructs a virtual image of the particle field. When looking or photographing through the hologram the particle configuration can be seen as it was during the recording of the hologram.

A very useful feature of holography is that also a real image of an object can be obtained without the need for a lens system. For this purpose, the hologram must be illuminated with the complex conjugate of the reference wave. For a divergent spherical reference wave this is a convergent wave from the opposite direction. In this case the hologram produces backward travelling object waves that form a real image of the light sheet. With a CCD-array, for example, a plane anywhere through the field could be examined for particles. If aberrating media like windows or low quality lenses have distorted the object wave on its way to the holographic recording plate, these distortions can be compensated during the backward path of the conjugate wave provided the distorting media are still in the old position.

It should be mentioned that holographic particle imaging in the off-axis configuration provides a solution of the ambiguity problem in double pulse particle velocimetry. When operated with a separate reference beam for each of the pulses the first and second image have been distinguished by Coupland (1987) for removing the directional ambiguity that is inherent in simple double pulse recordings. Hussain (1993) and Adrian (1993) have used a dual pulsed Nd:YAG laser system, for example, to supply suitable pulses that are aligned to travel identical object illumination paths but different reference paths. Upon reconstruction, first and second exposure can be viewed separately.

Off-axis holography is easily disturbed by vibrational noise, requires good coherence, and relies on an accurate reproduction of the reference wave for reconstruction. For small objects like the flow tracing particles, the so-called in-line arrangement is much simpler and provides better stability (Meng (1993)). In this case there is no extra reference wave, because the tiny particles leave enough of the illuminating plane wave light undisturbed so that it can serve as reference light. In-line holography relaxes many of the restrictions mentioned for off-axis holography and its application in particle studies has a long history (see, for example, Thompson (1989) and Vikram (1990)).

Recording

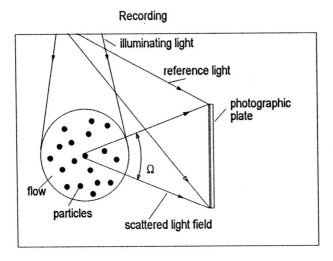

Reconstruction of virtual images

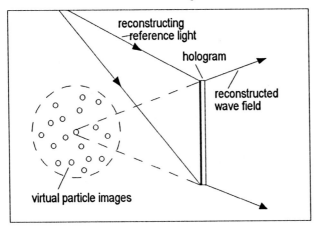

Fig. 1: Holography of a deep flow volume

2.2. Resolution in Holographic Particle Recording

Particle velocimetry calls for small and high quality particle images. Basically, the resolution in optical imaging is limited by diffraction and is determined by the angular spread Ω of the light that contributes to image formation. Omitting nonessential numerical factors, we obtain for the transversal resolution length (λ being the wavelength of light)

$$d_t \approx \lambda / \Omega \tag{1}$$

and for the longitudinal resolution length

$$d_l \approx \lambda / \Omega^2 \tag{2}$$

When a particle is smaller in size than the resolution length it will not be resolved in the image, when it is larger its geometrical image has to be convolved with a point spread function whose width is given by the resolution length.

Generally, in holography the angular spread is limited by the size of the holographic plate and its distance from the particle. This assumes that the particle scatters evenly over the range occupied by the plate. For the commonly used particle sizes (diameter d_p), this is the case for a large angle between illumination and observation direction (as for the 90° viewing in PIV). In the in-line arrangement, however, the pronounced contribution from forward scattered light is essential. The intensity in this angular region exceeds the average intensity in other directions by several orders of magnitude. It is concentrated in the central lobe of the diffraction pattern of the particle which is given approximately by $\Omega = \lambda / d_p$. Thus

$$d_t \approx d_p \tag{3}$$

$$d_l \approx d_p^2 / \lambda \tag{4}$$

In many practical cases where d_p is considerably larger than λ longitudinal focusing will be much lower than transverse focusing.

Various proposals have been made to improve the resolution in in-line holography: an auxiliary illumination beam has been introduced by Royer (1977) to enhance the scattered object light or two traditional in-line setups at right angles have been combined by Hussain (1993) or Bernal (1993). In another approach multibeam illumination was introduced by Hussain (1993) to increase the aperture by recording several scattered fields, each of extent Ω, on the same hologram. Thus highly effective forward scattering is combined with good axial focusing by a large aperture. Such setups, however, get rather complicated so that a great advantage of in-line holography is sacrificed. These considerations and the beneficial features mentioned earlier place a large bonus on off-axis holography for particle velocimetry. Despite the disadvantage that the overall intensity of the scattered light is much lower for sidewise observation we have therefore made off-axis holography our method of choice.

2.3 Noise in Holographic Particle Recording

The application of holographic particle velocimetry to a volume of large depth introduces a new problem. When the positions or displacements of a pair of particles must be evaluated from a reconstructed image (real or virtual) the light from additional particle images produces a disturbing noise background. This is especially pronounced in in-line holography, where light from both the virtual and real image overlaps and in addition is disturbed by the unused portion of the reconstructing wave. Off-axis holography deserves another credit in this respect, as the various wave components propagate in different directions. The contributions from out-of-focus particles, however, cannot be eliminated when a deep volume is illuminated as a whole. Here, the light sheet concept of PIV is superior, because only the particles of interest are illuminated.

For an evaluation of an ordinary deep-volume holographic recording the virtual image of the particle field can be photographed by focusing on different planes in depth, one after the other. When the planes are oriented parallel to the dominant flow direction these records provide the basis for interrogation as in PIV. In this case, what is the degradation due to the many out-of-focus particles?

We have carried out an analysis that is based on a numerical simulation of the image plane superposition of more or less focused images from particles that are distributed within a volume of given depth. A single plane is well in focus, the rest of the volume is sampled in depth by a series of other planes at equal 1mm intervals. All particles are assumed to scatter the same energy that is spread evenly over a disk-shaped image. Within the depth of focus, the size of the image is given by the diffraction limit (16 μm in the present case, set by the F-number of 8 of the imaging optics), outside by the appropriate geometric growth determined by the 4^0 cone of light limited by the aperture. Particle positions are placed at random, the average number per 1 mm^2 interrogation area of 1 mm depth equals 15. This has been found by Keane (1990) to be a proper choice for reliable evaluation. Particle pair separation in the focus plane was chosen to be 160 μm, the minimum value for good performance. Concerning sampling in depth we assume for the moment being that the characteristic spatial depth scale of the flow is smaller than the spacing between sheets. Velocity values in neighbouring sheets are thus independent, so that for each out-of-focus plane magnitude and direction of the displacement was varied randomly up to a maximum displacement of 500 μm. Under these conditions it is not possible to validate an evaluation by interpolation for which oversampling is required. This will be discussed once more at the end of this paragraph.

It must be mentioned that in the photography of particles from a

414

holographically reconstructed double exposure image field, the contributions from first and second exposure particles add up coherently. This differs from a directly obtained double exposure record, where they are superimposed one after the other and thus are summed incoherently. In sheet oriented PIV with low particle density this difference is of no importance because hardly any particle images overlap. In our case, however, the out-of-focus images of considerable size offer many possibilities for interference. Thus, in our simulation we attributed a random phase to each particle image. The constant phase approximation within a particle image is justified in the present geometry because the cone of light is restricted to less than the first Fresnel zone.

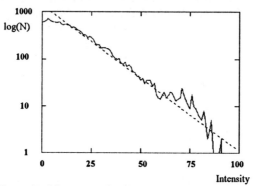

Fig. 2: Intensity histogram for 60mm volume

The image intensity calculated from this model environment was processed like a common PIV record. We calculated the power spectrum by a Fourier transformation, applied a zero order blocking filter and obtained the autocorrelation function from a second Fourier transformation. Finally, a search for the highest peak was to yield the particle displacement in the plane in focus. The value thus obtained was compared with the input displacement. An experiment was rated successful when both the values agreed.

With growing depth of the volume the image plane intensity showed an increasing amount of noise that deteriorated the in-focus images and spoiled the corresponding double structure. The intensity distribution gradually assumed speckle character which was proven by calculating the probability density function of the intensity. For a 60 mm deep volume Fig. 2 demonstrates the negative exponential dependence.

Accordingly, the autocorrelation peaks for the in-focus displacement gradually were buried in noise. This development is shown by Fig. 3 where we have plotted scans through the autocorrelation function along the direction of the displacement. A single plane particle field is compared with a volume of one centimeter depth (10 layers of model particles). The shown examples are from the outcomes that could be evaluated correctly.

Light sheet of 1mm thickness:

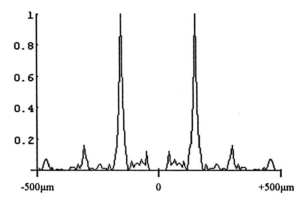

Volume of 10 mm depth:

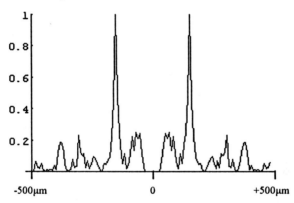

Fig. 3: Scans through autocorrelation functions (center is blocked)

Table 1. Results of simulation

depth [mm]	autocorrelation		crosscorrelation	
	failed evaluations (of 100)	confidence interval [%]	failed evaluations (of 100)	confidence interval [%]
1	4	[0,7]	0	[0,7]
10	14	[7,20]	0	[0,7]
20	56	[45,66]	16	[8,21]
40	72	[63,80]	58	[51,66]
80	97	[95,100]	-	-

The results of the experiments are summarized in the first row of Table 1. 100 realizations were calculated for each depth of volume and the number of erroneous evaluations determined. Furthermore, the 95% confidence interval is also given.

The table illustrates that a single plane is evaluated correctly in 96% of all cases, which compares favourably with the results obtained by Keane (1990). When the volume is expanded to a depth of 10 mm, already 14% of the evaluations are erroneous, for 80 mm depth this rate goes up to 98%. Our simulations thus suggest that the simple expansion of the depth of the inspected region finds a limit at about 10 mm.

A crosscorrelation evaluation instead of an autocorrelation provides two benefits. The first is that not only the value and orientation but also the sign of the velocity vector will be obtained. The second advantage - in this discussion the more interesting one - is the improved signal to noise ratio. This is due to the reduced number of possible particle pairings that produce the background noise. Furthermore, the crosscorrelation peak must not compete with a high central maximum as in autocorrelation. Therefore, the simulations are repeated with the same parameters but crosscorrelation. The second row in Table 1. shows that, indeed, a depth twice the size in autocorrelation is allowed. In a 20 mm deep volume, for example, 16% of the data will be incorrect. This is nearly the same value as obtained in the 10 mm deep volume interrogated by autocorrelation.

It is interesting to compare the results of our simulation with the performance of an experimental holographic system for flow velocimetry. Barnhart (1994) has evaluated phase conjugate real images of an off-axis hologram by crosscorrelating the light intensity from small interrogation areas picked up by a CCD-target. He was able to analyse a flow structure of 60mm depth with a validation rate of 85% which is much better than our predictions. His conditions, however, differ from our model in two aspects. Firstly, crosscorrelation operates with dynamically adjusted windows to reduce the loss of image pairs. Secondly and probably more important, the flow structure changes only slightly within the 1mm sampling interval in depth. Thus, the crosstalk from neighbouring planes does not contribute so much to noise. Rather the averaging over depth improves the rate of valid data. Our simulation, on the contrary, assumes independent data in adjoining sheets. The improvement is made possible because of oversampling in depth and must be paid for by handling a much larger amount of partly redundant data. A simulation of an accordingly modified flow field is planned to cover this situation.

2.4. Holography of multiple light sheets

The simulation has shown that the probability for correct evaluation of the particle separation from a plane of focus through a 3-D field of particle images decreases considerably due to the noisy influence of out-of-focus particles. We have therefore introduced an alternative approach in the holographic exploration of flows by recording a set of light sheets that sample the flow field in depth (Hinsch (1990), Hinsch (1993)). If the flow displays a predominant direction that determines the orientation of the light sheet, the velocity often changes only gradually in depth. When spacing the sheets according to the depth scale of the flow structures, much valuable information about the three-dimensionality of the flow can be obtained with a system of just a few light sheets. At the same time, favourable properties of ordinary PIV are preserved. First of all, optimum use is made of the available light by redirecting the sheet several times through the flow (Fig. 4). Because the particles scatter only little light, the quality of the light sheet stays sufficient for several passes. If needed, suitable beam shaping optics can improve the sheet dimensions after multiple passes through the measuring field. Secondly, each light sheet is available for any of the well-established or developing evaluation techniques including three-dimensional aspects like stereo interrogation. Similary to PIV, for example, a photo can be produced of any of the light sheets for further interrogation.

The holographic light sheet concept promises also signal-to-noise-ratio (SNR) improvement, because dark regions separate the illuminated sheets. In the evaluation of a specific plane in depth, however, noise from the other sheets may still be disturbing.

Again we have used a simulation as in the earlier case of a deep volume. We distribute ten sheets in depth each seperated by a given interval and focus on the closest of these planes. We then calculate the light distribution in the image plane and the autocorrelation function and determine the reliability for measuring the right particle separation for the in-focus plane.

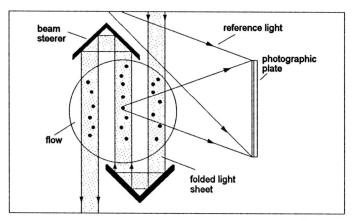

Fig. 4: Redirecting the light sheet with prisms to cover a flow volume

Let us first choose a separation of 1 mm between sheets. This configuration is equivalent to the particle arrangement that we have studied earlier in the 10 mm deep volume, i.e., the continuous illumination in depth. The simultaneous presence of particles in ten sheets thus raises the error to 14% as compared to 4% for the single sheet. The same type of calculations were done for sheet separations of 2, 4, and 8 mm. It was found that for ten sheets with 2 mm separation an error of 9% still must be accepted while at 4 mm spacing the error

418

has resumed the value for single sheet evaluation, i.e., 4%. The calculations thus suggest that single sheet evaluation improves the SNR over multiple sheet evaluation as long as the volume is densely sampled over a depth range of some 20 to 30 mm. These data are a good estimate for improvements by our technique of a separate reconstruction of light sheets which avoids crosstalk in densely sampled deep volumes by the reconstruction of single light sheets one after the other due to a sophisticated balance of the limited coherence of the laser.

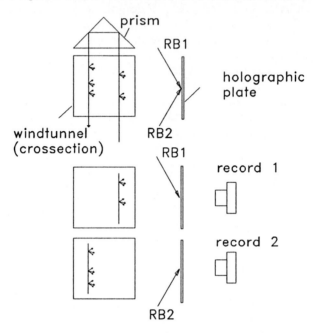

Fig. 5: Setup for holographic recording with limited coherence and selective reconstruction of single sheets.

The principle of this method is illustrated in the example of Fig. 5., showing the setup with two sheets used for an experiment on a wind tunnel flow. We are looking at the 0.25 m x 0.25 m cross section of the tunnel in the measuring region. A prism beam steering device redirects the sheet through the tunnel at a spacing that is smaller than the flow scales to be resolved. The scattered light from both the light sheets is incident on a holographic plate together with two reference beams from different directions in space. All of these waves are present at the same time, but provisions have been made that light from each sheet interferes only with one reference beam each. For this purpose the limited coherence length of the laser light is used. Mode selective devices in the laser produce a coherence length that is just sufficient for the holographic recording of a single light sheet together with a reference beam that matches in optical path length. Thus, each reference beam length is made to coincide with just one light sheet. By shifting through the set of reference beams during reconstruction, particles in one sheet can be evaluated without disturbing light from any other plane in depth.

There are some restrictions to this method. First, the intensity and quality of the reconstructed images decreases with the number of holograms on the plate. Second, a minimum angular separation between the reference beams must be observed. Finally, the coherence function of the laser repeats with a period of twice the resonator length. Because each new sheet calls for at least one coherence length in delay the resonator length divided by the coherence length gives the number of sheets that can be recorded. If more sheets are desired, a reference beam will reconstruct more than one sheet. In this case optical crosstalk is minimized when these sheets are spaced as far apart as possible.

To illustrate the principle of the method we would like to recall a result that we have presented earlier, a study of the oblique vortex shedding at a cylinder. The wake of circular cylinders at low Reynolds numbers is characterized by the well-known von Kármán vortex street. Ideally, this is a two-dimensional structure. Often, however, it is observed that the vortices are not shed parallel to the cylinder axis. The oblique shedding has been attributed to effects from the termination of the vortex axes at the ends. The slight three-dimensionality of this phenomenon is ideally suited to demonstrate the capability of our PIV method. Already two planes reveal important features of the flow structure.

The experiment was carried out in a wind tunnel at a speed of 0.47 m/s and with a cylinder of 6 mm diameter (Re = 190). Two light sheets were produced at 40 mm distance. 5 μm water droplets containing 10% glycerin served as tracer particles. The light originated from a 10-J Q-switched double pulse ruby laser (optical resonator length 0.7 m) that was modified for about 0.2-m coherence length. For this purpose one etalon in conjunction with the output mirror was removed. The original second etalon was exchanged for a model of specifically designed Q-factor to yield the desired coherence. A flow section of 50 mm x 85 mm about 55 mm downstream from the cylinder was observed. Holograms were made on Agfa Gevaert HOLOTEST 10E75 plates.

Virtual images of the double exposure particle fields were reconstructed from the hologram. To avoid distortions by a wavelength change, ruby laser pulses were also used for reconstruction, not changing the optical setup. Particle images were photographed with an 85-mm lens at F-number 4 and magnification 0.33. Identical magnification in both light sheets was guaranteed by translating the camera for optimum focus without changing its setting. For further processing, positive contact transparencies were made of the original photographs. Quantitative velocity data were obtained by automatic processing of the transparencies in a combination of Young's fringe generation and digital computation of the autocorrelation function. In the final plot the mean velocity is subtracted to emphasize the structures in the flow.

420

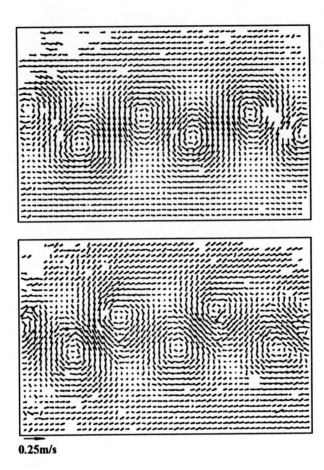

0.25m/s

Fig. 6: Velocity in two planes in the wake of a cylinder. The spatial shift between back and front plane indicates oblique vortex shedding. The mean velocity has been subtracted, flow from right to left.

Velocity vector plots of both the light sheets are shown in Fig. 6. The cylinder is positioned 55 mm upstream to the right and halfway up the frame. The vortex street is evident in both the planes. A clearly visible spatial shift in the periodic structure is evidence of the oblique vortex shedding. It amounts to about 9 mm which yields an angle of about 13°

3 Conclusions

Our analysis has shown that moderately three-dimensional flows can be sampled successfully by the holographic recording of a set of light sheets. First of all this method makes best use of the available light when a sheet is redirected through the flow repeatedly. Furthermore, a study of signal-to-noise conditions has shown that this is superior to the extraction of a single plane from a deep-volume holographic recording. We have also analyzed the noise produced by out-of-focus particles in the multiple light sheet concept and estimated the improvement that can be gained if individual light sheets can be reconstructed separately from the holographic record. For densely sampled volumes the reliability of the evaluation can thus be raised by our light sheet holography with controlled coherence. Since this requires more effort in the optical setup, an estimate of the gain can guide in system decisions.

References

Adrian, R.J. et al. (1993), An HPIV system for turbulence research, in: Holographic Particle Image Velocimetry, E.P. Rood, ed., ASME FED 148, 21.

Barnhart, D.H. et al. (1994), Phase-conjugate holographic system for high-resolution particle-image velocimetry, Appl. Opt. 33, 7159.

Bernal, L.P. and Scherer, J. (1993), HPIV measurements in vortical flows, in: Holographic Particle Image Velocimetry, E.P. Rood, ed., ASME FED 148, 43.

Coupland, J.M. et al. (1987), Particle image velocimetry: Theory of directional ambiguity removal using holographic image separation, Appl. Opt. 26, 1576.

Hinsch, K.D. et al. (1990), Holography with controlled Coherence for 3-D Particle Image velocimetry (PIV), in: HOLOGRAPHICS 1990, T. Tschudi, ed., Messago Publishing Company, Nürnberg.

Hinsch, K.D. et al. (1993), Holographic and stereoscopic advances in 3-D PIV, in: Holographic Particle Image Velocimetry, E.P. Rood, ed., ASME FED 148, 33.

Hinsch, K.D. (1993a), Particle Image Velocimetry, in: Speckle Metrology, R.S. Sirohi, ed., Marcel Dekker, New York.

Hussain, F. et al. (1993), Holographic particle velocimetry: prospects and limitations, in: Holographic Particle Image Velocimetry, E.P. Rood, ed., ASME FED 148, 1.

Keane, R.D. and Adrian, R.J. (1990), Optimization of particle image velocimeters. Part I. Double pulsed systems, Meas. Sci. Technol. 1, 1202.

Meng, H. et al. (1993), Intrinsic speckle noise in in-line particle holography, J. Opt. Soc. Am. A 10, 2046.

Royer, H. (1977), Holographic velocimetry of submicron particles, <u>Opt. Commun. 20</u>, 73.

Smigielski, P. et al. (1982), Visualisation of subsonic air flow section using high speed holography and measurement of the velocity field, <u>SPIE Vol. 348 High Speed Photography,</u> 608.

Thompson, B.J. (1989), Holographic methods for particle size and velocity measurement - recent advances, in: Holographic Optics II: Principles and Applications, G.M. Morris, ed., <u>Proc. SPIE 1136,</u> 308.

Vikram, C.S. (1990), "Holographic Particle Diagnostics," <u>SPIE Milestone Series,</u> Bellingham.

A COMPACT AND SIMPLE ALL OPTICAL EVALUATION METHOD FOR PIV RECORDINGS

Andreas Vogt, Frank Reichel[1], and Jürgen Kompenhans

Institut für Strömungsmechanik, Deutsche Forschungsanstalt für Luft- und Raumfahrt (DLR), Bunsenstraße 10, D-37073 Göttingen, Germany

[1]Jenoptik Technologie GmbH, Carl-Zeiss-Straße 1, D-07739 Jena, Germany

ABSTRACT

The Particle Image Velocimetry (PIV) is a technique which is utilized for measuring complete velocity vector fields in a few microseconds time. However, a lot of time is required during the evaluation of the PIV recordings. If the auto-correlation technique is used it is necessary to calculate a few thousand auto-correlation functions (ACFs) for each PIV recording. In this paper an evaluation system for photographical PIV recordings will be described, which allows a completely analog-optical determination of the ACFs of small sub areas of the recording. An optically addressable liquid crystal spatial light modulator was utilized to store the Young's fringes pattern appearing in the frequency plane at the output of a first optical Fourier processor and to provide them as the input of a second optical Fourier processor. Our analog optical autocorrelator offers a high processing speed and simultaneously a high resolution in the autocorrelation plane. The realized set up is compact and very easy to handle.

KEYWORDS: Particle Image Velocimetry (PIV), optical autocorrelator, Young's fringes, optically addressable spatial light modulator, liquid crystal spatial light modulator, optical Fourier transform

1 INTRODUCTION

Today Particle Image Velocimetry (PIV) is a quite well known measurement technique in fluid mechanical research. Although complete systems with satisfyingly performance are already commercially available, the development is still going on since PIV is still a relatively new method.

Some of the existing PIV systems utilize CCD cameras and video techniques for the recording of the PIV images and then perform a completely digital evaluation. However, for fluid mechanical problems with increased requirements concerning the space-bandwidth product (i. e. high resolution and a large observation field) the conventional photographic technique is still superior to the digitally working video systems. Photographic film (24×36 mm^2 format) can have a resolution of more than 100 line pairs per mm (lp/mm), whereas CCD chips with e. g. 2048 × 2048 pixels and 25 × 25 mm² sensor size achieve only 40 lp/mm. Because a huge amount of information contained in one PIV exposure has to be processed during the evaluation, systems utilizing the conventional photographic recording have a second major advantage: It is possible to perform a great part of the evaluation of the exposures analog optically so that the computation time can be reduced significantly. Furthermore, with digital systems there is still the problem of transferring the recorded data (4 MByte/recording for the above example) onto a recording device in a reasonable amount of time and also to supply the storage capacity for up to several hundred images during a measuring campaign.

For the evaluation of the PIV exposures a few thousand two-dimensional ACFs per exposure need to be computed. For each of these functions the local maximum indicating the mean particle displacement in a small subregion of the original image has to be found [1], [2]. The computation of the ACF, which is done indirectly by computing two two-dimensional Fourier transforms in most cases, and the peak detection in the autocorrelation plane are the critical procedures for the PIV evaluation with respect to time. The desired ACF can be determined digitally or, with relatively simple setups, optically. There have been quite a few suggestions how to realize such optical based techniques [3], [4], [5]. Although the digitally and optically based systems increased their processing speed considerably in the last years for the time consuming evaluation of the PIV recordings, a faster determination of each ACF is still desirable and the ultimate aim remains the evaluation in real time. Therefore the competition between the digitally and optically working techniques has not stopped yet. Digitally based systems increase their processing speed every few months by utilizing new and more powerful hardware components whereas on the optics side new optical components are developed and employed. Because in principle the optical systems can yield the ACF in real time (the determination of the involved Fourier transforms is carried out with the speed of light) the work in this field still continues.

As mentioned above two two-dimensional Fourier transforms have to be determined to get the desired ACF [1]. The input to the first optical Fourier processor is achieved by simply illuminating a small so called 'interrogation area' of the photographic negative of the PIV exposure with a laser light beam. The major problem for the existing optical systems is to get the output data of the first optical Fourier processor, the so called Young's fringes, into the second one. In principle the output of the first processor can be recorded with a video camera and subsequently be fed to a liquid crystal spatial light modulator (LC-SLM) [4], [5]. Another possibility is to write the fringe pattern optically to an appropriate light modulator. Optical set ups with electronically adressable liquid crystal light modulators are easy to handle and allow frame rates (data input) about 100 Hz. However, due to the finite size of their pixels (currently about 60 × 76 μm²) the spatial resolution is limited. Furthermore the input data to these devices, the fringes, need to be digitized before they are fed into this kind of light modulator. The optically adressable $Bi_{12}SiO_{20}$ (BSO) spatial light modulators on the other hand have a higher spatial resolution but at the same time require some effort to operate (high voltage, flash lamps for erasing the data, etc.).

The optically working evaluation system presented here combines the advantages of the above mentioned techniques by simultaneously avoiding their disadvantages. Our system utilizes an optically adressed liquid crystal spatial light modulator capable of frame rates in the order of the video frequency and with a high spatial resolution of 40 lp/mm and an active area of 10 mm in diameter (in the frequency domain). The device is extremely easy to handle (operating voltage ca. 5 V a. c., frequency 100 - 2000 Hz) so that the entire optical setup itself is also quite simple and compact.

2 EXPERIMENTAL SET UP

2.1 Optical part of the evaluation system

Currently two systems for the evaluation of photographical PIV recordings by means of the optically addressable SLM (type 300 p/01 of *Jenoptik Technologie*) have been realized. The one built at DLR makes use of already available hardware components from an existing analog-optical / digital evaluation system which involves the employment of a workstation for the peak detection and post processing of the data. The second system which was developed at *Jenoptik Technologie* is a PC based system. The optical set up for both systems is the same in principle. Figure 1 shows the optical part of the evaluation system as it was realized at DLR.

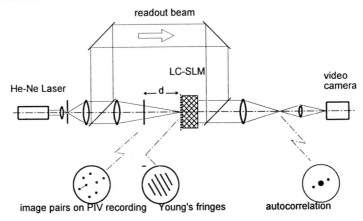

Figure 1: All optical autocorrelator for the evaluation of photographical PIV recordings

The light source, a He-Ne laser located on the left side of Figure 1, is followed by a spatial filter which removes the imperfections of the original laser beam. The spatial filter improves signal to noise ratio in the frequency plane as well as in the autocorrelation plane. After the spatial filter the beam is splitted. The rays passing the beam splitter (the write beam) are focused by the subsequent Fourier lens onto the light modulator. The PIV recording which is to be analyzed is placed between Fourier lens and Fourier plane, i. e. in the converging part of the beam [10]. (This arrangement allows an adjustable interrogation spot size on the PIV recording.) This way the Young's fringes pattern (i. e. the two dimensional spectrum) corresponding to the particle image displacement inside the present interrogation spot is imaged onto the light sensitive layer of the light modulator.

The readout beam is reflected onto the back plane of the SLM by means of two mirrors and a second beam splitter. This beam impinges onto the right hand side of the SLM and is phase modulated by the device according to the Young's fringes pattern. (The working principle of the optically addressable spatial light modulator is explained in detail below.) Due to a mirror inside the SLM the readout beam is reflected back again. The part of the now modulated readout beam passing the beam splitter is then also Fourier transformed by another Fourier lens thus yielding the two dimensional spectrum of the Young's fringes pattern, i. e. the autocorrelation of the greyvalue distribution inside the inter-rogation spot. The ACF is recorded by a video camera.

2.1.1 Optically addressable LC spatial light modulator

The optically adressed LC-SLM (see also [6], [7], [8]) is a sandwich system consisting of a photoconductor, a liquid crystal layer, a dielectric mirror, alignment layers, and transparent electrodes on glass substrates (figure 2). The alternating voltage applied to the transparent electrodes is divided according to the impedance of the different layers. As light impinges onto the photoconductive material, the conductivity of the latter increases so that the corresponding increase in the voltage difference across the liquid crystal results in a variation of the orientation of the molecules. This SLM may therefore be used to transform two-dimensional intensity distributions into suitable distributions of refractive index on the output side, provided the selection and control of suitable photoconductors does not produce any noticeable lateral charge balancing.

In the SLMs Plasma-CVD-deposited a-Si:H on indium-tin-oxide (ITO) - coated glass substrates is used as a photoconductive layer. To reflect the light, a multilayer dielectric mirror is produced between a-Si:H and LC by electron-beam evaporation of TiO_2 and SiO_2. With planar, untwisted alignment a nematic LC is a uniaxial birefringent medium. Alignment is achieved by a polyimide layer.

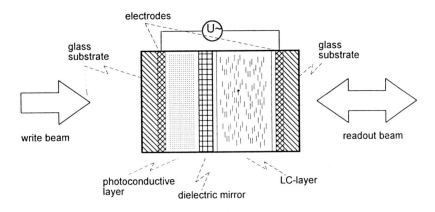

Figure 2: Principle of the optically adressable liquid crystal light modulator

Due to the birefringent properties of the molecules light passing through the liquid crystal is phase modulated proportional to the local tilt of the molecules. The write beam with the input pattern incidents from the left onto the device whereas the readout beam falls onto the device from the right. The readout beam passes the liquid crystal and is reflected by the dielectric mirror so that it passes the crystal twice. Leaving the device the readout beam is thus phase modulated corresponding to the input pattern.

The scope of possible SLM applications in Fourier processors is determined, among other things, by the absolute and spectral sensitivity of the LC-SLM [9]. Figure 3 shows the illumination dependence of the phase shift $\Delta\varphi$ of the readout light with different drive frequencies. For this measurement the readout wavelength is given by $\lambda_{out} = 551$ nm and the write wavelength by $\lambda_{in} = 633$ nm. The direction of polarization of the readout light is parallel to n_e, the extraordinary refractive index. At a drive frequency of $\omega = 100$ Hz and an effective voltage of $U_{rms} = 3{,}25$ V an illumination intensity of $I_\pi \approx 10$ µW/cm^2 is required to generate a phase shift of $\Delta\varphi = \pi$. The higher the selected drive frequency, the higher the illumination intensity must be chosen ($I_\pi \sim \omega$). The stabilisation of the phase shift on the readout side is guaranteed by controlling the SLM drive frequency. The maximum spectral sensitivity for typical a-Si:H layer thicknesses of about 3 µm ranges between 600 and 630 nm.

In another test rectangular gratings were projected onto the a-Si:H-layer with a line-to-space ratio of 1:1 to determine the resolution of the device. From the intensity of the zeroth order of diffraction I_0, the diffraction efficiency η can be determined as a function of the spatial frequency ε of the grating. Figure 4 shows the dependence measured for a LC layer thickness $d_{LC} = 6$ µm. A diffraction efficiency of max. $\eta = 100$ % is obtained for rectangular phase profiles with a line-to-space ratio of 1:1 and a phase deviation of $\Delta\varphi = \pi$. The drop in diffraction efficiency with increasing spatial frequency is due to the change in phase form caused by a spread of the field distribution in the LC and a lateral charge balance in the a-Si:H or the mirror system, respectively. With the usual LC thicknesses of $d_{LC} = 6$ μm, the resolution limit is around 35 - 40 lp/mm.

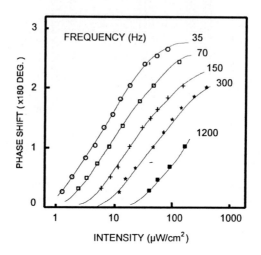

Figure 3: Dependence of phase shift on write light intensity for different drive frequencies

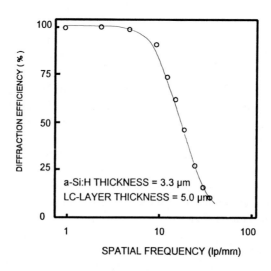

Figure 4: Modulation transfer function of the LC-SLM

The dynamic behaviour of the LC-SLM, i.e. the optical response to the illumination pulses, is determined by the speed of the voltage change between the a-Si:H and LC-layers as well as by the dynamics of the LC molecule distribution of the LC-layer. Figure 5 shows an oscillogram with the switching behaviour of the SLM. The illumination pulse was attained with a mechanical shutter, the exposure intensity I_π giving a phase difference of the readout light of $\Delta\varphi = \pi$.

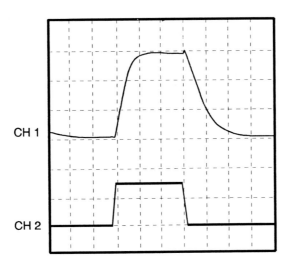

Figure 5: Time response of the LC-SLM (CH1: output light response; CH2: write light, horizontal scale 20 ms div^{-1})

2.2 Digital part of the evaluation system

After the ACF is recorded the coordinates of the peaks in the autocorrelation plane representing the particle image displacement are still to be determined. This is usually done digitally. The ACF is therefore transferred to the memory of a computer by means of a frame grabber. In our case a workstation (SUN sparc station 10/41), or a 486-PC, are utilized for this purpose.

With no a priori knowledge about the position of the correlation peaks the whole autocorrelation plane has to be searched for these peaks. The algorithm implemented on the workstation clips the greyvalue distribution of the ACF at an automatically calculated threshold and connected parts of the ACF with grey-values bigger than the threshold are labeled. In a second step the centre of these labelled areas is determined considering the grevalues of the individual pixels and their location. A curve fit of the peaks with e. g. a two dimensional Gaussian function is also implemented. The peak finding algorithm is quite time consuming and together with the digitizing process requires on the order of 1 second for each interrogation spot with the workstation based system. Since the ACF can be provided at nearly the video frequency a dedicated processor (pipelined digitizer and array processor) could significantly improve the interrogation speed. This has not been implemented up to now, as the fully auto-matic DLR PIV system is rather aimed at high quality (high spatial resolution, small fluctuations, no loss of data) than at high speed evaluation.

The PC based system of *Jenoptik Technologie* comprises an IBM compatible 486-PC equipped with slots for a x-y stage controller, the SLM controller and the frame grabber. The CCD camera is a Sony XC-77 CE (756 × 581 pixels), working on 2/3" interline-transfer-module basis. The x-y stage can be operated with a maximum speed of 52400 steps /second (minimum step size 1 μm). An alternating (square-wave) voltage programmable in amplitude and frequency is generated by the SLM controller and this a.c. voltage is selected according to the PIV film contrast. Using this system components in combination with a plain peak detection algorithm in an interactively defined region of interest a processing speed of 165 ms per velocity vector can be achieved. Therefore an evaluation of a PIV recording consisting of approx. 3500 interrogation spots as e. g. in Figure 9, 10 or 12 lasts less than 10 minutes. It is expected that this speed can even be increased by a factor of two through the use of PC systems with higher bus speeds (e. g. local bus, PCI bus).

3 TEST OF THE NEW EVALUATION SYSTEM

Some basic tests were performed before the first PIV recordings were analyzed by the all optical autocorrelator. First of all the linearity of the system was checked. For this purpose we utilized some 'artificial' PIV exposures, so called specklegrams. The utilized specklegrams were achieved with a double exposure of a speckle pattern before and after this pattern was shifted by a certain amount. Therefore they show a noise pattern of constant displacement over the whole image. The displacement of the different specklegrams is determined by a visual measurement using a microscope.

The evaluation of specklegrams with three different displacements (152 μm, 217 μm, and 406 μm, resp.) confirmed the linear relationship between the displacement on the analyzed images and the location of the correlation peaks in the ACF plane.

To determine the scale in the ACF plane further tests were performed by evaluating three different specklegrams at different positions with respect to the distance between PIV recording and SLM in the optical set up [5]. With these tests also a linear dependency between the scale in the ACF plane and the distance d between the recording and the corresponding first Fourier plane (see figure 1) was checked and proven. The recordings with the nominal displacement of 151 μm and 406 μm were evaluated at two different distances d and the recording with the nominal displacement of 217 μm at three different distances.

These measurements were simultaneously utilized to calibrate the entire evaluation system, i. e. the optical and the digital part of the system as a whole. This is necessary because the result of the peak determination is related to the pixel grid of the frame grabber that recorded the ACF. Thus for calculating the displacement on the analyzed PIV recording a relation between these pixel coordinates and the original displacement with the unit 'meter' has to be established.

Figure 6 shows the results in form of the determined calibration factor. The expected linear relationship between displacement on the recordings and the distance d on the one hand, and the displacements given by the peaks in the ACF plane on the other hand is obvious. The standard deviation error bars are considerably smaller ($\approx 10^{-2}$ µm/pixel) than the symbols utilized for the measurement points and therefore not plotted in this figure.

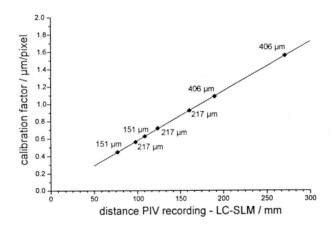

Figure 6: Linearity and calibration curve of the all optical autocorrelator
(see text for details)

4 COMPARISON WITH OTHER EVALUATION SYSTEMS

To compare the performance of the new all optical evaluation system with a conventional analog-optical / digital system as described by Höcker [11] and with the analog-optical / analog-optical system presented by Vogt and Kompenhans in [5] three different PIV recordings of the kind described in the previous chapter were analyzed with the different systems. The computer programs utilized for the processing of the ACFs were the same in all three cases.

The displacements of the three analyzed patterns were determined by eye under a microscope and were found to be 151 µm, 217 µm and 433 µm. A displacement of 200 µm between the image pairs on a PIV negative is a value which is typical for the DLR-PIV system. The measurement of the displacements by eye might not be the best way to proceed, but for the purpose of comparing the evaluation systems the constancy of the displacement is the only important feature, not the displacement value itself. The fact that the displacement is constant all over the recording is inherent, to the way the specklegrams were created.

Each of the images was evaluated at 400 different interrogation spots within an area in the centre of the recordings. The results are given in the tables 1, 2 and 3. The first column indicates the measured shift on the recording, the second the result of the evaluations, the third the standard deviation of the determined displacement and the fourth the standard deviation with respect to the absolute value of the determined shift.

The last columns of the tables indicate that the new all optical evaluation method yields the smallest *relative* error for the determined displacements. This is mainly due to the fact that the optics of the evaluation system allow an individual and continuous adjusting of the distance of the correlation peaks with respect to the center in the ACF plane. Such an adaptation can not be carried out with fully digitally working systems. The possibility to adapt the scaling within the evaluation system also explains that after a rearrangement of the optical set up a larger displacement in the specklegrams can yield a smaller value for the corresponding displacement in the ACF plane.

Table 1: Analog-optical / digital evaluation

Δs_{ref} / µm	$\bar{r}$ / Pixel	$\bar{\sigma}_r$ / Pixel	$\bar{\sigma}_r$ / $\bar{r}$
151	44,55	0,0192	$4,31 \cdot 10^{-4}$
217	40,74	0,0138	$3,39 \cdot 10^{-4}$
433	34,18	0,0110	$3,22 \cdot 10^{-4}$

Table 2: Analog-optical / analog-optical evaluation with electronically addressable LCD-SLM

Δs_{ref} / µm	$\bar{r}$ / Pixel	$\bar{\sigma}_r$ / Pixel	$\bar{\sigma}_r$ / $\bar{r}$
151	198,85	0,0609	$3,06 \cdot 10^{-4}$
217	196,56	0,0422	$2,15 \cdot 10^{-4}$
433	168,78	0,0375	$2,22 \cdot 10^{-4}$

Table 3: Analog-optical / analog-optical evaluation with optically addressable LC-SLM

Δs_{ref} / µm	$\bar{r}$ / Pixel	$\bar{\sigma}_r$ / Pixel	$\bar{\sigma}_r$ / $\bar{r}$
151	339,70	0,082	$2,41 \cdot 10^{-4}$
217	314,59	0,044	$1,40 \cdot 10^{-4}$
433	380,22	0,044	$1,16 \cdot 10^{-4}$

In another more qualitative comparison the system utilizing the electronically addressable SLM and the optically addressable SLM were compared with PIV recordings of a turbulent flow field. These recordings were made 1.3 m downstream of a regular grid with a rod diameter of 1.5 mm and a mesh size of 19.0 mm. This grid produced an additional turbulence of 0.46 % (measured with a hot wire anemometer) in an otherwise laminar wind tunnel flow. The figures 7 and 8 show the results of this comparison in a small sector of the recorded flow field. (Figure 9 shows the result of the whole evaluation.) From these figures the enhanced resolution of the new all optical autocorrelator becomes evident.

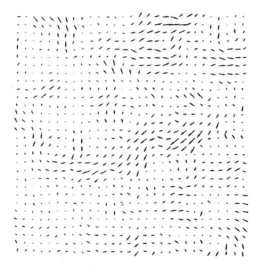

Figure 7: Sector of a flow field behind a grid, determined with an evaluation system utilizing an electronically addressable SLM

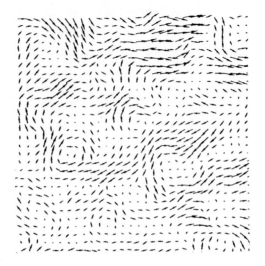

Figure 8: Same sector as in figure 7, but now evaluated with an evaluation system utilizing the optically addressable SLM

5 EXAMPLES OF DIFFERENT APPLICATIONS

The new evaluation system was utilized to evaluate PIV recordings of different measurement campaigns performed at DLR Göttingen. The following figures show the results of a flow field behind a grid, behind a circular cylinder in cross flow, above a NACA 0012 airfoil, and the flow field behind a car model. It is emphasized that the displayed results are raw data, i. e. they are not filtered or smoothed in the postprocessing.

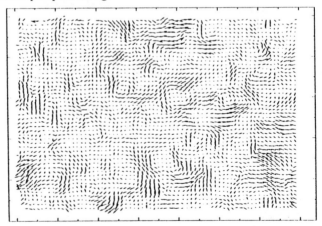

Figure 9: Turbulent flow field behind a grid, $U_x = 10$ m/s, mean flow velocity subtracted. Turbulence level 0.46%, size of observation field 10×15 cm^2, (peak detection and post processing with the workstation based system of DLR)

Reference Vector: 20.00 m/s

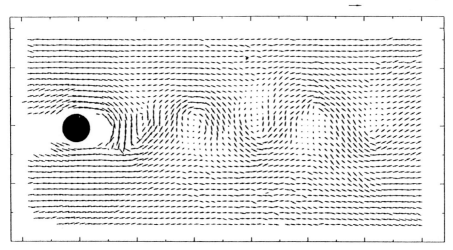

Figure 10: Wake behind a circular cylinder in cross flow; Re = 20000, U_∞ = 15 m/s, u_{ref} = 8.7 m/s subtracted, diameter of cylinder 2 cm, (peak detection and post processing with the workstation based system of DLR)

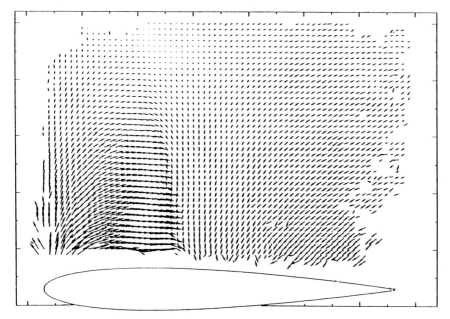

Figure 11: Flow field above a NACA0012 airfoil (from [12]), Ma = 0.75, angle of attack: 5° u_{ref} = 311 m/s subtracted, chord length of airfoil 20 cm, (peak detection and post processing with the workstation based system of DLR)

436

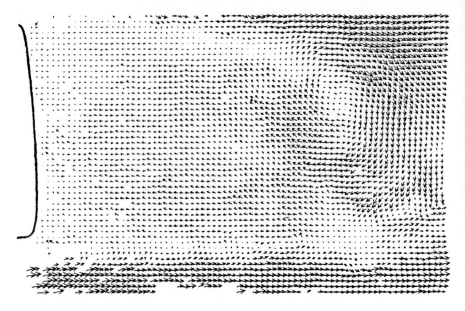

Figure 12: Flow field behind a car model (from [13]), $U_\infty = 50$ m/s, $u_{ref} = 38$ m/s subtracted, size of observation field 20×31 cm^2, (peak detection and post processing with the PC based system of Jenoptik, evaluation time ≈ 10 min)

6 CONCLUSION

The results presented in this paper show that the all optical autocorrelator with the optically addressable spatial light modulator is a very suitable tool for the evaluation of photographic PIV recordings.

The first major advantage of the system is the short time which is necessary to determine the ACF. Due to the fact that the ACF can be determined nearly at video frequency, the need for faster but nevertheless exact and general peak searching algorithms in the ACF plane now becomes increasingly important. Sophisticated peak detection algorithms require up to one second on a workstation based system to determine the correct correlation peaks. Only when this speed is increased into the range of the video frequency full advantage of the high processing speed of the all optical autocorrelator can be taken.

The second advantage of the described set up is the enhanced displacement and therefore velocity resolution compared to conventional evaluation systems utilizing the Young's fringes method and a digital Fourier transform. This improvement allows the detection of even small velocity fluctuations (turbulence of 0.46 % in the presented example) and very weakly pronounced structures in the flow.

REFERENCES

[1] Keane, R. D., Adrian, R. J., "Optimization of Particle Image Velocimeters", L.I.A., Vol 68, ICALEO (1989), pp. 141 - 160

[2] Prasad, A. K., Adrian, R. J., Landreth, C. C., Offutt, P. W. "Effect of resolution on the speed and accuracy of particle image velocimetry interrogation", Experiments in Fluids, 13, (1992), pp. 105 - 116

[3] Lee, J., Farrel, P. V., "Particle Image Velocimetry measurements of IC engine valve flows", pp. 25.4.1, Conference Proceedings of the Sixth International Symposium on Application of Laser Techniques to Fluid Mechanics, July 20th - 23rd, 1992, Lisbon

[4] Morck, T., Andersen, P. E., Westergaard, C. H., "Processing speed of photorefractive optical correlators in PIV-processing", pp. 27.2.1., Conference Proceedings of the Sixth International Symposium on Application of Laser Techniques to Fluid Mechanics, July 20th - 23rd, 1992, Lisbon

[5] Vogt, A., Raffel, M., Kompenhans, J., "A comparison of optical and digital evaluation of photographic PIV recordings"; pp. 27.4.1., Conference Proceedings of the Sixth International Symposium on Application of Laser Techniques to Fluid Mechanics, July 20th - 23rd, 1992, Lisbon

[6] Efron, U., Grinberg, J., Braatz, P. O., Little, M. J., Reif, P. G., Schwartz, R. N., "The silicon liquid-crystal light valve", J. Appl. Phys. 57 (1985), pp. 1356 - 1368

[7] Ashley, P. R., Davis, J. H., "Amorphous Silicon Photoconductor in a Liquid Crystal Spatial Light Modulator", Appl. Opt. 26 (1987), pp. 241 - 246

[8] Gabor, A. M., Landreth, B., Moddel, G., "Integrating mode for an optically addressed spatial light modulator", Appl. Opt. 37 (1993), pp. 3064 - 3067

[9] Reichel, F., Löffler, W. "Optical space frequency analysis for real-time pattern recognition", J. of Optoelectronics, 1994 (in press).

[10] Goodman, J. W., "Introduction to Fourier Optics", Mc Graw Hill, 1986

[11] Höcker, R., Kompenhans, J., "Some technical improvements of Particle Image Velocimetry with regard to its application in wind tunnels", Proceedings of the International Congress on Instrumentation in Aerospace Simulation Facilities-ICIASF '89, Sept. 18 - 21, 1989, pp. 545-554

[12] Raffel, M., Kompenhans, J., "PIV mesurements of unsteady transonic flow fields above a NACA 0012 airfoil", Proceedings of the 5th International Conference on Laser Anemometry - Advances and Applications, Veldhoven, The Netherlands, 23rd - 27th August 1993, SPIE Proceedings Volume 1987, pp. 527 - 534

[13] Vogt, A., Raffel, M., Kompenhans, J., Hack, A., "Particle Image Velocimetry für instationäre Strömungen in großen Windkanälen", Tagungsband Lasermethoden in der Strömungsmeßtechnik, 2. Fachtagung 1993, 14. - 16.9.93, PTB Braunschweig, pp. 33.1-33.3

A new paradigm for Particle Tracking Velocimetry, based on graph-theory and pulsed neural networks

Dominique DEROU and Laurent HERAULT

LETI - (CEA-Technologies Avancées)
DSYS / SCSI
CENG-17 rue des Martyrs
38054 - Grenoble Cedex 9 - France

1. Introduction

The Particle Tracking Velocimetry (PTV) technique consists in recording, at different instances in time, positions of small tracers particles following a fluid flow and illuminated by a sheet, or pseudo sheet, of light. It aims to recognize each particle trajectory, constituted of n different spots and thus to determine each particle velocity vector. In the present paper, we devise a new method, taking into account a notion of global consistency between the trajectories to be extracted, in terms of visual perception and physical properties. It is based on *a graph-theoretic formulation of the particle tracking problem* and on the use of *original neural networks, called pulsed neural networks*.

1.1 Background

Early studies in fluid mechanics were forced to use intrusive, imprecise and complex techniques, rarely giving full-field measurements. Now, with the outstanding recent progress in computer technology and image processing, significant improvements can be obtained by using optical methods and image analysis. Today, particle tracking is one of the simplest and most powerful methods of quantitative flow visualization. It is among the few ones capable of providing instantaneous maps of magnitude, over extended areas, such as velocity or vorticity.

Some reviews on particle tracking velocimetry (PTV) describe the principles and applications of many types of PTV ([ADR86], [ADR91], [AGU87], [BUC92], [HES88]), which are applicable from slow to supersonic flows. Various methods have been developped for PTV, using classical image processing techniques : estimation of the particle's mean displacement between two images and search of the new position of the particle on the following image thanks to the estimated velocity [NIS89], minimization of an objective function [CHA85], point-by-point matching by statistical analysis of displacement vectors ([GRA89], [HAS91]). It appears that those current methods present several drawbacks :

- the algorithms use image processing techniques like correlation or search-tree, which imply a high sensitivity to noise and a low speed of computation,
- they provide degraded results. As a consequence, a post-processing step is necessary to suppress errors.

Obviously, improvements can be obtained by using more sophisticated image processing techniques, so as to use the information present in the original image, with a higher efficiency. Several papers express some trajectory grouping problem as combinatorial optimization problems and try to minimize a global cost function including local constraints ([DEN89], [GYU91], [MOH92], [PAR89], [PET91], [SHA88], [YUI91]). A major drawback is that all these methods need to be parameterized. Moreover, their computational time is so large that they have been applied only onto small size problems.

1.2 A new formulation for the particle tracking velocimetry problem

We suggest an approach using the notion of perceptual grouping and some physical properties of the flow. This approach provides a generalized formulation of the particle tracking problem as a global optimization problem. We propose a new neural network concept, named pulsed neural network, to deal with this optimization problem. Our algorithm consists in three distinct processing stages, as illustrated by figure 1.1 :

1. Creation of potential trajectories extracted from the original image, thanks to metric constraints imposed by the image acquisition process.
2. Determination of a matrix representing mutual consistency and another one representing incompatibility constraints between potential trajectories, by use of perceptual grouping notions and physical properties of the studied phenomenon.
3. Extraction of a set of trajectories satisfying each incompatibility constraint in terms of global consistency. This stage is formulated as a global optimization problem and is solved through two pulsed neural networks.

The method we propose can be also used for detecting false velocity vectors provided by particle image velocimetry (PIV) algorithm based on correlation. In this case, the stage of generation of potential trajectories is replaced by a PIV algorithm generating some regularly distributed velocity vectors.

1.3 Paper organization

The remainder of this paper is organized as follows. Section 2. briefly describes the pre-processing used to obtain the potential trajectories from the

440

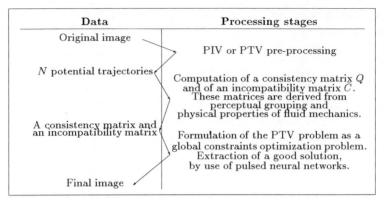

Fig. 1.1. Synoptic diagram of our algorithm.

original image. Then it introduces the perceptual notion of consistency between trajectories and expresses the particle tracking problem in terms of a global optimization problem. Section 3. outlines two neural network methods used to solve this optimization problem. In section 3.2, we present the case in which the cost function to be maximised is a binary function. The particle tracking problem can then be formulated as a maximum independent subset problem, in a graph. This section details the original network method used to solve the maximum independent subset problem. Section 3.3 is devoted to the resolution of the particle tracking problem as a global optimization problem. Some experimental results, obtained with real images, are presented in section 4., followed by a general discussion giving some directions for future work.

2. A combinatorial optimization formulation

2.1 Extraction of potential trajectories

The basic information that our particle tracking algorithm needs is a set of potential trajectories extracted from the original image.

In case of PTV, in a first module, our analysis starts with a classical point detection process. It involves the recognition of each spot constituting a virtual particle trajectory by applying computation of local histograms, determination of thresholds thanks to histograms, thresholding of the original image, connexity analysis, and computation of the centroïd of each particle spot.

The second module receives as input a list of points and their respective attributes (here, the grey level). Its purpose is to determine a list of potential trajectories. It uses *a priori* knowledge (metric constraints) coming from the experimental process. This knowledge can be expressed by : "a potential trajectory is made up of n distinct and consecutive spots, laying on a parameterized curve and verifying some metric relations". For instance, in our fluid mechanics application, trajectories are composed of three aligned spots such that the length between the first two points is twice the length between the two other points. Fig.2.1 shows an inset to Fig.2.2 showing a blow-up of a small area containing some complete particle trajectories.

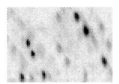

Fig. 2.1. Example of particle trajectory, made up of three points, aligned and in a constant length ratio.

The method uses a local search tree for each point. This tree is defined by the possible matches between this point and the others. It is pruned by using local constraints corresponding to the *a priori* above-mentionned knowledge. By this way, a list of potential trajectories satisfying the experimental conditions is obtained. These potential trajectories represent the potential instantaneous velocity vectors of particles. Obviously, local constraints can lead to select erroneous trajectories.

In case of PIV, the set of potential trajectories is provided by any PIV algorithm. Whichever PIV method may be used, some erroneous velocity vectors will exist.

What follows seeks to detect and suppress those erroneous trajectories. Fig.2.3 depicts all potential trajectories created after the processing stage of the original image presented in Fig.2.2.

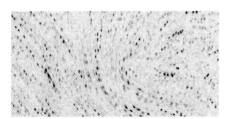

Fig. 2.2. A particle tracking image. From this image (440 ∗ 220 pixels), 590 spots are extracted.

Fig. 2.3. Potential trajectories extracted after the pre-processing stage. 140 potential trajectories have been created.

2.2 Perceptual grouping and particle image velocimetry

Salient structures can often be perceived in an image at a glance. They appear to attract our attention without the need to scan the entire image in a systematic manner, and without prior expectations regarding their shape. The processes involved in the perception of salient structures appear to play a useful role in segmentation and recognition, since they allow us to immediately concentrate our attention on objects of interest in the image ([LOW85], [SHA88]). Figure 2.4a depicts an example of perceptual grouping easily experienced by the human visual system. The geometric shapes are immediately perceived although their contours are fragmented, implying that the gaps do not hinder the immediate perception of such objects. In this case, one must group together several line segments of the shapes to distinguish them from the noisy background. A more striking example of illusory contours is found in the Kanizsa illusion presented in figure 2.4b. Here, one generally perceives the structure of a square lying with its corners on four circles rather than four independent and partial circles. The perception of the square largely consists in contours which do not physically exist but which nevertheless are perceived.

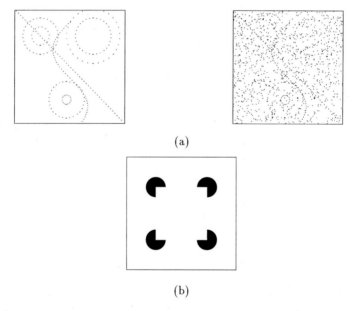

(a)

(b)

Fig. 2.4. (a) Two instances of perceptual arrangements. (b) The Kanizsa square illusion.

Many authors have emphasized the capacity of the human visual system to complete some components of an image, in order to get trajectories, edges, regions, by using local spatial constraints and satisfying a global consistency in the image, like in the illusory contours examples. We intend to use this observation in order to detect and suppress some of the trajectories, which are radically inconsistent with the flow motion. Furthermore, some physical properties of the observed phenomenon can be used (like fluid viscosity, velocity, Reynolds number).

By use of these neuropsychology considerations and these fluid mechanics properties, we define consistency and incompatibility coefficients between any two potential trajectories. The consistency coefficient between two potential trajectories is the higher as trajectory pair is consistent with the fluid motion in its local environment. The binary incompatibility coefficient between two potential trajectories indicates the strict incompatibility between trajectories. Our goal is to extract and quantify, through these coefficients, the trajectories ability to induce a continuity impression beyond their physical limits.

2.3 The consistency coefficient

Let i and j be any two potential trajectories. We want that the interaction between these trajectories satisfies the perceptual knowledge on the flow. If the pair of trajectories (i, j) contributes to the perception of the global movement, the coefficient q_{ij} must be high. Otherwise their interaction must be small. This coefficient is a consistency measure. We use some Gestallt principles [WER58] so as to express the components of a consistency coefficient (our underlying goal is to keep the interpretation as simple as possible in a Gestallt sense) :

- *Similarity* : similar elements are grouped together. In fluid dynamics, at a certain scale, neighboring particles have about the same velocity. This observation is the higher as the fluid viscosity is high.
- *Proximity* : elements that are close together tend to be grouped together. Only a neighborhood of each potential trajectory must be taken into account. Otherwise, the influence of any particle over another particle decays with distance : closer particles will tend to influence each other more than distant ones.
- *Good continuation* : elements that lie in a common channel or along a common line or smooth curve are grouped together. A fluid mechanics flow characteristic is that all neighboring particles seem to be on similar streamline. In a certain neighborhood, all pairs of potential trajectories have a constant curvature radius (with a high value for a laminar flow, and a small one for a vortex). Thus, if a potential trajectory is erroneous, the radius of the virtual circle, tangent to this trajectory and to a correct trajectory of its neighborhood, will be very different from the mean value of the radius calculated in its neighborhood.

Coherence is defined by a combination of similarity, proximity and good continuation notions. Let us define precisely each component of the consistency

444

coefficient.

2.3.1 Similarity. The similarity coefficient $q_{ij}^{similarity}$ constrains particles i and j to have a similar velocity. We devise the following expression :

$$\forall j \in \mathcal{V}(i), \ q_{ij}^{similarity} = (1 - \frac{\Theta_{ij}}{\pi/2}).(1 - \frac{|\ l_i - l_j\ |}{\max(l_i, l_j)})$$

where : l_i (resp. l_j) denotes the modulus of the vector corresponding to the potential trajectory i (resp. j), Θ_{ij} is the angle between vectors corresponding to potential trajectories i and j (Fig. 2.5) and $\mathcal{V}(i)$ defines the neighborhood around potential trajectory i. The similarity coefficient quantifies the

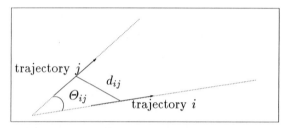

Fig. 2.5. Notations for the consistency coefficient.

similarity in direction and modulus of two particles. Its value belongs to $[0, 1]$.

2.3.2 Proximity. The proximity coefficient must vanish smoothly as the two trajectories are farther from each other. We choose :

$$\forall j \in \mathcal{V}(i), q_{ij}^{proximity} = \exp -\frac{d_{ij}^2}{2.\sigma_d^2}$$

where d_{ij} is the distance between the trajectories i and j, and σ_d is a fraction of the standard deviation of all the distances over the image.

2.3.3 Good continuation.

Principle. The aim of this coefficient is to favour trajectories which either have a tendency to be perceived as elements of greater forms like streamlines or contribute to the visual perception of these streamlines. This capacity to induce a continuity impression is translated in two ways.

Firstly, it appears in the type of neighborhood used. Psycho-experimental studies ([GAI93], [LIN88]) have demonstrated that, in seventy percent of cases, human subject perceive a discontinuity in an extension of two segments

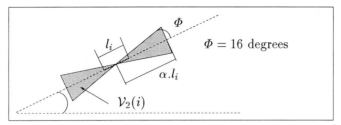

Fig. 2.6. Presentation of the neighborhood $\mathcal{V}_2$.

as soon as the angle between those two segments is more than nearly sixteen degrees. Thanks to this data, a search neighborhood $\mathcal{V}_2(i)$ can be defined, as represented in figure 2.6.

Secondly, in this neighborhood, particle trajectories must contribute to infer streamlines. Trajectories belonging to a common channel of streamlines have the same curvature radius. We consider a potential trajectory i. Then we consider successively all the potential trajectories j belonging to the just defined neighborhood. We define a generalized curvature radius of the virtual circle approximately tangent to i and j, as well as the more frequent (prevailing) generalized curvature radius. The good continuation principle implies that the potential trajectories j whose generalized curvature radius with potential trajectory i is about the prevailing curvature radius will be favoured.

Definition of the generalized curvature radius. There does not always exist a circle tangent to any two segments. Parent and Zucker [PAR89] define a radius of curvature for a circle tangent to two segments. But their definition supposes the existence of such a real circle. The formulation of Sha'ashua and Ullman [SHA88] requires the same condition. In our case, we need to have an approximation of the radius of the circle the best fitted to the two given segments. Thus we propose a generalized curvature radius ρ_{ij}. The generalized curcature radius needs the computation of the intersection point of the two median lines passing respectively through potential trajectories i and j (see figure 2.7). Finally the generalized curvature radius is defined as the mean of the distances d_i and d_j between the intersection point C and the features i and j :

$$\rho_{ij} = \frac{d_i + d_j}{2}$$

This definition of the generalized curvature radius has been chosen as the formulation the best accounting for the curvature of the virtual circle approximately tangent to the potential trajectories i and j. In the special case where there exists a real circle tangent to i and j, then ρ_{ij} is the classical curvature radius of the circle.

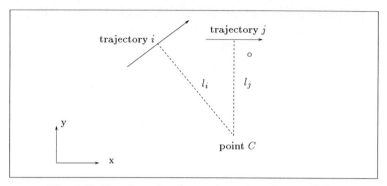

Fig. 2.7. Notations for the good continuation coefficient.

Good continuation coefficient. Let i be a potential trajectory. All the potential trajectories j present in the neighborhood $\mathcal{V}_2(i)$ of i are selected. For each pair of potential trajectories (i, j), we define the generalized curvature radius ρ_{ij} of the virtual circle tangent to potential trajectories i and j.

$$\rho_{prevailing}(i) = \arg\left(\max_a(\text{card}(E_a(i)))\right)$$

$$\text{with } E_a(i) = \{j \mid j \in \mathcal{V}_2(i), \rho_{ij} = a\}$$

The prevailing curvature radius gives the most frequent curvature radius and indicates the form of the general flow motion around i. In a fluid mechanics application, if $\rho_{prevailing}(i)$ is high, i lies in a laminar zone. If it is low, the area around i presents a vortex aspect.

In order to satisfy the good continuation principle, the pair of potential trajectories (i, j) must induce a streamline following the general flow motion. It must then have a generalized curvature radius close to $\rho_{prevailing}(i)$. The higher the difference between ρ_{ij} and $\rho_{prevailing}(i)$ is, the more the pair (i, j) is inconsistent with the global flow motion. The smaller it is, the more the pair of potential trajectories (i, j) contributes to the visual perception of the streamlines by eyes. We choose for the definition of the good continuation coefficient :

$$q_{ij}^{good\ continuation} = \exp\left(-\delta.\frac{\mid \rho_{ij} - \rho_{prevailing}(i) \mid}{\rho_{prevailing}(i)}\right)$$

where δ is a constant parameter. We choose such an equation, in order to take into account the proportionality ratio between ρ_{ij} and $\rho_{prevailing}(i)$. Thus a percentage of the difference between ρ_{ij} and $\rho_{prevailing}(i)$ is used, in order to normalize its influence over the entire image.

The good continuation coefficient will be equal to 1 if ρ_{ij} is equal to $\rho_{prevailing}(i)$, i.e. if i and j belong to a prevailing streamline. It will be the smaller as i and j are inconsistent with the flow motion.

2.3.4 Formulation of the consistency coefficient. We define the consistency coefficient as a combination of the coefficients just defined :

$$q_{ij} = \mathcal{H}(q_{ij}^{similarity}, q_{ij}^{proximity}, q_{ij}^{good\ continuation})$$

The function $\mathcal{H}$ can be a linear combination of $q_{ij}^{similarity}$, $q_{ij}^{proximity}$ and $q_{ij}^{good\ continuation}$ or another more complex function. It gives a weighted measure of the consistency of trajectories pair. The more two trajectories have high consistency, the more salient they are. Consistency coefficients define a consistency matrix $Q = (q_{ij})_{(i,j)\in<1,N>^2}$, with $q_{ij} \in [0,1]$.

2.4 The incompatibility coefficient

The incompatibility coefficient establishes hard constraints between potential trajectories. Indeed, incompatibilities can exist between some trajectories. Two strict conditions induce incompatibility :

- *Sense* : two neighboring trajectories cannot have opposite senses.
- *Common point* : two trajectories cannot share a common spot.

2.4.1 Sense. The sense coefficient imposes that two neighboring particles i and j must have strictly the same sense. This coefficient is calculated by :

$$\bar{c}_{ij}^{sense} = D(i,j)$$

where: $D(i,j) = \begin{cases} 0 & \text{if } i.j \geq 0 \\ 1 & \text{if } i.j < 0 \end{cases}$

2.4.2 Common point. Since the flow seeded with particles is illuminated by a thin sheet of light, particles cannot be superposed and particles trajectories cannot cross and share a common spot.

$$\bar{c}_{ij}^{common\ point} = \begin{cases} 1 & \text{if i and j share a common point} \\ 0 & \text{otherwise} \end{cases}$$

2.4.3 Formulation of the incompatibility coefficient. We define the incompatibiliy coefficient as the result of the boolean OR operation between the sense coefficient and the common point coefficient. Thus, the incompatibility coefficient will belong to $\{0,1\}$ and will be equal to 0 only if both hard constraints are satisfied and 1 otherwise :

$$\bar{c}_{ij} = \bar{c}_{ij}^{sense} \text{ OR } \bar{c}_{ij}^{common\ point}$$

Incompatibility coefficients define an incompatibility binary matrix :

$$\bar{C} = (\bar{c}_{ij})_{(i,j)\in<1,N>^2}, \text{ with } \bar{c}_{ij} \in \{0,1\}.$$

2.5 A new combinatorial optimization statement of the PTV problem

A consistency matrix Q, defining the coherence between each pair of potential trajectories, and an incompatibility matrix $\bar{C}$, representing all hard constraints between potential trajectories, have been generated. At this stage, we intend to select the most consistent trajectories which are mutually compatible, and to remove all erroneous trajectories among the N potential trajectories provided by the first stage of the algorithm. The solution of the particle tracking problem is consequently a subset of the potential trajectories, in which all "erroneous trajectories" have been detected and suppressed. Those "erroneous trajectories" are potential trajectories which display a low consistency with their local environment or which do not satisfy the constraints inherent in the problem. For this reason, the problem can be expressed as finding a subset of potential trajectories maximizing a quality function representing the global consistency of the image, and satisfying all hard constraints (i.e. no elements of the subset are incompatible).

Let $\boldsymbol{p} = (p_1, \ldots, p_N)$ be a vector, in which $p_i = 1$ if the trajectory i is labelled as a good one (corresponding to a real particle trajectory), or $p_i = 0$ if it is labelled as an erroneous one. From a mathematical point of view, the particle tracking problem can be formulated as a global constraints optimization problem :

$$\text{finding } \boldsymbol{p} \text{ maximizing } E(\boldsymbol{p}) = \sum_{i=1}^{N} \sum_{j=1}^{N} q_{ij} \cdot p_i \cdot p_j$$
$$\text{while satisfying : } \begin{cases} \forall i, \ p_i \in \{0, 1\} \\ \forall i, j, \ p_i = p_j = 1 \Longrightarrow \bar{c}_{ij} = 0 \end{cases} \quad (2.1)$$

This framework represents a new unified formulation of the particle tracking problem and the selection of false velocity vectors in PIV problem, as a global optimization problem with constraints. We will devise two approaches to solve the problem. We will first present the case in which the consistency matrix is binarized in order to reduce the computational time of the further optimization process. We will show that this binary case is strictly equivalent to the problem of extracting a maximum independent subset in a graph. We will devise a pulsed neural network algorithm to solve it. We have developed another pulsed neural network in order to solve the general formulation of the particle tracking problem.

2.5.1 Binary case : Formulation as a problem of extracting a maximum independent subset in a graph.

Model. Let us define an incompatibility graph $\bar{\mathcal{G}} = (V, E)$, with a set of vertices $V = \{0, 1, \ldots, N\}$ where each vertex is a potential trajectory and an edge set $E = \{ij \mid (i, j) \in < 1, N >^2, i \neq j\}$. The adjacency matrix of graph $\bar{\mathcal{G}}$, $\bar{G} = (\bar{g}_{ij})_{(i,j) \in <1,N>^2}$, is a binary matrix and represents another notion of compatibility ($\bar{g}_{ij} = 0$) or incompatibility ($\bar{g}_{ij} = 1$) between potential trajectories i and j. This matrix is made up of coefficients resulting from the

OR operator between the incompatibility coefficients $\bar{c}_{ij}$ previously described and the complement of binarized consistency coefficients $\mathcal{B}(q_{ij})$ defined by :

$$\bar{g}_{ij} = \bar{c}_{ij} \text{ OR } \overline{\mathcal{B}(q_{ij})}$$

where the operator $\mathcal{B}$ is a binarization function, described in the following (cf. table 2.8), with $\mathcal{B}(q_{ij}) \in \{0, 1\}$ and $\overline{\mathcal{B}(q_{ij})} = 1 - \mathcal{B}(q_{ij})$.

Two potential trajectories are connected by an edge in graph $\bar{\mathcal{G}}$ if they are incompatible. In this case, we show (in [DER94a]) that the global optimization problem described by equation (2.1) reduces to the search of the largest independent subset of the incompatibility graph $\bar{\mathcal{G}}$. Therefore, we speak of formulation as a maximum independent subset problem. The maximum independent subset problem is a hard graph-theoretic problem. In section 3.2, we present a new neural algorithm to solve it.

Binarization of the consistency matrix. We define the notion of consistency support :

$$\Omega_i = \frac{\sum_{j \in \mathcal{V}_i} q_{ij}}{\text{card}(\mathcal{V}_i)}$$

where $\mathcal{V}_i$ is the neighborhood of potential trajectory i. The consistency support Ω_i should be the higher as the potential trajectory i is consistent with the global flow motion. Moreover, potential trajectories which are radically inconsistent and evidently erroneous should have minimal support. We can consider that the potential trajectories, which have a consistency support less than the mean consistency support among all potential trajectories, are incompatible with all the potential trajectories belonging to their neighborhood. Then a second threshold can be achieved on the consistency coefficient. For a given i, binary values $\mathcal{B}(q_{ij})$ of the edges of the consistency matrix Q are defined thanks to table 2.8.

In this table, Ω_{mean} is defined by : $\Omega_{mean} = \frac{\sum_i \Omega_i}{N}$ with N the number of potential trajectories, and $\alpha_{\mathcal{B}}$ is a constant parameter, experimentally fixed to $0, 5$.

$\forall j \notin \mathcal{V}_i$	$\forall j \in \mathcal{V}_i$		
	$\Omega_i < \Omega_{mean}$	$\Omega_i \geq \Omega_{mean}$	
		$q_{ij} \geq \alpha_{\mathcal{B}}$	$q_{ij} < \alpha_{\mathcal{B}}$
$\mathcal{B}(q_{ij}) = 1$	$\mathcal{B}(q_{ij}) = 0$	$\mathcal{B}(q_{ij}) = 1$	$\mathcal{B}(q_{ij}) = 0$

Fig. 2.8. Calcul of the compatibility coefficient $\mathcal{B}(q_{ij})$.

2.5.2 General case. The problem is to find a subgraph of the graph $\bar{\mathcal{C}}$, whose adjacency matrix is the matrix $\bar{C}$, while maximizing the global consistency of the solution, i.e. maximizing the function :

$$E(\boldsymbol{p}) = \boldsymbol{p}^T.Q.\boldsymbol{p}$$

The algorithm and the pulsed neural network related to this approach are presented in section 3.3.

3. Pulsed neural network

3.1 The pulsed discrete time neural network model

The basic artificial neuron can be modelled as a multi-input nonlinear device with weighted interconnections, also called synaptic weights. The cell body (soma) is represented by a threshold function H which is a hard limiter with hysteresis (Schmitt Trigger). If u_i is the input (potential) and p_i the output of a neuron i, then the H function is defined as :

$$p_i^{(z+1)} = H\left(u_i^{(z+1)}\right) = \begin{cases} 1 & \text{if } u_i^{(z+1)} > \alpha \\ 0 & \text{if } u_i^{(z+1)} < \beta \\ p_i & \text{otherwise} \end{cases}$$

where

$$u_i^{(z)} = u_i(z\tau)$$

with the sampling period τ. Parameters α and β are thresholds that describe the ability of a neuron to modify its output (see figure 3.1).

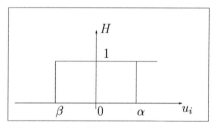

Fig. 3.1. Hard limiter with hysteresis.

The vector $\boldsymbol{p}^{(z)} = (p_1^{(z)}, \ldots, p_N^{(z)})$ represents the state of the neural network at time $z\tau$. If $p_i^{(z)} = 1$, then the neuron i is activated at time $z\tau$. Otherwise, it is disactivated.

The artificial neural network we use consists of recursive neurons which are potentially fully interconnected. At any time $z\tau$, the input of the i-th neuron is not a weighted mean of the outputs of the other neurons, as in a classical Hopfield neural network. In the general case, the input of the i-th neuron is a function of the outputs of the others neurons.

The temporal evolution of the i-th neuron input is given by the following evolution equation :

$$u_i^{(z+1)} = u_i^{(z)} + \Delta u_i^{(z+1)}$$

where

$$\Delta u_i^{(z+1)} = F_i\left(p_1^{(z)}, \ldots, p_{i-1}^{(z)}, p_i^{(z)}, p_{i+1}^{(z)}, \ldots, p_N^{(z)}\right) = F_i(\boldsymbol{p}^{(z)})$$

When $\Delta u_i^{(z+1)} > 0$, the neuron i is excited at time $(z+1)\tau$. Otherwise, it is inhibited.

In a discrete Hopfield neural network [HOP82], the F_i function is defined by :

$$F_i(\boldsymbol{p}^{(z)}) = \sum_{j,j\neq i} T_{ij} p_j^{(z)} + I_i$$

where T_{ij} is the weight of the synaptic link between neuron j and neuron i and I_i is an external input on neuron i. In terms of dynamics, in an Hopfield neural network, the potential of a neuron is updated after its output is updated. When a neuron is updated, its input is calculated as the sum of a weighted mean of the other neuron outputs and of an offset. Moreover, one can associate to an Hopfield neural network a Lyapunov funtion, called energy, that is minimized through the neural network dynamics. It is given by:

$$E = -\frac{1}{2}\sum_i\sum_j T_{ij} p_i p_j - \sum_i I_i$$

Then, obviously, the evolution equation of the i-th neuron is given by:

$$\Delta u_i^{(z+1)} = -\Delta E/\Delta p_i$$

In a pulsed neural network, the F_i function is more general :

$$F_i(\boldsymbol{p}^{(z)}) = \Lambda\left(\delta_1^{(z)} f_1^i\left(\boldsymbol{p}^{(z)}, I_i\right), \ldots, \delta_c^{(z)} f_c^i\left(\boldsymbol{p}^{(z)}, I_i\right), \ldots\right) \qquad (3.1)$$

where:

- $\forall c, \delta_c^{(z)} \in \{0, 1\}$. These functions are defined by the neural network designer.
- $f_1^i, \cdots, f_C^i$ are functions of the neuron outputs and of the external input.

Practically, we choose :

$$F_i(\boldsymbol{p}^{(z)}) = \sum_{c=1}^{C} \lambda_c \delta_c^{(z)} f_c^i \left(\boldsymbol{p}^{(z)}, I_i \right)$$

At any time $z\tau$, a neuron update consists in calculating a linear combination of the functions $f_1^i, \cdots, f_C^i$ of the neuron outputs and of the external input. Depending on the time $z\tau$ of updating, the function $\delta_c^{(z)}$ is equal to 0 or 1. Thus, the contribution of a function f_c^i will be taken into account or will not.

The discrete Hopfield neural network has a structure of a specific pulsed neural network where :

- $C = 1$
- for any neuron i, $f_1^i \left(\boldsymbol{p}^{(z)}, I_i \right) = \sum_{j, j \neq i} T_{ij} p_j^{(z)} + I_i$
- $\forall z, \delta_1^{(z)} = 1$

In the general case, neurons are self-connected, i.e. the binary output $p_i^{(z)}$ of a neuron i at time $z\tau$ is used at time $(z+1)\tau$ to calculate $u_i^{(z+1)}$. It has been demonstrated, in [DER94b] that, if binary neurons are self-connected, then there is no function $E(\boldsymbol{p})$ verifying : $\forall z, \forall i, \Delta u_i^{(z+1)} = -\Delta E / \Delta p_i$. The pulsed neural networks is then radically different from the Hopfield one.

By using such kind of recursive pulsed neural networks, we can apply the following methodology to solve any discrete combinatorial optimization problem:

1. Represent the problem from the neural network standpoint. Determine the input representation in such a way that the final solution is indicated by the neuron outputs. Each neuron output is associated with a variable of the problem. For many combinatorial problems, the binary output state of the neurons is used.

2. Express the cost function and the constraints as functions f_c^i of the neuron outputs $\boldsymbol{p}$. One can define several cost functions if the problem is a multi-criteria one. A function f_c^i associated with a constraint returns 0 if the constraint is satisfied.

3. Infer the evolution equation of any neuron i that is associated with the functions f_c^i. In this step, one defines the δ_c functions. For example, we can take $\delta_c^{(z)} = 1$ during ω complete updates of the neurons after the convergence of the network. When the δ_c^i functions associated with the cost functions are 0, the network evolves in a relaxation mode. Otherwise, the network evolves in a pulsation mode, where it tries to escape from a local minimum in order to minimize the cost function.

4. Randomly start the network. In relaxation mode, the evolution equations will drive the network to a feasible solution if it exists. Then, by alternating pulsation and relaxation modes, the network will return admissible

solutions minimizing the cost function. The neural network jumps among feasible solutions. The best one will be recorded.

3.2 A neural network for the PTV maximum independent subset problem

Given the previously described graph of incompatibility $\bar{\mathcal{G}}$, the particle tracking problem can be solved by finding the largest independent subset of the graph $\bar{\mathcal{G}}$ (see Section 2.5.1). An independent subset is a subgraph such that there is no edge between any pair of vertices in that subset. The maximum independent subset problem is to find the largest independent subset of a given graph.

The neural network devised for the maximum independent subset problem is a pulsed neural network, as the one described previously.

Initially all inputs and outputs are randomly generated. In order to speed up the convergence, the potential u_i are bounded. The convergence mode used is randomly asynchronous mode. At each step of the iterative process, a unique neuron is randomly selected and updated using the evolution equation.

Definition 3.1. *Let the following functions be defined by:*

$$deg \quad : \quad <1, N> \to \mathbb{N}$$
$$x \mapsto card\{j \in <1, N >| \ \bar{g}_{xj} = 0\}$$

$$h \quad : \quad \mathbb{R} \to \{0,1\} \qquad\qquad l \quad : \quad \mathbb{R} \times \{0,1\}^N \times \mathbb{N} \to \{0,1\}$$

$$x \mapsto \begin{cases} 1 & if \ x = 0 \\ 0 & if \ x \neq 0 \end{cases} \qquad (x, p, z) \mapsto \begin{cases} 1 & if \ x > R(p^{(z)}, z) \\ 0 & otherwise \end{cases}$$

where $R(p^{(z)}, z)$ is the size of the largest independent subset found at z.

The evolution equation is defined by the expression of the function F_i :

$$F_i(p^{(z)}) \;=\; p_i^{(z)} \cdot \left\{ -A \cdot \left[1 - h\left(\sum_{j \in V(i)\backslash\{i\}} \bar{g}_{ij} \cdot p_j^{(z)} \right) \right] \right\} +$$

$$(1 - p_i^{(z)}) \cdot \left[B \cdot h\left(\sum_{j \in V(i)\backslash\{i\}} \bar{g}_{ij} \cdot p_j^{(z)} \right) + C \cdot \delta^{(z)} \cdot l(deg(i), p^{(z)}, z) \right]$$

The function $\delta^{(z)}$ returns 0 while the network has not converged ($\exists i \mid \Delta u_i^{(z+1)} \neq 0$), and returns 1 during several complete updatings of the network (here equal to 4) as soon as the network has converged. This defines a pulsation. When $\delta^{(z)} = 0$, the network evolves in a relaxation mode. When $\delta^{(z)} = 1$, it evolves in a pulsation mode. The parameters A, B and C are stricly positive coefficients, whose choice is experimentally indifferent.

During a relaxation, the networks evolves freely towards an attractor which encodes a feasible solution of the problem. During a pulsation phase, the network tries to escape from the attractor in which it has fallen during the previous relaxation phase.

Let us consider that the network is in a relaxation phase. Only the first two terms will operate. The first two terms of the evolution equation ensure that the solution satisfies all binary constraints, i.e. that the solution is an independent subset of the graph $\bar{\mathcal{G}}$.

If neuron i is activated $(p_i^{(z)} = 1)$, only the first term of the equation is used. If none of the neurons j with which neuron i is incompatible $(\bar{g}_{ij} = 1)$ is activated $(p_j^{(z)} = 0)$, then $\Delta u_i^{(z+1)}$ will be null. Neuron i will remain activated. If some of these incompatible neurons are already activated $(p_j^{(z)} = 1$ and $\bar{g}_{ij} = 1)$, neuron i must tend to be inhibited. The potential variation will be negative, since A is strictly positive. Therefore the first term is an inhibitory term. It verifies that every activated neuron (belonging to the selected subset) is incompatible with none of the other activated neurons.

If neuron i is not activated $(p_i^{(z)} = 0)$, the second term will operate. This term has an exciting effect. If none of the neurons j with which neuron i is connected $(\bar{g}_{ij} = 1)$ is activated $(p_j^{(z)} = 0)$, then neuron i has a high probability to belong to a good solution. Therefore neuron i is excited. If it is not the case, the state of neuron i does not change. As the previous term, this term ensures that the solution satisfies the constraints.

The use of these two terms would be sufficient to provide a solution satisfying the constraints, i.e. an independent subset. Indeed, we demonstrated in [DER94a], that this network converges on a solution satisfying all constraints in a limited computational time. At convergence, for any neuron i, $\Delta u_i^{(z+1)} = 0$. However the problem is to find a near optimal solution, that we define to be large enough. The network must have the ability to find several solutions all satisfying the constraints, in order to choose the best one. It must be able to leave a local "minimum" and to converge towards a new solution.

The pulsation phase will enable the network to converge to a new feasible solution, since the pulsation phase provides a new initial point far enough from the local minimum found during the previous relaxation phase. The idea is to excite some neurons in a right and clever manner. Between two pulsation phases, the network converges and provides a solution. Since the desired solution is a large one, from convergence to convergence, we stored the best solution ever found. Its size is called $R(\boldsymbol{p}^{(z)}, z)$. If a neuron i has a degree in the compatibility $\mathcal{G}$ smaller than the $R(\boldsymbol{p}^{(z)}, z)$, size of the largest solution already found, it cannot belong to a new independent subset solution with higher size. So it will not be excited. On the other hand, if its degree in the compatibility graph is higher than $R(\boldsymbol{p}^{(z)}, z)$, we can assess that it can belong to another largest feasible solution. All neurons, whose degree in compatibility graph is higher than $R(\boldsymbol{p}^{(z)}, z)$ and which are inactivated, are excited during the pulsation phase. After the pulsation, $\delta^{(z)}$ is fixed to 0 again. The network, evolving in a relaxation phase (with the first two terms), then converges once again to another feasible solution.

The stopping criterion is related to the test made during pulsation phases. If none neuron has a degree higher than $R(p^{(z)}, z)$, then the solution found cannot be improved : it is optimal. The final solution is the largest subset among the feasible solutions proposed by the network.

We notice that this network is powerful to solve any maximum clique problem. We will not present any results obtained with this neural network, because results are better with the algorithm for non binarized matrix. The interest of this method is that :

- It provides a new method to solve the maximum independent subset problem, that guarantees that all constraints are satisfied.
- It can be easily implemented on specialized chips.

3.3 A pulsed neural network for the global optimization problem

In section 2.5.2, the particle tracking problem has been formulated as a global optimization problem. Our goal is to devise a neural network capable of giving an optimal solution for the general problem described by equation (2.1).

The evolution equation is defined so as to maximize the quality function $E(p)$, while satisfying all hard constraints. The problem is to find an independent subset of the incompatibility graph $\bar{C}$ that maximizes the quality function $E(p) = \sum_{i=1}^{N} \sum_{j=1}^{N} q_{ij} . p_i . p_j$. For this reason, an updated neuron must have a transformation that increases the quality function. A neuron, whose state is inactivated ($p_i^{(z)} = 0$) just before its updating, must be shifted to $p_i^{(z+1)} = 1$ only if its new state contributes towards an increase of quality function. Furthermore, so as to obtain an independent subset of the incompatibility graph, we impose that, if an inactivated neuron i is updated to $p_i^{(z+1)} = 1$, then all the activated neurons j incompatible with it (such that $\bar{c}_{ij} = 1$ and $p_j^{(z)} = 1$) will be disactivated. Indeed, if a neuron is updated to 1, it infers that this neuron is regarded as belonging to the final solution. In order to obtain an independent subset for the solution, all neurons already activated and incompatible with this neuron must be disactivated at the same time.

Consider that neuron i, such that $p_i^{(z)} = 0$, is updated. It will be shifted to 1, and the activated neurons incompatible with it will be shifted to 0, if these transformations induce an increase of the quality function :

Before the transformations : $p_i^{(z)} = 0$

After the transformations :

$$\begin{cases} p_i^{(z+1)} = 1 \\ \forall j \text{ such that } p_j^{(z)} = 1 \text{ and } c_{ij} = 1, \ p_j^{(z+1)} = 0 \end{cases}$$

Now we determine the expression of the quality function variation, when these transformations of neuron states are performed :

$$\Delta E_i^{(z+1)} = -2 \sum_{\{s|\bar{c}_{is}=1\}} \sum_{j\neq s\,,\,j\neq i} q_{js}.p_j^{(z)}.p_s^{(z)}$$

$$+2 \sum_{j\neq s\,,\,j\neq i} q_{ij}.p_j^{(z)} - \sum_{\{s|\bar{c}_{is}=1\}} p_s^{(z)} + 1$$

The evolution equation will use this relation. The evolution equation of a neuron i is given by :

$$\begin{cases} F_i(\boldsymbol{p}^{(z)}) = (1 - p_i^{(z)}).\max(0, \Delta E_i^{(z+1)}) - p_i^{(z)}.\delta^{(z)}.k(\boldsymbol{p}^{(z)}, i) \\ \forall j \neq i\,,\; F_j(\boldsymbol{p}^{(z)}) = -\bar{c}_{ij}.(1 - p_i^{(z)}).\max(0, \Delta E_i^{(z+1)}) \end{cases}$$

The function $k(\boldsymbol{p}^{(z)}, i)$ used in the previous relation is an inhibitory function. Its purpose is to inhibit some activated neurons, in an attempt to test another solution. Two possibilities are :

- Random selection of X activated neurons ($p_i^{(z)} = 1$), which are disactivated at time $t + 1$ ($k(\boldsymbol{p}^{(z)}, i) = \infty$ and so $p_i^{(z+1)} = 0$). This number can decrease from high value at the beginning till 0.
- Determination of the mean degree of the incompatibility graph $\bar{C}$. Disactivation of all nodes whose degree is higher than this mean degree. Thus, the inhibitory function intends to inhibit all nodes which display an "a priori bad" degree.

The dynamic system evolution devised consists, in the first term, in updating neurons whose output is equal to 0. An inactivated neuron will be activated if its state shift increases the quality function. Since, at the same time, all activated neurons incompatible with the updated neuron are shifted to 0, hard constraints are thus imperatively guaranteed. The use of the first term would be sufficient to provide a solution satisfying all hard constraints, i.e. an independent subset. This network converges towards a solution satisfying all constraints in a limited computational time. At convergence, for all neurons i, $\Delta u_i^{(z+1)} = 0$. At this time, the network has reached a local "maximum".

Notwithstanding, the problem is to find a near optimal solution, that we define as maximizing the quality function $E(\boldsymbol{p})$. The network must have the ability to find several solutions all satisfying the constraints, in order to choose the one with the highest quality function value. The network must be able to leave a local "maximum" and converge towards a new solution. The pulsation function and the inhibitory function (second term of the evolution equation) will enable the network to converge to a new feasible solution. Initially the pulsation function $\delta^{(z)}$ is set equal to 0, until each neuron potential variation $\Delta u_i^{(z+1)}$ becomes null. Then $\delta^{(z)}$ is activated to 1 during one complete updating of the N neurons. During this pulsation phase, the network modifies completely its state vector. The second term of the equation comes in. The idea is to inhibit some of the activated neurons. Previously, we present two possibilities for the inhibitory function. After the pulsation, $\delta^{(z)}$ is set to 0 again. The network, conducted by the first term of the evolution equation, can converge once again to another feasible solution.

The network can be viewed as a network composed of Potts model neurons. A Potts model is a generalization of the two-state Ising model to an arbitrary number of discrete states [FIS91]. In our case, a Potts neuron, i.e. a vector, corresponds to each variable p_i. This vector has a components number equal to the degree of vertex i in the incompatibility graph $\bar{C}$ plus one. Since the running mode chosen is an asynchronous by block mode, at each step of the iterative process, all components of Potts neuron i are simultaneously updated.

4. Experimental results

The algorithm described in the previous section has been applied to many series of fluid mechanics images. The experimentation that has provided these images was carried in a wave tank. A uniform intensity incoherent white light source was used. A mechanical chopper has enabled to pulse the continuous light to obtain stroboscopic illumination. The fluid has previously been sowed with tracer particles, which scatter light into a numerical photographic system. A trajectory is made up of three consecutive and aligned spots. The directional ambiguity is overcome by using particles tagging pulses : the duration between the first two pulses is twice the duration between the two following pulses. The problem of particle tracking is formulated as a global optimization problem. We present the results of the application of the second pulsed neural network.

The original image (1500*900pixels) is presented in Fig.4.1. Fig.4.2 shows potential trajectories. Fig. 4.3 presents the final result, obtained after neural network processing. Experiments were performed on a standard SUN SPARC 10 workstation. Applying the grouping strategy directly on the entire image is not realistic, since there exists relations on local neirghborhood. Instead, we apply the algorithm locally to windows. Windows of size 300*300 pixels have been scanned one by one. To avoid side effects, these windows overlap. In this example, starting with a total of 10005 spots extracted from the original image, the pre-processing step generates 4051 potential trajectories. The pulsed neural network extracts 2306 particle traejctories. On image 4.2, we observe that many potential trajetories are radically erroneous. On image 4.3, despite of the complexity of the task, the pulsed neural network has converged to a visually correct result.

The algorithm we propose is quite rapid : the processing has required about 3 minutes on a SUN SPARC 10 workstation. The result proposed in Fig. 4.3 testifies to the great visual quality of the algorithm. Regardless

458

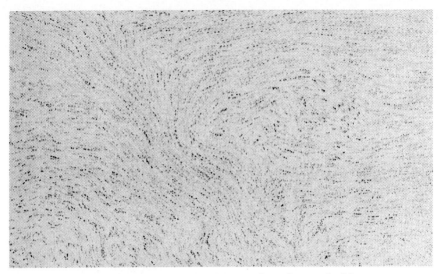

Fig. 4.1. Example of real image (1500*1000 pixels).

Fig. 4.2. Image of potential trajectories.

Fig. 4.3. Final result.

of the image tested, the algorithm recognizes most of the particles seeded in the fluid. A statistical description of the recognized particles has showed that, among all particles recognized, about 97% were right particles. The quality of the results provided by our algorithm is undoubted. However, we must observe that some particles have not been recognized, more particularly in vortex center. The main reason is that particles velocity is so low in vortex center that corresponding positions spots are superposed and cannot be identified as individual spots. Improvements have to be done on the pre-processing algorithm, in order to be able to discriminate superposed and connected spots.

It is very difficult to quantify the performance of a particle tracking algorithm on real images. The experiment presented here is therefore qualitative. A quantitative comparison is under studies with numerical simulations provided by finite-elements based methods. Such comparisons are currently being done. Thanks to them it will be possible to quantify the performance and limitations of our algorithm.

5. Conclusion

In this paper, we have proposed a new paradigm for the trajectory grouping problem, with special emphasis on the problem of particle tracking.

First and foremost, we suggested a mathematical encoding of the problem, which takes into account metric constraints specific to the problem, perceptual properties of the image and physics properties of the studied phenomena. Second, we proposed a new statement of the particle tracking problem as a global optimization problem. This new formulation has the advantage to take into account :

- available information on the physical nature of the experiment and of the studied phenomena,
- perceptual information on the original numerical image.

Endly, in order to solve this combinatorial optimization problem, we devised a new kind of neural networks, named pulsed neural networks. The advantages of these new neural networks are :

- The pulsed neural networks need no coefficient. Accordingly they have a completely black-box behaviour from a user point of view.
- When the network has converged, the solution found is guaranteed to be a feasible solution, satisfying all constraints.
- The networks regularly put forward a feasible solution. It is possible to impose a computational time limit.
- The iterations number necessary to converge is much smaller than in most of other methods.
- The algorithm can be directly implemented on a parallel multiprocessor or on a specialized chip, with a high improvement of performances.

The first fundamental contribution of our work is a formulation of the particle traking problem as a graph-theoretic problem and the formal formulation of this problem as a global optimization problem, which uses metric relations, specific to the experimental process and to the fluid, and some perceptual information included in the image. The second main contribution relies on the neural networks devised in order to solve this constraint satisfaction problem. These neural networks guarantee the optimization.

It is clear from the real experimental results proposed that our algorithm provides very good solution. Nevertheless realistic simulations must be investigated in order to quantify the performances and limitations of our algorithm. In the future, we plan to improve our method so as to be able to treat any type of fluid flows (laminar, vortex). Moreover we consider essential to quantify the quality of our results. This work will be done shortly.

6. Acknowledgements

The authors would like to thank J.M. Dinten for many valuable discussions, as well as J.J. Niez (CEA-CESTA) and P. Puget. They are grateful to Y. Dolias (CEA-DRN), who provided them with fluid mechanics images, and G. Cognet (CEA-DRN) for helpful conversations on fluid mechanics theory.

References

[ADR86] R.J. ADRIAN. Multi-point optical measurements of simultaneous vectors in unsteady flow–A review. *Int. J. Heat Fluid Flow*, 23:261–304, 1986.

[ADR91] R.J. ADRIAN. Particle-imaging techniques for experimental fluid mechanics. *Annual Review of Fluid Mechanics*, 23:261–304, 1991.

[AGU87] J.C. AGUI and J. JIMENEZ. On the performance of particle tracking. *Journal of Fluid Mechanics*, 185:447–468, 1987.

[BUC92] P. BUCHHAVE. Particle image velocimetry–Status and trends. *Experimental Thermal and Fluid Science*, 5:586–604, 1992.

[CHA85] T.P.K CHANG, A.T. WATSON and G.B. TATTERSON. Image processing of tracer particle motions as applied to mixing and turbulent flow. *Chemical Engineering Science*, 40(2):269–285, 1985.

[DEN89] B. DENBY and S.L. LINN. Status of HEP net research in the USA. *Computer Physics Communications*, 57:297–300, 1989.

[DER94a] D. DEROU and L. HERAULT. Perceptive grouping and pulsed neural network. Application to particle tracking velocimetry. Technical report, Commissariat a l'Energie Atomique-LETI-Departement Systemes, February 1994. LETI/DSYS/SCSI/MV-25.

[DER94b] D. DEROU and L. HERAULT. Pulsed neural networks and perceptive grouping. In *Fourth European Conference on Computer Vision*, pages 521–526, Stockholm, Sweden, May 1994.

[FIS91] K.H. FISHER and J.A. HERTZ. *Spin glasses*. Cambridge University Press, 1991. ed. D. Edwards.

[GAI93] M. GAIO, J. MADELAINE and B. VICTORRI. Extraction de structures linéaires globales dans des images télédétectées. *Les cahiers du LAIAC*, (9), 1993.

[GRA89] I. GRANT and A. LIU. Method for the efficient incoherent analysis of particle image velocimetry images. *Applied Optics*, 28(10):1745–1747, 1989.

[GYU91] M. GYULASSI and M. HARLANDER. Elastic tracking and neural network algorithms for complex pattern recognition. *Computer Physics Communications*, 66:31–46, 1991.

[HAS91] Y. HASSAN and R. CANAAN. Full-field bubbly flow velocity measurements using a multi-frame particle tracking technique. *Experiments in Fluids*, 12:49–60, 1991.

[HES88] L. HESSELINK. Digital image processing in flow visualization. *Annual Review of Fluid Mechanics*, 20:421–485, 1988.

[HOP82] J. HOPFIELD. Neural networks and physical systems with emergent collective computationnal abilities. *Proc. of National Academy of Science of USA*, 79:2554–2558, 1982.

[LIN88] N. LINK and S. ZUCKER. Corner detection in curvilinear dot grouping. *Biological Cybernetics*, 59:247–256, 1988.

[LOW85] D.G. LOWE. *Perceptual organization and visual recognition*. Kluwer Academic Publishers, Boston, 1985.

[MOH92] R. MOHAN and R. NEVATIA. Perceptual organization for scene segmentation and description. *IEEE Transactions on Pattern Analysis and Machine Intelligence*, 14(6):616–635, 1992.

[NIS89] K. NISHINO, N. KASAGI and M. HIRATA. Three-dimensional particle tracking velocimetry based on automated digital image processing. *Transactions of the ASME*, 111:384–391, 1989.

462

[PAR89] P. PARENT and S.W. ZUCKER. Trace inference, curvature consistency and curve detection. *IEEE Transactions on Pattern Analysis and Machine Intelligence*, 11(8):823–839, 1989.

[PET91] C. PETERSON and T.ROGNVALDSSON. An introduction to artificial neural networks. Lectures given at the 1991 CERN School of Computing, Ystad, Sweden, Juillet 1991.

[SHA88] A. SHA'ASHUA. Structural saliency: the detection of globally salient structures using a locally connected network. Master's thesis, dept. of Applied Math., Weizmann Institute of Science, Rehovot, Israel, 1988.

[WER58] M. WERTHEIMER. *Readings in Perception*, chapter Principles of perceptual organization, pages 115–135. Princeton University Press, 1958. ed. D. Beardslee and M. Wertheimer.

[YUI91] A.L. YUILLE, K. HONDA and C. PETERSON. Particle tracking by deformable templates. In *Int. Joint Conference on Neural Networks, IJCNN-91*, volume 1, pages 7–12, Seattle, WA, 1991.

Precision Whole-Field Velocity Measurements with Frequency-Scanned Filtered Rayleigh Scattering

Richard B. Miles, Joseph N. Forkey, Noah Finkelstein, and
Walter R. Lempert

Department of Mechanical & Aerospace Engineering
Princeton University, Princeton, New Jersey 08544 U.S.A.

Abstract. We present an analysis of Filtered Rayleigh Scattering, a diagnostic technique for precision measurements of velocity, temperature, and density in unseeded air flows. Precision measurements require that the filter characteristics be accurately known and that the geometrical factors be properly modeled. Failure to account for these effects in low F number optical systems will lead to large errors in both the velocity and temperature measurements. The best signal-to-noise is achieved with a sharp cut-off blocking filter whose width exceeds the linewidth of the Rayleigh scattered light. The signal-to-noise can be further enhanced by operating in the ultraviolet region of the spectrum where large enhancements in the scattering cross section and better optical filters are available.

Keywords. Rayleigh scattering, flow measurement diagnostics, filters, lasers

1 Introduction

1.1 Motivation

Filtered Rayleigh Scattering is a diagnostic tool which has been recently developed for imaging high-speed flow fields [Miles and Lempert (1990)]. The approach relies on the utilization of a narrow linewidth laser source, which illuminates the flow, and an atomic or molecular blocking filter which is placed in front of the camera. The filter has a very sharp cut-off (from 10% transmission to 90% transmission in a few hundred MHz) and an attenuation of many orders of magnitude on line center. Filtered Rayleigh Scattering has two major applications. The first is the observation of weak scattering by blocking strong background light which is only a few hundred MHz away. The second is the spectral characterization of the scattered light which yields spatially resolved measurements of velocity, temperature, and density.

Both of these applications can be achieved using either molecular scattering from species in the flow, or particle scattering from naturally occurring or artificially introduced seed particles. While temperature and density cannot be extracted from particle scattering, the use of seed particles is attractive since particle scattering is generally stronger, and therefore, yields a greater signal-to-noise ratio than does molecular scattering. The spectral profile of particle scattering is much narrower than the absorption band of the filter. So, when particles are used, the amount of light transmitted is approximately equal to the filter transmission value at the scattering frequency, which is a function of the Doppler shift due to the velocity of the particles. The filter then acts to convert velocity into transmission percentage. When used with particles, Filtered Rayleigh Scattering is similar to a technique developed by J. Meyers and co-workers, called Doppler Global Velocimetry [Meyers and Komine (1991)]. That technique uses an optically thin iodine absorption line, and two simultaneously acquired images of the particles: one with the filter, and one without the filter. The accuracy of Doppler Global Velocimetry has not yet been established, but involves issues similar to those discussed here, plus image registration and absorption path length considerations.

The drawback of using particles is that they only yield information about particle velocity and particle density. In order to extract information about the flow itself, one must assume that the particles are uniformly seeded and that they accurately follow the flow. In cases where these assumptions can be made with a high degree of certainty, the use of seed particles is appropriate. In other cases, however (when these assumptions are of questionable validity, or when the particular facility being used cannot be seeded, due to either technical or practical impediments), Rayleigh scattering from molecules can be used to extract the same information, albeit with lower signal levels. In this case, the scattering comes directly from the constituents of the flow, so the true flow velocity and density are measured directly. Furthermore, because the spectral shape of molecular Rayleigh scattering is dependent on the thermal motion of the scattering molecules, these experiments also yield a measure of static temperature--a parameter which cannot be determined from particle scattering methods.

In order to avoid interference from particle scattering, a typical molecular Rayleigh scattering experiment requires that the flow be free of particles. However, since Filtered Rayleigh Scattering makes use of a sharp cut-off, blocking filter, it is possible to design an experiment which will reject the narrower linewidth particle scattering while simultaneously transmitting the thermally broadened molecular Rayleigh scattering. In experiments where the laser frequency is scanned, the velocity of the particles can be measured along with the molecular velocity and temperature. Large particle loading may overwhelm the filter's extinction capability, seriously limiting the measurement of molecular properties.

The robustness of Filtered Rayleigh Scattering lies in its ability to make use of either particle or molecular scattering, and in its ability to suppress background from particle contamination when molecular scattering is of interest.

In this paper, we discuss theoretical models and experimental results of Filtered Rayleigh Scattering in an unseeded flow. Many of our comments, however, may be extended to the case in which particle scattering is used.

1.2 Previous Work

The blocking of strong background light and the spectral characterization of molecular Rayleigh scattering have both been demonstrated in recent laboratory experiments. The blocking capability of the filter can be used to remove light scattered from windows, walls, and models in order to permit the observation of weak Rayleigh scattering from the flow field itself. For example, a series of two-dimensional cross-sectional slices of the flow field within a Mach 3 inlet duct were imaged using this technique, leading to a volumetric picture of the crossing shock and boundary layer structure [Forkey, et. al. (1994)]. Without the background suppression feature, the light scattered from the air would have been obscured by the light scattered from windows and walls. In this experiment, the flow cross sections were observed from downstream through a mirror placed on a sting in the flow, and the resulting Doppler effect provided enough frequency shift to achieve strong background suppression. This feature has also been used to observe compressible mixing layers [Elliot, et. al. (1992)].

The ability of the filter to resolve the spectral character of the light scattered from a flow was demonstrated in a Mach 5 free jet, which was illuminated at 45° by a laser sheet [Miles, et. al. (1992)]. As the laser was frequency tuned, the light scattered from the Mach 5 jet was swept in frequency through the absorption band of the filter. In this case, the light that was scattered was not only shifted in frequency due to the Doppler effect, but it was also broadened by thermal motion of the air molecules. By observing the Rayleigh scattering from air at a particular location in the flow field, as the laser frequency was swept, the flow velocity, temperature, and relative density were determined. Other researchers have also addressed this aspect of Filtered Rayleigh Scattering, exploring the velocity profile of mixing layers [Elliott, et. al. (1994)] and examining the potential of Filtered Rayleigh Scattering for measurement of mass flow [Winter and Shirley (1993)].

The utility of the Filtered Rayleigh Scattering technique for the measurement of velocity, temperature, and density, relies on the accuracy with which these measurements can be made. That accuracy depends on numerous factors. The fundamental noise process is shot noise associated with the number of photoelectrons collected from each resolvable image element. To reach this limit, however, many other parameters must be precisely known. These include precise tracking of the laser frequency, a precise characterization of the cell transmission versus frequency, and a precise understanding of the complexity introduced by the collection optics. Two paradigms will be used in this discussion. The first is the iodine molecular vapor filter used in conjunction with a frequency-doubled Nd:YAG laser, and the second is the mercury atomic vapor filter used in conjunction with a frequency-tripled Titanium:Sapphire laser. All Filtered Rayleigh scattering experiments, to date, have made use of the first paradigm, the iodine molecular vapor filter. For future measurements, however, the mercury vapor filter shows great promise. Preliminary mercury filter characterizations will be shown together with a discussion of the potential impact of the mercury vapor/Titanium:Sapphire system.

466

2 Iodine Filter Paradigm

The operation of the iodine molecular filter relies on specific absorption lines associated with the B-X electronic band of iodine. These transitions have been studied before, and current literature exists which delineates line positions and line strengths [Gersternkorn and Luc; Tellinghuisen (1982); Glaser (1985)]. In addition, the line broadening can be modeled assuming a combination of thermal and collisional processes. While the accuracy of the literature values is good, it is not sufficient for precision measurements. For example, relative line positions must be known to an accuracy of a few MHz, while literature values have quoted uncertainties of a few hundred MHz. The line broadening parameter must be verified to the same accuracy as line positions, and the absorption coefficients must be known to within a few percent. These are all important since the measurement of the velocity has an error of approximately 1 m/sec if the line position is off by 1 MHz (or 1 part in 10^8). Similarly, the temperature measured depends critically on the accuracy with which the slope of the filter edge is known. This slope is a function of the broadening parameters and the extinction coefficients. Generally, the iodine molecular absorption cell is operated at high enough vapor density so that the absorption line is optically thick. In that case, the maximum slope of the cut-off scales as the square root of the log of the product of the line center extinction coefficient, α_0, times the path length, l, divided by the thermal linewidth, Δf [Miles, et. al (1991)]:

$$\frac{d\left(\frac{I}{I_0}\right)}{df} = \frac{0.74}{\Delta f}\sqrt{ln\left(\alpha_0 l\right)} \tag{1}$$

In order to achieve sharp cut-off, the thermal broadening must be small (i.e., the temperature low, and the mass of the molecule high), and the extinction must be large (i.e., high vapor pressure or long path length). The accuracy of the measurement improves with a sharp cut-off filter. This suggests that a heavy gas vapor with a high vapor pressure such as iodine or mercury, is desirable. It also suggests that the cell operate with a large extinction coefficient. As a practical matter, this is limited by the out-of band absorption, normally associated with very weak absorption phenomena. In the iodine cell there is a background continuum absorption due to the C-X transition in iodine. This continuum absorption leads to a maximum contrast before the off-line transmission begins to drop significantly. In iodine, this maximum contrast is approximately 5 orders of magnitude.

A series of experiments have been conducted in order to establish with precision the location of the absorption lines, the behavior of those lines as a function of cell temperature and partial pressure, the maximum extinction, and the effect of weak background absorption. These experiments used a pair of continuous wave Nd:YAG lasers. The first of these was frequency-doubled, frequency-modulated, and locked to a nearby optically thin iodine absorption line using a well stabilized reference cell and a standard first derivative nulling technique. A portion of the infrared beam from this reference laser was passed into an optical fiber and onto a detector. The second laser was frequency-doubled and passed through the iodine cell to be characterized. A portion of the infrared light from this laser was combined with the infrared light from the first laser in the optical fiber and passed onto the detector. The beat frequency between the two lasers was measured using a high frequency counter which was accurate to better than 1 MHz. The transmission of the iodine cell was recorded while the frequency of the second laser was scanned and precisely monitored. These transmission versus frequency measurements were repeated for various different iodine cell vapor pressures, leading to optically thick absorption profiles measured with an accuracy on the order of 1 MHz.

A model of the iodine absorption profile was initially constructed using literature values for line positions and extinction coefficients, and the classical expression for thermal broadening. This model was then corrected using the accurately measured line position. Figure 1 shows the measurement of the iodine absorption spectrum and the corrected model fit to that measurement. Note the correction has been made for the optically thick line of interest, the residual error is still apparent on other lines shown. Once this model has been developed and validated, it can be used as a tool to deconvolve from the experimental data the lineshape, position, and strength of the scattering from the particular observed element in air.

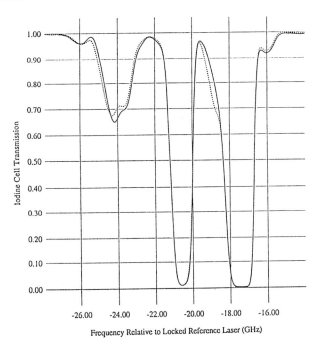

Fig. 1. Plots show comparison between I_2 cell data (solid line) and theoretical model of iodine absorption (dotted line). Data is renormalized to 100% transmission away from absorption lines, since background absorption and absorption due to windows is not included in the model.

In the case where the scattering light level is low, fast (i.e., low F number) optics are desirable in order to collect the most light possible. In this case, additional corrections must be made to the model to account for the subtended angle of the collection optics and the variation of scattering angle across the field-of-view. The corrections for the subtended collection angle must take into account two important phenomena. The first is that the Doppler shift associated with light collected through one portion of the optics is slightly different from that associated with light collected through another portion of the optics. This occurs because the Doppler shift observed is related not only to the angle of the incident laser beam, but also to the angle at which the scattering molecule is observed. The second correction is due to the varying path lengths of the light rays as they pass through the iodine cell. In the case of an optically thick transition, these varying path lengths do not introduce a significant change in the data, but should be taken into account for the highest precision measurements.

For example, the cell transmission of light scattered from an idealized Mach 5 jet (velocity of 721 m/sec, static temperature of 50 K, and static pressure of 40 torr) illuminated at 45°, and with scattered light collected normal to the jet (see Fig. 2), is shown in Figs. 3 and 4 for collection optics with a lens diameter, d_o, of 1/5 of the distance from the lens to the jet, D (Fig. 3), and with a lens diameter equal to the distance from the lens to the jet (Fig. 4). Each figure shows three curves. The solid curve corresponds to the response of the system neglecting the geometric factors and neglecting broadening of the light scattered from the air (i.e., assuming a temperature of 0°K or a mass of ∞). This curve just gives the filter response. The dashed line shows the system response to the point in the center of the field-of-view and includes the geometric factors associated with light collection. The dotted curve incorporates the thermal broadening of the air and represents the modeled response of the filter and optical collection system to the 50 K air in the Mach 5 flow. The curves all cross at the same point since the broadening in all cases is symmetric. This crossing point is a measure of the velocity of the flow. It is evident that the $d_o/D = 1$ collection optics lead to an effective broadening of the filter cut-off, which would, in the absence of proper modeling, be misinterpreted as a higher temperature of the gas flow. These geometrical factors are different for different points in the image due to the different scattering angles of the light collected. For example, Fig. 5 shows the effect on the observed element at the farthest edge of the field-of-view. This figure corresponds to Fig. 4 (i.e., $d_o/D = 1$), but note that the frequency is offset somewhat (about 350 MHz) because the element observed is seen from a different angle, so it gives a significantly different apparent velocity. This shift is still apparent with a $d_o/D = 5$ collection geometry (Fig. 6), but is smaller (about 70 MHz). Nevertheless, failure to account for this shift will lead to a significant error (approximately 50 m/sec or 7%) in the velocity measurement of the Mach 5 jet.

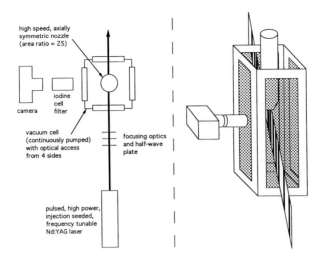

Fig. 2. Experimental geometry for Mach 5 jet calculations and measurements. The laser is incident at 45° to the downward directed jet. Flow conditions: Mach 5 with T_0=300 K, P_0=400 psi (velocity: 721 m/s; static temperature: 50 K; static pressure: 40 torr).

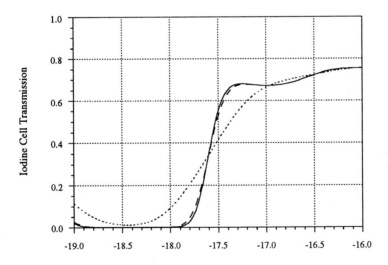

Frequency Relative to Locked Reference Laser (GHz)

Fig. 3. Response of the I_2 filter to the Mach 5 flow with a collection lens placed 5 times its diameter from the sample volume. The solid curve is the ideal response assuming no frequency broadening of the scattered light, the dashed curve is the same with geometrical collection factors added in, and the dotted curve is the predicted response including broadening, geometry, and the filter profile.

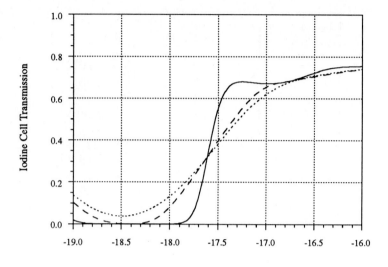

Fig. 4. Response of the I_2 filter to a Mach 5 flow with the collection lens placed a distance equal to its diameter from the observed sample volume. Curves are as identified on Fig. 3.

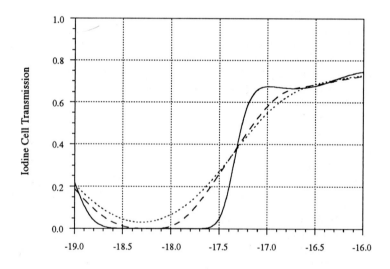

Fig. 5. Response of the I_2 filter to the sample volume element at the edge of the field-of-view for the same collection geometry as in Fig. 4. Note that the apparent response of the filter has been shifted by approximately 350 MHz.

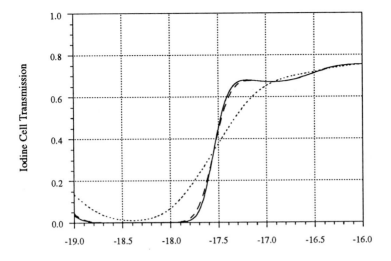

Frequency Relative to Locked Reference Laser (GHz)

Fig. 6. Response of the I_2 filter to a sample volume element at the edge of the field-of-view for the same collection geometry as in Fig. 3. Note that the frequency offset seen in Fig. 5 is reduced, but is still present and must be taken into account.

In addition to the fundamental filter/collector image transfer characteristics, one must also add the laser linewidth to fully model the image transfer function. Once this modeling is complete, then the data associated with any point in the image can be deconvolved so that the frequency shift and line profile associated with the scattering at that point can be determined. A least-squares fit to that data then yields the temperature, velocity, and density at that particular point in the flow field. For example, data from one resolution element in a real Mach 5 jet set-up as shown in Fig. 2, is shown in Fig. 7 together with the fit. Here, direct (unshifted) background scattering of the laser is also present and included in the fit. The data is noisy, largely because of jet fluctuations rather than shot noise. Even with this large amount of noise, the fit gives the velocity and density with a statistical uncertainty of 4%, and temperature with a statistical uncertainty of 8%.

Once the transfer function of the laser, filter, and collection optics is determined, the limitations of the measurement accuracy are, primarily, the signal-to-noise ratio of each resolvable element in the image, and the experimental factors, such as the precise measurement of the laser incidence angle, the measurement of the camera angle, and the stability of the high power laser. Stability of the high power laser is important because it is this laser which is used for the scattering measurements. The frequency of this high power laser, however, cannot be monitored directly with sufficient accuracy. Instead, this laser is frequency locked to a cw seed laser, whose frequency is tuned and precisely

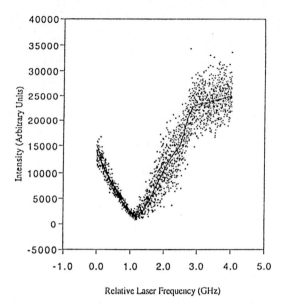

Fig. 7. Demonstration of data fitting routine. Data from overexpanded Mach 5 nozzle flow is represented by dots. Fit to data is represented by the solid line. The scatter in the data is largely due to fluctuations in the flow.

monitored. Stability error can arise due to: (1) the inability of the high power laser to instantaneously track the frequency of the cw seed laser, or (2) the finite stabilization time of the cw seed laser as its frequency is scanned. If the scan is too rapid, the higher power laser cannot track, and the injection locking becomes intermittent. With designs currently under development, the high power laser will track the injection laser on a pulse-by-pulse basis, so intermittent locking will not be present. There may, however, be some residual frequency pulling that has yet to be examined.

In the absence of experimental factors, the fundamental limitation is the shot noise associated with the number of photoelectrons collected from each resolved element of the flow. For example, assuming: a laser with an energy, E_L of 100 mJ at 0.532 micron (frequency-doubled Nd:YAG laser); expanded with a height, H, of 2 cm, and a thickness, δ, of 50μ; air at standard temperature and pressure $\left(N = 2.69 \times 10^{19} \, / cc \right)$ with a differential Rayleigh cross section, $\partial\sigma/\partial\Omega$, of $6 \times 10^{-28} \, cm^2 \, / sr$; a resolved volume element, V, of 50μ x 50μ x 50μ; a collection system consisting of an F5 lens, imaging a 2 x 2 cm cross-sectional area of the flow onto a 1 cm^2 CCD device with 25 x 25μ elements (a magnification factor, m, of 0.5); and an additional 10% loss due to optical transmission $\left(\eta = 90\% \right)$, the predicted number of photons, n, present at the detector per resolution element (Eq. 2) is 160 for each pulse of the laser.

$$n = \left(\frac{E_L}{\hbar w}\frac{1}{H\bullet\delta}\right)(N)\left(\frac{\partial\sigma}{\partial\Omega}\right)(v)(\eta)\left[\frac{\pi}{4F^2}\left(\frac{1}{1+\frac{1}{m}}\right)^2\right] \tag{2}$$

With a 40% quantum efficiency detector (e.g., a high efficiency CCD), this corresponds to 64 photoelectrons collected. The shot noise is equal to the square root of the number of photoelectrons, so with a single shot, the density can only be measured to one part in eight, or 12.5%. To achieve a density measurement of better than 1%, at least 10,000 photons per resolution element must be accumulated. This corresponds to approximately 160 laser pulses, or 16 sec at 10 pulses/sec. For more accurate single pulse measurements, faster collection optics are required. For example, an F=1 lens will collect 25 times as many photons, leading to an accuracy of approximately 2.5% for the measurement of density in a single shot.

The measurement of velocity and temperature requires one of the following: scanning of the laser frequency, observing the scattering volume through filters from several different angles, using multiple laser illumination angles, or using filters with slightly different cut-off frequencies. The most straight-forward approach is to scan the laser frequency, which leads to a time-averaged velocity and temperature at each point. The velocity and temperature are then found by fitting the intensity versus frequency curve, as discussed previously. Essentially, the velocity is determined from the laser frequency at which the filter attenuates the signal by 50%, and the temperature is determined by the rate at which the signal is attenuated as the laser is swept in frequency. The accuracy of both these measurements depends on the quality of the fit. By tuning the laser far enough so that the scattered light has a frequency such that it is fully transmitted through the molecular filter at the beginning of the scan, and again is fully transmitted at the end of the scan, then the velocity can be determined quite accurately from the symmetry of the attenuation profile. In this case, the accuracy of the velocity measurement is similar to that of the density. The temperature measurement, on the other hand, requires that the slope of the curve be fitted. This means that the data associated with low transmissivity is important. This data corresponds to a reduced number of photons collected, and, consequently, higher shot noise. For this reason, the temperature measurement is intrinsically less precise than velocity or density.

An estimation of the filter performance can be made by recognizing that an ideal notch blocking filter can be visualized as a low pass step filter with a cut-off frequency, f_{-c}, plus a high pass step filter with a cut-off frequency of f_{+c}. The signal passing through the low pass filter can be expressed as the integral of the scattering frequency profile R(f) up to the cut-off frequency.:

$$S_- = \int_{-\infty}^{f_{-c}} R\left(f - f_D - f_L\right)df \tag{3}$$

474

where the argument $f - f_D - f_L$ includes the Doppler shift, f_D, associated with the flow velocity and the frequency, f_L, of the illuminating laser. The signal passing through the high pass filter is the integral beginning with the upper cut-off frequency and continuing to infinity:

$$S_+ = \int_{f+c}^{\infty} R\left(f - f_D - f_L\right) df \tag{4}$$

The overall signal recorded is just the sum of these. As the laser frequency, f_L, is scanned, the signal is swept through the filter profile, and the light collected represents a convolution of the filter characteristic with the signal. For example, if the scattered signal, R(f), has a Gaussian profile, 10% of the light will be attenuated when the lower frequency cut-off, f_{-c}, is $\sqrt{2} \times \Delta f$ above the peak scattering frequency. As the laser is swept, 50% attenuation will occur when the filter cut-off is exactly at the peak frequency, and 90% will occur when the scattering peak frequency is $\sqrt{2} \times \Delta f$ above the lower cut-off. There will be very little light transmitted until the upper cut-off is reached, and then the signal returns to full strength following the same curve. In order to achieve the maximum contrast, the filter width should be much greater than the linewidth of the scattered light. If, for example, the filter width is just equal to the Gaussian scattering linewidth, Δf, then even at the minimum, the filter transmission will still be 16%, since some light will be transmitted corresponding to those portions of the scattering falling above the upper cut-off and below the lower cut-off. For molecular blocking filters much narrower than that, the temperature becomes difficult to measure because the contrast between full transmission and maximum attenuation is reduced. The iodine filter has a linewidth of approximately 1.8 GHz, which matches well the 1.1 GHz Gaussian linewidth of light scattered from room temperature air.

3 Mercury Vapor Filter Paradigm

An attractive approach to reducing the shot noise is to move into the ultraviolet portion of the spectrum. The Rayleigh scattering cross section scales as the fourth power of the frequency times the square of the polarizability. In the ultraviolet portion of the spectrum the frequency is approximately a factor of two higher, and the polarizability is increased by 10% or so. Many different materials can be used for optical filters in the ultraviolet, but a very attractive candidate is mercury vapor. Mercury absorbs strongly at 0.2537 microns, and also has the desirable characteristics of high mass and high vapor pressure. The fact that it is an atomic vapor leads to a very simple absorption spectrum consisting of nine spectral lines which come from the six naturally occurring isotopes of mercury (196, 198, 200, 201, 202, and 204), and associated hyperfine splitting. This region of the spectrum is accessible using frequency-tripled Titanium:Sapphire lasers, or frequency-tripled Alexandrite lasers. An example of an experimental scan of a 5 cm long mercury cell with a side-arm temperature of 23°C is shown in Fig. 8. A close-up of the Hg_{202} transition is shown in Fig. 9, along with the thermally and collisionally broadened model. The maximum transmission of 70% is a consequence of window reflectivity and does not correspond to background absorption in the mercury vapor. The predicted contrast of this filter is 31 orders of magnitude.

An experiment set-up with the same collection optics as described in the previous section, and, again, using a single 100 mJ laser pulse expanded to 2 cm sheet, gives 1.55×10^3 photons collected onto a 25 micron x 25 micron detector array. Using a thinned CCD device, 40% collection efficiency can be achieved, leading to more than 600 photoelectrons collected per pulse. This corresponds to approximately 10 times the number of photoelectrons collected at 0.532 microns in association with the iodine filter. This increased count reduces the shot noise significantly. For example, with this collection geometry the shot noise decreases from approximately 12% to 4% per laser shot. Decreasing the F number to an F1 system increases the number of photons collected per resolution element to 15,000, which leads to a shot noise of less than 1% per pulse. The 2.76 GHz bandwidth of the filter matches well the approximately 2 GHz width which is characteristic of the Rayleigh scattered light at this wavelength.

The mercury filter/Titanium:Sapphire laser system will lead to much higher accuracy in the measurements of density, temperature, and velocity, than would be possible for the iodine filter/frequency-doubled Nd:YAG laser configuration. Several other features of the ultraviolet approach are worth mentioning. One is that the on-line center extinction can be much greater than for iodine vapor, due to the absence of background absorption bands in mercury, so that light scattered from boundary layers close to surfaces can be directly observed. In addition, in the ultraviolet, background scattering from windows and walls is reduced due to the low reflectivity of metallic surfaces at this wavelength. This leads to an additional enhancement in the practical signal-to-noise limit of the experiment.

4 Conclusions

Filtered Rayleigh Scattering promises to be a highly accurate measurement tool for air flows. In order to achieve high accuracy, the cell must be properly characterized to take into account the filter transmission and collection geometry. Once these have been properly accounted for, the measurement of velocity, temperature, and density is limited primarily by the shot noise associated with photon detection. This shot noise can be reduced by using fast collection optics and high efficiency detectors. Due to the fact that Rayleigh scattering does not saturate, signal levels may also be increased by using higher power laser sources. A particularly attractive approach to reducing the noise level is to move into the ultraviolet portion of the spectrum and take advantage of large enhancement in the scattering cross section. In this case, a mercury atomic vapor filter is used and there is a significant reduction in shot noise.

476

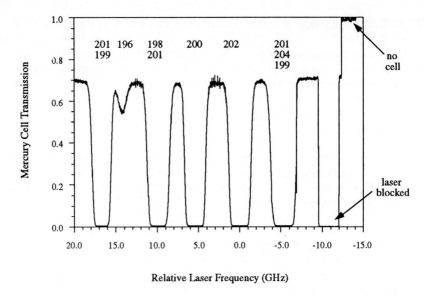

Fig. 8. Experimental scan of mercury vapor absorption in a 5 cm long cell with a side arm temperature of 23°C and a cell temperature of 48°C.

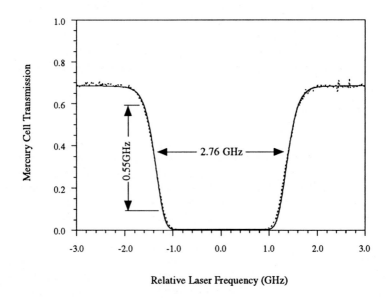

Fig. 9. Blow-up of experimental data (dotted) from Fig. 8 in the vicinity of Hg_{202} transition, along with modeling prediction (solid line).

5 Acknowledgments

This work was conducted under the support of the Air Force Office of Scientific Research, NASA-Langley, NASA-Lewis, and Small Business Innovative Research Programs under M.L. Energia, Inc. and Schwartz Electro-Optics, Inc.

6 References

Elliott, G.S.; Samimy, M.; and Arnette, S.A. (1992) A study of compressible mixing layers using Filtered Rayleigh Scattering, AIAA-92-0175.

Elliott, G.S.; Samimy, M.; and Arnette, S.A. (1994 Details of a molecular filter based velocimetry technique, AIAA-94-0490.

Forkey, J.N.; Lempert, W.R.; Bogdonoff, S.M.; Miles, R.B.; and Russell, G. (1994) Volumetric imaging of supersonic boundary layers using filtered Rayleigh Scattering background suppression. AIAA 94-0491, 32nd Aerospace Sciences Meeting & Exhibit, Reno, Nevada.

Gersternkorn, S. and Luc, P., Atlas du Spectre d'Absorption de la Molecule d'Iode, (Editions du CNRS).

Glaser, M. (1985) Identification of hyperfine structure components of the iodine molecule at 640 nm wavelength, Opt. Comm., Vol. 54 (No. 6), p. 335.

Meyers, J.F. and Komine, H. (1991) Doppler Global Velocimetry: A new way to look at velocity. Laser Anemometry--Vol. 1, ASME, p. 289.

Miles, R.B. and Lempert, W.R. (1990) Two-dimensional measurement of density, velocity, and temperature of turbulent air flows from UV Rayleigh scattering, Appl. Phys. B, Vol. B51, p.1.

Miles, R.B.; Lempert, W.R.; and Forkey, J.N. (1991) Instantaneous velocity fields and background suppression by Filtered Rayleigh Scattering, AIAA-91-0357, 29th Aerospace Sciences Meeting, Reno, NV

Miles, R.B.; Forkey, J.N.; and Lempert, W.R. (1992) Filtered Rayleigh Scattering measurements in supersonic/hypersonic facilities. AIAA-92-3894, 17th Aerospace Ground Testing Conference, Nashville, TN.

Tellinghuisen, J. (1982) Transition strengths in the visible-infrared absorption spectrum of I_2, J. Chem. Phys., Vol. 76 (10), p. 4736.

Winter, M. and Shirley, J.A. (1993) Air mass flux measurement system using Doppler shifted Filtered Rayleigh Scattering, AIAA-93-0513.

Author Index

Springer-Verlag
and the Environment

We at Springer-Verlag firmly believe that an international science publisher has a special obligation to the environment, and our corporate policies consistently reflect this conviction.

We also expect our business partners – paper mills, printers, packaging manufacturers, etc. – to commit themselves to using environmentally friendly materials and production processes.

The paper in this book is made from low- or no-chlorine pulp and is acid free, in conformance with international standards for paper permanency.

Printing: Mercedesdruck, Berlin
Binding: Buchbinderei Lüderitz & Bauer, Berlin